INTRODUCTION TO SYNTHETIC ELECTRICAL CONDUCTORS

INTRODUCTION TO SYNTHETIC ELECTRICAL CONDUCTORS

John R. Ferraro

Department of Chemistry
Loyola University
Chicago, Illinois

Jack M. Williams

Chemistry and Materials
   Science Divisions
Argonne National
   Laboratory
Argonne, Illinois

1987

ACADEMIC PRESS, INC.
Harcourt Brace Jovanovich, Publishers
Orlando   San Diego   New York   Austin
Boston   London   Sydney   Tokyo   Toronto

ACADEMIC PRESS, INC.
Orlando, Florida 32887

United Kingdom Edition published by
ACADEMIC PRESS INC. (LONDON) LTD.
24–28 Oval Road, London NW1 7DX

Library of Congress Cataloging in Publication Data

Ferraro, John R., Date
    Introduction to synthetic electrical conductors.

    Includes index.
    1. Organic conductors.   2. Electric conductors.
I. Williams, Jack M. (Jack Marvin), Date
II. Title.   III. Title: Synthetic electrical conductors.
QD382.C66F47   1987   620.1'1297   86-22165
ISBN 0–12–254120–0 (alk. paper)

PRINTED IN THE UNITED STATES OF AMERICA

87 88 89 90   9 8 7 6 5 4 3 2 1

This book is dedicated to our families, who so generously relinquished family time so that we might work on this book. JMW especially wishes to thank Joan, Kelly, and Hillary for their continuing interest and support.

CONTENTS

5   **TRANSITION ELEMENT-MACROCYCLIC LIGAND
         COMPLEXES**   205

6   **MISCELLANEOUS CONDUCTORS**   244

7   **INTRODUCTION TO THE PHYSICS OF LOW-
         DIMENSIONAL SYSTEMS**   263

PREFACE

In this book the authors have attempted to include in one volume an introdution to synthetic electrical conductors, which have fascinated scientists of various disciplines for the past 15 or so years. Most of the subjects covered in this book have appeared separately as symposium compendiums in the past.

We emphasize the words in the title of the book—*introduction* and *synthetic*. The book is intended to serve as an introduction to the various types of electrical conductors. However, references to reviews and books appear after each chapter, and the bibliography has been included for the reader who wishes to pursue the subject matter further. Only electrically conductive materials that must be synthesized in the laboratory are discussed. Additionally, the book is written by chemists, and the approach is obviously different from that which might be written by a representative of another discipline.

Chapter 1 is introductory in nature and presents historical information concerning the alloy superconductors (e.g., Nb_3Ge). Chapter 2 discusses the organic charge-transfer conductors, which have been the source of the only organic superconductors. Chapter 3 concerns itself with the polymeric conductors (organic and inorganic). Chapters 4 and 5 include material on the Krogmann salts and the metallo-macrocyclic metals. Chapter 6 covers the poly(ethylene oxide) polymer-ion and graphite conductors. In Chapter 7 a general discussion is provided on the physics of ID systems, which are so pertinent to many of the organic conductors. Chapter 8 compares pressure and temperature effects on electrical conductors. Chapter 9 is a summary chapter, and attempts to provide possible directions for future research. The appendix includes a recent bibliography, some of which have appeared after the book was submitted to the publishers. This appendix attempts to keep the book current up to publication time, which is difficult to achieve in such fast-moving research areas as those included in this book.

Finally, the thrust and depth given to various topics in the chapters reflect the interests and prejudices of the authors. We apologize for the possible omission of some topics, whether on purpose or inadvertently. Any omissions are based on several factors: (1) the profusion of publications that are accumulating, (2) the effort to keep the size of the book at a reasonable level, and (3) the individual prejudices of the authors in regard to choice of topics. We also recognize that we have omitted certain references. We hope that this book will provide readers from various disciplines with insight into the rapidly growing field of synthetic metals.

ACKNOWLEDGMENTS

One author (JRF) wishes to acknowledge the Searle Foundation for support of some of this work. Both authors express their thanks to Mrs. V. R. Bowman and Mrs. C. M. Cervenka of Argonne National Laboratory for their efforts in the preparation of this manuscript. JMW wishes to acknowledge the support of the U.S. Department of Energy, Office of Basic Energy Sciences, Division of Materials Sciences, under Contract W-31-109-ENG-38.

1 INTRODUCTION

In the past 20 or more years there has been a rapid development in the syntheses and characterizations of unusual new materials, which have high or metallic-like electrical conductivities. These materials, often by virtue of their crystalline packing arrangements, exhibit anisotropic behavior in such properties as electrical, optical, and magnetic, and are considered to be one-dimensional compounds at room temperature in most instances.

A primary motivation behind these research activities has been based on the potential technological applications that these substances might possess. The primary goal in this increased research has been the search for a high-temperature superconductor. The search for superconductivity has long motivated scientists of several different physical disciplines. Superconductivity is defined, in a broad sense, as the complete absence of electrical resistance. The great industrial potential of materials possessing superconductivity justifies the effort being expended in this direction. Electrical power transmission without resistance would increase the efficiency of the nation's national grid and more efficient electrical motors and extremely powerful electromagnets would become possible. The absence of resistance with a consequent decrease in heat generation would also make possible larger, more powerful and efficient computers. The ability of a superconductor to expel an external magnetic field (Meissner effect) would allow a vehicle to float above a superconducting roadway for a rapid, smooth, and efficient mode of transportation. Military and space applications are also possible.

In the instances where superconductivity has been discovered, extreme

1

cryogenic temperatures are necessary. Consequently, costs of obtaining and maintaining superconductive conditions have been prohibitive.

The early research in this area was devoted to metal alloys. In 1973, Nb_3Ge was found to possess the highest recorded superconducting transition temperature (T_c) for any material at 23.2 K (Testardi, 1973a; Testardi *et al.,* 1974; Gavaler, 1973; Gavaler *et al.,* 1974). At these temperatures one still is borderline in the liquid helium cryogenic domain. A rise in T_c of 10 K or more degrees would bring one easily into the hydrogen domain (bp 20.4 K) or, neon (bp 27.1 K) and materially reduce cooling costs. T_c should be in the range 20–40 K for any practical application [see Laverick (1969)]. †

The superconductive alloy materials with the highest T_c have all come from the A-15 family of alloys (Dew-Hughes, 1975). The A-15 structure, illustrated in Fig. 1, is one involving close-packed structures, with a primitive cubic cell of 8 atoms and belonging to the space group O_h^3-$Pm3m$. The stoichiometry is A_3B, with the A element derived from Groups IVA, VA, and VIA, while the B element is from the right side of the periodic table, up to Group VB. A most important structural feature of the A-15 materials are the orthogonal chains of A atoms (linear and one dimensional) with metal–metal distances *less* than in pure A. The discovery of numerous

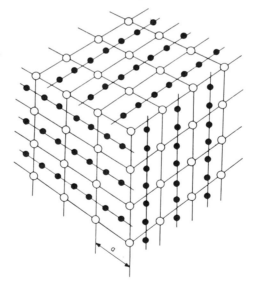

Fig. 1.   The A-15 structure. Extended arrangement of atoms in A_3B compounds; (●) A atoms and (○) B atoms. The structure shows the three orthogonal sets of linear chains formed by the A (transition metal) atoms. [Reprinted from article by Testardi 1973b.]

† A major breakthrough has recently occurred in this area (see p. 323).

chemical systems possessing similar metal atom chains, and often having metal–metal distances *less than* those in the parent metal, provided great impetus toward the study of the electrical and magnetic properties of these novel systems. Such is the case for the platinum atom chain systems discussed in Chapter 4.

Figure 2 shows that elements in the periodic chart relative to their possible superconductivity. At atmospheric pressures fewer than 30 elements are recognized as superconductors and 14 elements become superconductive at elevated pressures. Examination of Table I shows that the A-15 alloys are generally made up of a transition element superconductor in combination with another element. There are 66 compounds with an A-15 structure and 46 are known to be superconductors as tabulated in Table I (Dew-Hughes, 1975).

An empirical relationship for determining the T_c for A-15 superconductors has been derived by Dew-Hughes and Rivlin (1974) and follows

$$T_c = 27.5(T_A - 2)M_B^{-1/2} \qquad (1)$$

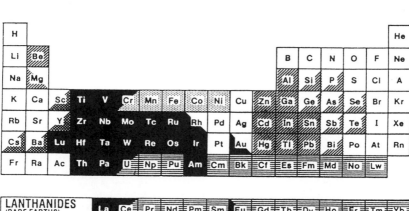

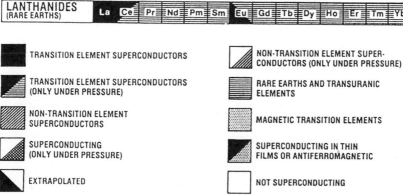

Fig. 2. Superconducting elements in the periodic table.

TABLE I

Critical Temperature of all A-15 Compounds Known to be Superconductive[a]

Compound	T_c	Compound	T_c	Compound	T_c
Ti$_3$Ir	4.6	V$_3$Si	17.1	Ta$_3$Ge	8
Ti$_3$Pt	0.49	V$_3$Ge	7	Ta$_3$Sn	6.4
Ti$_3$Sb	5.8	V$_3$Sn	4.3	Ta$_3$Sb	0.72
Zr$_3$Au	0.92	V$_3$Sb	0.8	Cr$_3$Ru	3.43
Zr$_4$Sn	0.92	Nb$_3$Os	0.94	Cr$_3$Os	4.03
Zr$_3$Pb	0.76	Nb$_3$Rh	2.5	Cr$_3$Rh	0.07
V$_3$Os	5.15	Nb$_3$Ir	1.76	Cr$_3$Ir	0.17
V$_3$Rh	0.38	Nb$_3$Pt	10	Mo$_3$Os	11.68
V$_3$Ir	1.39	Nb$_3$Au	11	Mo$_3$Ir	8.1
V$_3$Ni	0.57	Nb$_3$Al	18.9	Mo$_3$Pt	4.56
V$_3$Pd	0.08	Nb$_3$Ga	20.3	Mo$_3$Al	0.58
V$_3$Pb	3.7	Nb$_3$In	8	Mo$_3$Ga	0.76
V$_3$Au	3.2	Nb$_3$Ge	23.2[b]	Mo$_3$Si	1.3
V$_3$Al	9.6	Nb$_3$Sn	18.3	Mo$_3$Ge	1.4
V$_3$Ga	15.4	Nb$_3$Bi	2.25		
V$_3$In	13.9	Ta$_{4.3}$Au	0.58		

[a] Dew-Hughes (1975).
[b] $T_c = 23.6$ K value for Nb$_3$Ge prepared by chemical vapor deposition [see Fujiura *et al.* (1985)].

where T_A is the critical temperature of pure A and M_B the atomic mass of element B.

Figure 3 shows a plot of T_c versus $M_B^{-1/2}$. Extrapolation of the plot for Nb$_3$Sn and Nb$_3$Ge to Nb$_3$Si produces a T_c of ~38 K. New phases of Nb$_3$Si have been synthesized at elevated temperature and pressure, but, unfortunately, to date, A-15 type Nb$_3$Si has not been prepared (see Waterstrat *et al.*. unpublished data).

Ultimately, the goal of much of the present research is to obtain better electrical conductors and, hopefully, a high-temperature superconductor. The quest for metallic alloys with higher T_c values continues, and scientists have turned their attention to other materials, such as organic solids. Almost 75 years ago it was suggested that organic solids might exhibit electrical conductivities comparable to metals (Kraus, 1913; McCoy and Moore, 1911). Little (1964, 1965), in two classic publications, discussed the possibility of creating a structure involving a linear hydrocarbon chain with alternating single and double bonds along the chain and surrounded by highly polarizable molecules like diethylcyanine iodide. He predicted that such a molecule would provide free passage of electrons along the chain from one site to another, and could become superconductive even at room temperature. This "excitonic" model for superconductivity stim-

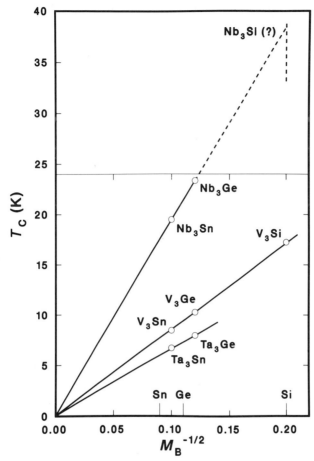

Fig. 3. Plot of T_c versus $M_B^{-1/2}$. [Taken in part from Dew-Hughes and Rivlin (1974). Reprinted from *Nature* **250**, 723. Copyright © 1974 Macmillan Journals Limited.]

ulated the search for syntheses of many new organic and metal chain-forming molecules which might behave as metals.

Other significant contributions were those of Krogmann and Hausen (1968) and Krogmann (1969), who reinvestigated and began characterizing the platinum chain-forming tetracyanoplatinate salts. This regenerated interest in these salts (Krogmann salts), and thus, another area of research in the search for new electrical conductors has developed. The syntheses of TCNQ [(7,7,8,8-tetracyano-*p*-quinodimethane)] by Melby *et al.* (1962), TTF (tetrathiafulvalene) prepared by Wudl *et al.* (1970), TTF-TCNQ (possibly the first "organic metal") synthesized by Ferraris *et al.* (1973) and

Coleman *et al.* (1973) were major contributions in the developments of the organic charge-transfer salt synthetic metals (synmetals). A chronological history of events in the saga of synthetic electrical conductors is listed in Table II.

Even if all the goals of a room temperature superconductor are not soon realized, the new synthetic metals offer numerous potential technological applications. In fact, a number of these have already surfaced in the area of energy conservation, medical electronics, sensing, display electrode materials, high-powered batteries, and thermoelectric uses.

Additionally, this research has brought together scientists from many disciplines. These include electrical engineers, chemists, physicists, materials scientists, organic and inorganic synthetic chemists, as well as theorists. These interactions have led to the creation of new areas of scientific study in the chemistry and physics of novel electrical conductors.

In this text we have arbitrarily chosen to classify and discuss the synthetic electrical conductors in four major areas. These include the organic

TABLE II

Chronological History of Electrical Conductors

Year	Discovery
1911–1913	First suggestion that organic solids could be electrical conductors
1911	H. K. Onnes discovers superconductivity in metallic Hg
1954	First reported conducting molecular compound—bromine salt of perylene
1957	Bardeen, Cooper, and Schrieffer (1957) theory of superconductivity developed
1962	Synthesis of TCNQ
1963	$Cs_2(TCNQ)_3$ type salts synthesized
1964	Little suggests superconductivity at room temperature
1968	Krogmann salts reinvestigated [related salts discovered by Knop (1848)]
1970	Synthesis of TTF
1973	Energy crisis in world
1973	Nb_3Ge superconductor, $T_c = 23.2$ K
1973	TTF-TCNQ synthesized—first organic metal (metallic to 54 K)
1975	$(SN)_x$ found to be conductor and superconductor at 0.3 K (first synthesized in 1910)
1977	$(SNBr_{0.4})_x$ superconductor at 0.36 K
1977	Transition element macrocyclic ligand conductors synthesized
1979	TMTSF-DMTCNQ synthesized, $\sigma = 10^5$ (ohm cm)$^{-1}$ at 10 kbar and 1 K
1979	Selenium-based Bechgaard salts synthesized-$(TMTSF)_2X$, (X = monovalent anion); superconductors under pressure at low temperature (~ 1 K)
1981	$(TMTSF)_2ClO_4$; first organic superconductor at *ambient pressure* and 1.4 K
1983	$(BEDT-TTF)_2ReO_4$ superconductor under pressure and low temperature ($T_c \simeq 2$ K)—first sulfur-based organic superconductor
1984–1986	β-$(BEDT-TTF)_2I_3$, β-$(BEDT-TTF)_2IBr_2$, and β-$(BEDT-TTF)_2AuI_2$ superconductors at ambient pressure with T_c's of 1.4, 2.8, and 5.0 K, respectively
1986–1987	La, Ba, Cu, O systems synthesized (e.g., $La_{1.85}Ba_{0.15}CuO_4$, $La_{2-x}Sr_xCuO_4$) with T_c's of = 30–37 K

charge-transfer metals (containing the new "organic" superconductors), the organic polymeric metals, the Krogmann salts (tetracyanoplatinates), and the transition element–macrocylic ligand metals.

From the data presently available, it has become clear that the research area of synthetic electrical conductors will provide a great potential for future developments, both in the fundamental as well as in the application sense.

LIST OF SYMBOLS AND ABBREVIATIONS

T_c	Transition temperature to superconducting state
T_A	Critical temperature of A element in A_3B, A-15 alloys
M_B	Atomic mass of element B in A_3B alloys
TCNQ	7,7,8,8-Tetracyano-p-quinodimethane
TTF	Tetrathiafulvalene
TMTSF	Tetramethyltetraselenafulvalene
DMTCNQ	(2,4-Dimethyl-TCNQ)
BEDT–TTF or "ET"	Bis(ethylenedithio)tetrathiafulvalene
σ	Conductivity in (ohm cm)$^{-1}$ units
$(SN)_x$	Polythiazyl (conductive inorganic polymer)

REFERENCES

Bardeen, J., Cooper, L. N., and Schrieffer, J. R. (1957), *Phys. Rev.* **108**, 1175.

Coleman, L. B., Cohen, M. J., Sandman, D. J., Yamagishi, F. G., Garito, A. F., and Heeger, A. J. (1973), *Solid State Commun.* **12**, 1125.

Dew-Hughes, D. (1975), *Cryogenics* **15**, 435.

Dew-Hughes, D., and Rivlin, V. G. (1974), *Nature* **250**, 723.

Ferraris, J., Cowan, D. O., Walatka, V. V., and Perlstein, J. H. (1973), *J. Am. Chem. Soc.* **95**, 948.

Fujiura, K., Watari, T., and Kato, A. (1985), *J. Less-Common Met.* **113**, L13.

Gavaler, J. R. (1973), *Appl. Phys. Lett.* **23**, 480.

Gavaler, J. R., Janocko, M. A., and Jones, C. K. (1974), *J. Appl. Phys.* **45**, 3009.

Kraus, H. J. (1913), *J. Am. Chem. Soc.* **34**, 1732.

Krogmann, K. (1969), *Angew. Chem. Int. Ed. Engl.* **8**, 35.

Krogmann, K., and Hausen, H. D. (1968), *Z. Anorg. Allg. Chem.* **358**, 67.

Laverick, C. (1969) *in* "A Guide to Superconductivity" (D. Fishlock, ed.), Chapter 8, pp. 128–144, Elsevier, New York.

Little, W. A. (1964), *Phys. Rev. A* **134**, 1416.

Little, W. A. (1965), *Sci. Am.* **212**, 21.

McCoy, H. N., and Moore, W. C. (1911), *J. Am. Chem. Soc.* **33**, 273.

Melby, L. R., Harder, R. J., Hertler, W. R., Mahler, W., Benson, R. E., and Mochel, W. E. (1962), *J. Am. Chem. Soc.* **84**, 3374.

Testardi, L. R. (1973a), *Phys. Today* **26**, 17.

Testardi, L. R. (1973b), *in* "Physical Acoustics—Principles and Methods" (R. N. Thurston, ed.), Chapter 4, pp. 194–292, Academic Press, New York.

Testardi, L. R., Wernick, J. H., and Royer, W. A. (1974), *Solid State Commun.* **15**, 1.

Waterstrat, R. M., Ferraro, J. R., and Basile, L. J., unpublished data.

Wudl, F., Smith, G. M., and Hufnagel, E. J. (1970), *J. Chem. Soc., Chem. Commun.*, p. 1453.

2 ORGANIC CHARGE-TRANSFER METALS INCLUDING NEW SUPERCONDUCTORS

I. INTRODUCTION

The first experimental indication that molecular compounds could exhibit interesting electrical properties, apart from those of an insulator, was made by Akamatu *et al.* (1954), who reported a resistivity of $\rho \cong 10$ ohm cm for a bromine salt of perylene. Perylene itself is an insulator with a resistivity $\rho = 10^{14}-10^{16}$ ohm cm. This observation was a verification of earlier predictions by McCoy and Moore (1911) in 1911, the same year superconductivity was discovered in mercury, that certain organic solids might become electrically conductive. In undergoing this dramatic transformation from an insulator to a conductor, the perylene–bromine solid must change significantly from the starting materials. Thus, for the first time metallic properties were observed in materials which did not contain metal atoms!

The first stable, highly conducting organic molecule was synthesized by Melby *et al.* (1962). The compound was TCNQ (7,7,8,8-tetracyano-*p*-quinodimethane) and its structure is depicted in Fig. 1. Thereafter, interest in conducting organic solids developed rapidly when it was discovered that many salts of TCNQ were electrically conductive (Siemons *et al.*, 1963; Acker *et al.*, 1960). For example, $Cs_2(TCNQ)_3$ was conductive at room temperature, but became an insulator when the temperature was lowered (Siemons *et al.*, 1963). For the salts $M(TCNQ)_2$, in which M is a monovalent cation (alkali metal or organic), extensive synthesis and properties investigations were undertaken (Acker *et al.*, 1960). Most of these salts were found to be semiconductors (i.e., having thermally activated

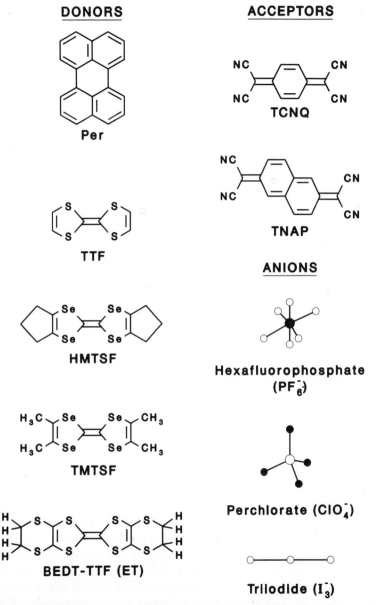

Fig. 1. Molecules that form conducting crystals described in this chapter. The left column contains the electron donors: Perylene (Per), tetrathiafulvalene (TTF), hexamethylene–TSF (HMTSF), tetramethyltetraselenafulvane (TMTSF), and bis(ethylenedithio)–TTF (BEDT–TTF or ET). The right column contains the acceptor molecules tetracyanoquinodimethane (TCNQ) and tetracyanonaphthalene (TNAP). [From Williams and Carneiro (1985).]

electrical conduction) with activation energies of 50–200 meV, but having possible metallic behavior (nonthermally activated conduction) at high temperature.

The TCNQ anions in the metallic salts are packed in pancake-like molecular stacks with extended π-electronic systems above and below the molecular planes, utilizing p_z orbitals of the acceptor. Electrons are delocalized in these systems and move from plane to plane along the TCNQ stacks. The conductivities are highly anisotropic, being $500 \times$ greater in the direction parallel to the stacks than in the perpendicular direction. This conduction anisotropy is an indication of the lack of interstack electronic communication in these systems. Although not impressive as conductors, these salts stimulated considerable thinking and interest in the synmetal field and research with related molecules commenced.

As interest in synmetals expanded it was found that a large number of salts could be synthesized involving charge transfer between a wide variety of organic donors and acceptor (organic and inorganic) species. More than 400 salts of TCNQ have been prepared (Scott *et al.*, 1978). These materials demonstrated a wide range of novel magnetic, electrical, and structural properties. Room temperature conductivities ranged from insulator values to $\sim 10^3$ (ohm cm)$^{-1}$. Simple salts often have the stoichiometry of 1:1 (cation donor–TCNQ acceptor), but can reach values of 1:2 or 1:4 in more complex salts. The reaction between donor (D) and acceptor (A) molecules in these salts can be illustrated as follows in which charge-transfer salts are possible:

$$DA, \quad [D^+ \cdot][A \cdot^-]$$
Charge-transfer compound

During the same period, the understanding of superconductivity in elemental metals underwent dramatic improvements with the development of the now well-known theory of Bardeen, Cooper, and Schrieffer (BCS) (Bardeen *et al.*, 1957). According to this theory, electrons form bound pairs (Cooper pairs) as a result of their interactions with the lattice vibrations (the phonons). In this theory the upper limit for the superconducting transition temperature is determined by the maximum phonon frequency, the Debye frequency. This understanding led Little (1964) to propose that high transition temperatures could be achieved in *molecular* metals (if these could be made) since the high frequencies of their internal vibrations might play the same role as phonons in ordinary superconductors. Little's proposal sparked considerable interest and marked the onset of an intensive search for superconductivity in molecular materials. The promise of possible high temperature superconductivity has remained a goal since that time.

Wudl *et al.* (1970) synthesized the organic electron-donor molecule TTF (tetrathiafulvalene) (see Fig. 1 for its structure) and observed that highly conductive materials could be prepared when it reacted with halogens and pseudo-halogens. Table I summarizes the electrical conductivities of these salts (Isett and Perez-Albuerne, 1978; Miller, 1978). Tetrathiafulvalene contains four sulfur atoms and is a good electron donor, readily giving up an electron to form a stable, positively charged cation. In combination with TCNQ a 1:1 salt crystallizes, and this was the first molecular crystal exhibiting genuine metallic behavior. Between the temperatures of 298 and 54 K, TTF–TCNQ possesses the characteristics of a metal (decreasing electrical resistance with decreasing temperature), although one must modify the concept of a metal to account for the complicated electronic structure of a two-component molecular compound. Superconductivity was never attained in TTF–TCNQ due to low temperature (54 and 38 K) metal–insulator transitions.

TTF and TCNQ are found to form independent molecular stacks in the structure of the TTF–TCNQ compound. Figure 2 illustrates the structure of the one-dimensional organic conductor of TTF–TCNQ. Extensive π-interaction and electron delocalization occurs along the stacks but not between them and this leads to structural transitions in each separate stack (TTF and TCNQ) and concomitant changes in electrical properties. Below 54 K the material exhibits a semiconducting state similar to some of the alkali metal salts of TCNQ, previously discussed. The state is often referred to as the Peierls insulator state, and was predicted by Peierls (1955) to exist in one-dimensional conductors (lacking significant interchain interaction). The transition to an insulator in a one-dimensional conductor

TABLE I

Conductivities of Several TTF Salts[a]

Compound	σ_{RT} (ohm cm)$^{-1}$	ID chain of cations
TTF Cl$_{0.92}$	3.7	TTF
TTF Br$_{0.7}$	300	TTF
TTF I$_{0.7}$	300	TTF
TTF (SCN)$_{0.54}$	550	TTF
TTF (SeCN)	~650	TTF
TSF Cl$_{0.5}$[b]	2000	TSF
TTF	100–540	
TCNQ	100	
TTF–TCNQ	~500	
TSF–TCNQ	~700–800	

[a] Taken in part from Miller (1978).
[b] TSF is the selenium analog of TTF.

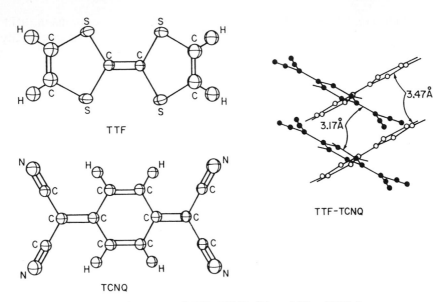

Fig. 2.   Structure of TTF–TCNQ. [From Miller (1978).]

arises from an electron–phonon instability in the lattice, and must be prevented if superconductivity is to occur in molecular systems.

A preponderance of molecular stacking occurs in these charge-transfer conductors, and appears to be an important requisite for the formation of a conducting solid. Stacking can be of two types (Torrance, 1979):

(a)   mixed stacks, in which the donors (D) and acceptors (A) stack alternately face-to-face such as [D–A–D–A–D–A · · ·], and

(b)   segregated stacks, in which the donors and acceptors form separate donor stacks [D–D–D–D · · ·] and acceptor stacks [A–A–A–A · · ·].

Molecular compounds with mixed stacks are not highly conductive because of electron localization on the acceptor species. However, a majority of the charge-transfer compounds that form segregated stacks are highly conductive (Torrance, 1979), as illustrated by the segregated stacks of TTF–TCNQ in Fig. 2. The π-overlap and charge-transfer interaction between adjacent molecules in the stacking directions are strong, causing the unpaired electrons to partially delocalize along these one-dimensional molecular stacks thereby enhancing conduction in that direction.

II.   TTF AND TCNQ VARIANTS

In an attempt to improve the electron-donating properties of TTF, syntheses of numerous TTF variants were performed. The earliest modification was the synthesis of TSF, the selenium analogue of TTF (Engler and Patel, 1974). The corresponding TSF–TCNQ salt was found to be isomorphous with TTF–TCNQ. The replacement of selenium for sulfur caused the electrical conductivity to double (see Table I). The conductivity also increases to a maximum as the temperature is lowered until a metal–insulator transition occurs. Whereas TTF–TCNQ shows a $T_{M \to I}$ at 54 K, TSF–TCNQ demonstrated a lower $T_{M \to I}$ at 40 K.

Bechgaard *et al.* (1976) further modified TTF by synthesizing TMTSF (tetramethyltetraselenafulvalene) (see Fig. 1 for its structure). As was previously observed for TTF–TCNQ versus TSF–TCNQ, TMTSF–TCNQ was also found to have an increased room temperature conductivity and lower $T_{M \to I}$ when compared with TTF–TCNQ. In charge-transfer compounds involving TMTSF, especially those containing monovalent anions, it is the donor stack of TMTSF which dominates the transport properties.

Other substituted donors and acceptors of TTF and TCNQ were also synthesized. For example, DMTCNQ (2,4-dimethyl-TCNQ), TMTTF (tetramethyltetrathiafulvalene), HMTTF (hexamethylenetetrathiafulvalene), and HMTSF (hexamethylenetetraselenafulvalene) were prominent among the new variants. Table II shows the conductivities for several of these salts. Figure 3 shows additional modifications of TTF and the salts formed with TCNQ together with conductivity values at room temperature.

Torrance (1979) arranged various cations (donors) used with TCNQ (acceptor) in terms of four classes. Class I (1:1 TCNQ inorganic salts, e.g., KTCNQ) and class II (1:1 TCNQ organic salts, e.g., TTF–TCNQ) are illustrated in Fig. 4. For the 1:2 salts classes III and IV result, as illustrated in Figs. 5 and 6. The basis for this classification was made on the differences between conductivities, which appear to group into the four classes, as illustrated in Figs. 7 and 8. In Table III the conductivity, energy of activation, and charge-transfer status for the four classes are tabulated. Class

TABLE II

Conductivities of Several TCNQ Salts

Salt	σ_{RT} (ohm cm)$^{-1}$
TTF–TCNQ	~500
TSF–TCNQ	~700–800
TMTTF–TCNQ	350
TMTSF–TCNQ	800
HMTTF–TCNQ	500
HMTSF–TCNQ	2000
TMTSF–DMTCNQ	400–$600/2 \times 10^5$ ($p = 10$ kbar at ~1 K)
TMTSF–TCNQ	1200

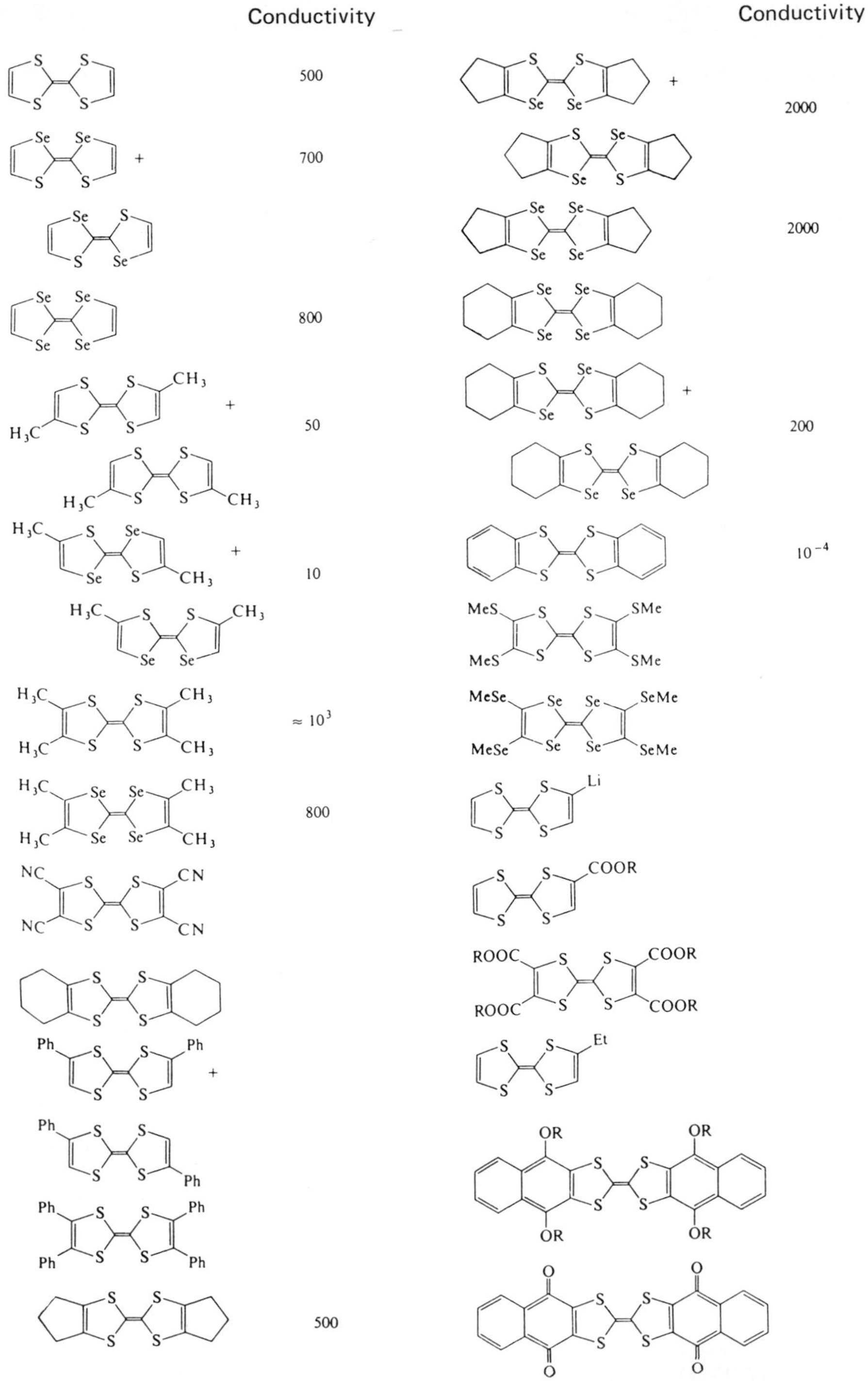

Fig. 3. Selected modifications of tetrathiafulvalene. Conductivity values of the TCNQ salt (ohm^{-1} cm^{-1}) at room temperature are given. [From Miller (1978).]

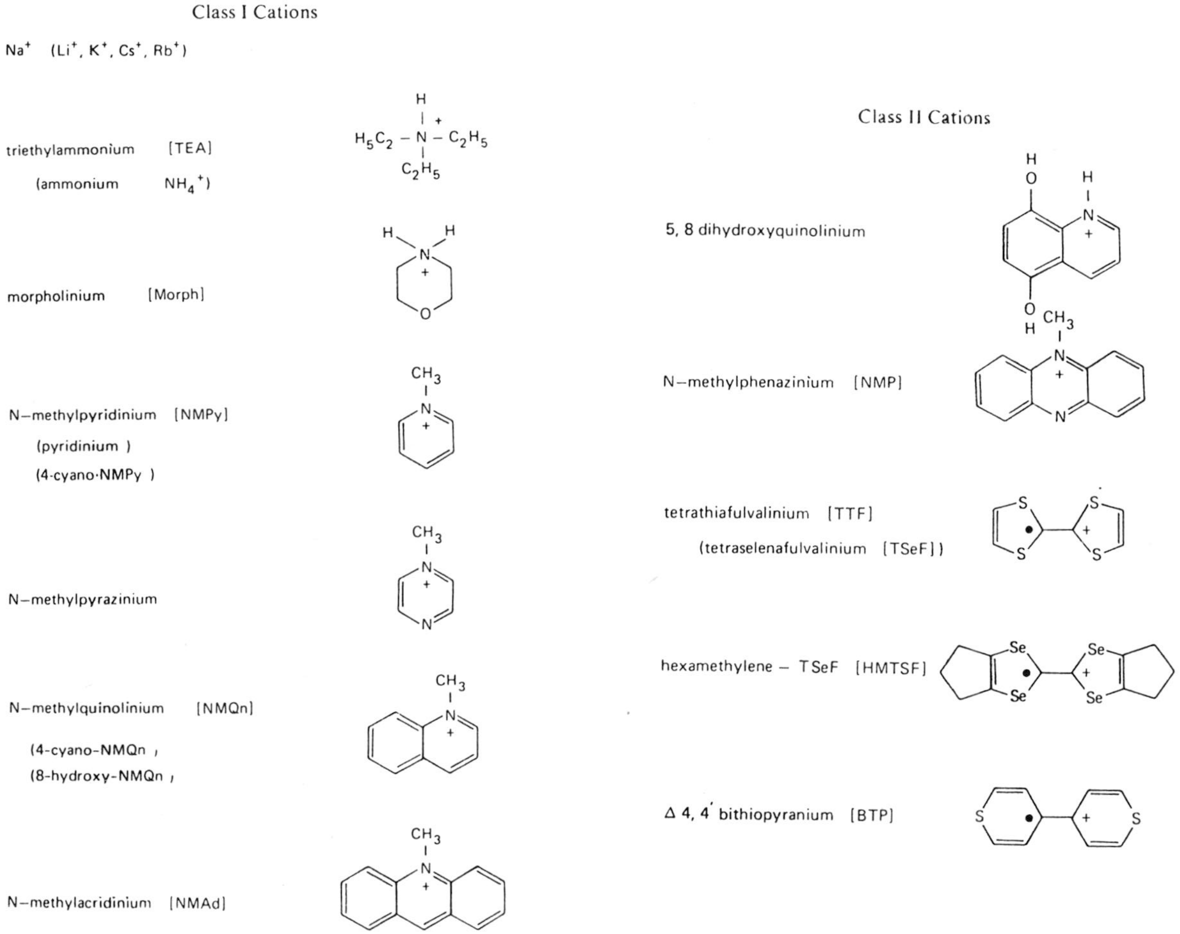

Class I Cations

Na^+ (Li^+, K^+, Cs^+, Rb^+)

triethylammonium   [TEA]

(ammonium NH_4^+)

morpholinium   [Morph]

N−methylpyridinium   [NMPy]

(pyridinium)

(4-cyano-**NMPy**)

N−methylpyrazinium

N−methylquinolinium   [NMQn]

(4-cyano-**NMQn** ,

(8-hydroxy-**NMQn** ,

N−methylacridinium   [NMAd]

Class II Cations

5, 8 dihydroxyquinolinium

N−methylphenazinium   [NMP]

tetrathiafulvalinium   [TTF]

(tetraselenafulvalinium   [TSeF])

hexamethylene − TSeF   [HMTSF]

Δ 4, 4´ bithiopyranium   [BTP]

Fig. 4. Classes I and II cations of simple 1:1 TCNQ Salts. [From Torrance (1978).]

Class III Cations
(1:2 salts unless noted)

Cs$^+$ (2:3)
triethylammonium [TEA]

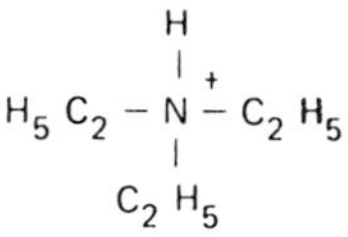

$$H_5 C_2 - \overset{\displaystyle \overset{H}{|} \ +}{\underset{\displaystyle \underset{C_2 H_5}{|}}{N}} - C_2 H_5$$

N-propylquinolinium [NPQn]
(4-cyano-N-methyl Qn [CNNMQn])

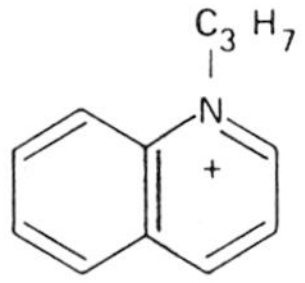

methyl-triphenylarsonium [M ϕ_3 As]
(Me ϕ_3 P)

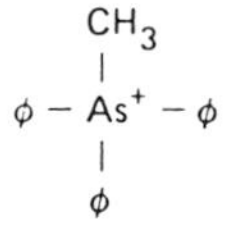

$$\overset{\displaystyle \overset{CH_3}{|}}{\phi - \underset{\displaystyle \underset{\phi}{|}}{As^+} - \phi}$$

N,N,N',N'-tetramethyl-p-phenylenediamine [TMPD]

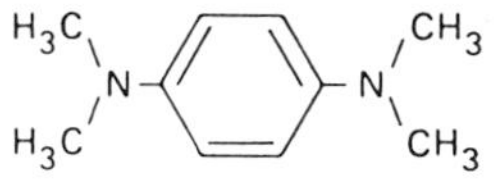

3,3-diethylthiazolinocarbocyaninium [Et$_2$ Tz CC]
(intermediate conductivity form)

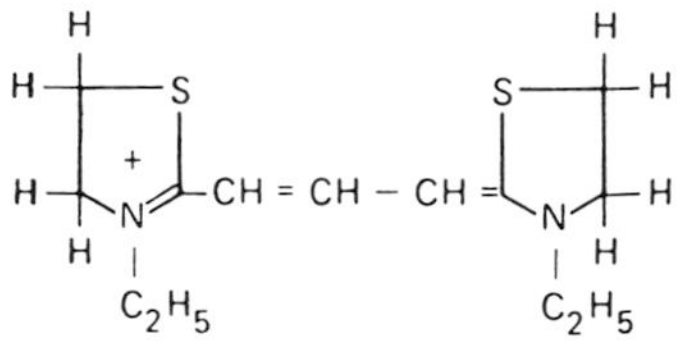

4-4' Bipyridinium

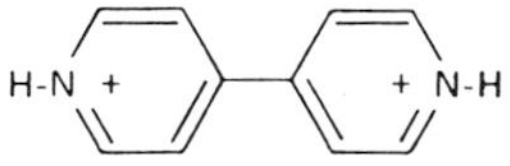

1,2-di (N-ethyl-4-pyridinium) ethylene [DEPE] (1:4)
(form II)

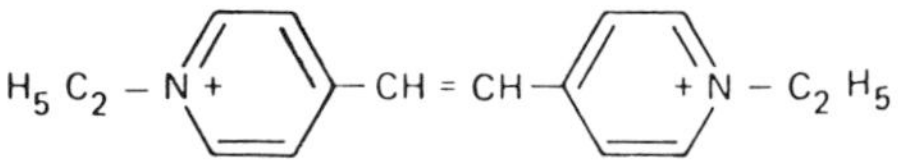

$$H_5 C_2 - N^+ \!\!\!\!\bigcirc\!\!\!\! - CH = CH - \!\!\!\!\bigcirc\!\!\!\! ^+ N - C_2 H_5$$

Fig. 5.   Class III cations. [From Torrance (1978).]

Class IV Cations
(1:2 salts unless noted)

quinolinium [Qn]
  (N-methyl Qn, phase A)

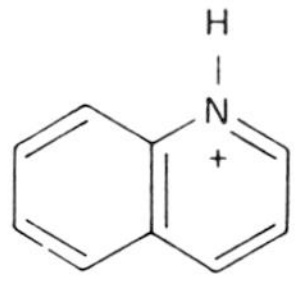

acridinium [Ad]
  (N-methyl Ad)

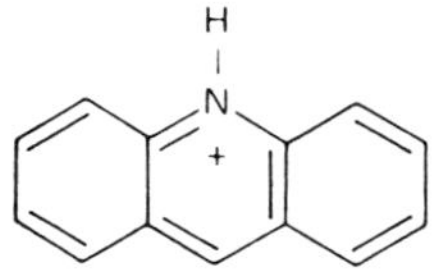

tetrathiotetracene [TTT]

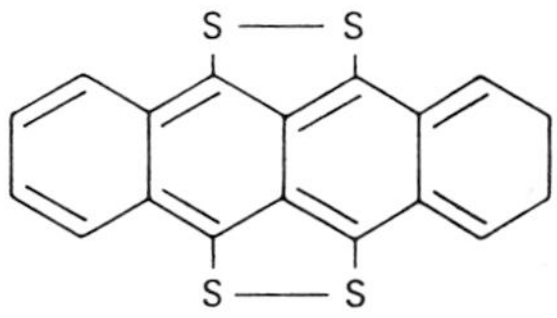

3,3-diethylthiazolinocarbocyaninium [Et$_2$ Tz CC]
  (high conductivity form)

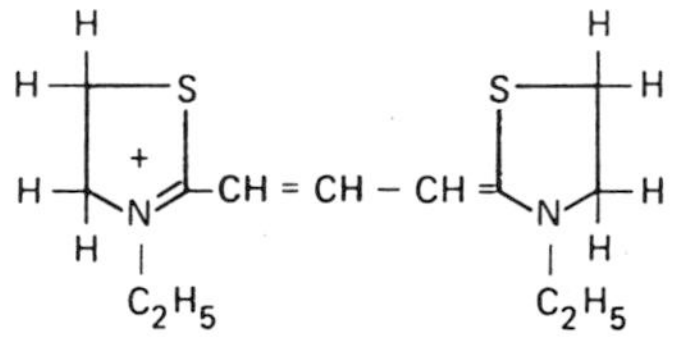

2-2′ Bipyridinium

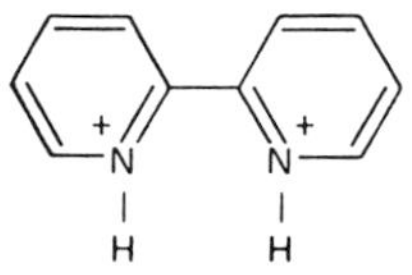

1,2-di (N-ethyl-4-pyridinium) ethylene [DEPE] (1:4)
  (form I)

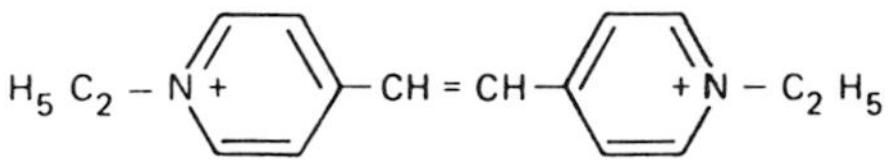

Fig. 6.   Class IV cations. [From Torrance (1978).]

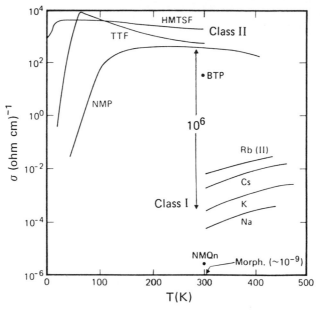

Fig. 7.   Temperature-dependent conductivity for a number of simple 1:1 TCNQ salts illustrating how they can be divided into classes I and II. [From Torrance (1978).]

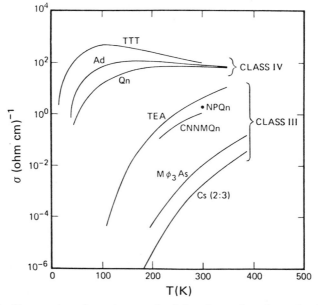

Fig. 8.   Temperature-dependent conductivity of a number of complex 1:2 TCNQ salts illustrating how they can be divided into classes III and IV. [From Torrance (1978).]

TABLE III

Various Physical Properties for Four Classes of TCNQ Salts

Class	Conductivity	Activation energy	CT[a] status
I	Insulators	Large	Complete CT reaction
II	σ high	Low	Incomplete
III	$\sigma < 10$ (ohm cm)$^{-1}$	High	Complete CT reaction
IV	$\sigma > 50$ (ohm cm)$^{-1}$	Low	CT incomplete

[a] CT is charge transfer.

I salts are insulators due to a complete charge transfer and a concurrent large activation energy; the class II and IV systems are better conductors because of incomplete charge transfer and lower energy of activation.

Torrance (1979) also considered numerous intermolecular effects, and interactions, which are believed to have important influences on the properties of charge-transfer salts of TCNQ. Among these are

(1) U_0, the Coulombic repulsion energy between two electrons on the same molecule;

(2) V_1, the Coulombic repulsion between electrons on neighboring molecules;

(3) $4t$, the bandwidth arising from the strong π-molecular overlap between molecules along the stack;

(4) J, the exchange interactions between neighboring spins;

(5) p, the number of unpaired electrons per TCNQ;

(6)   stoichiometry;

(7)   polarizability and size of the cations;

(8)   interactions between stacks (both overlap and Coulombic);

(9)   nature of cations (symmetric or asymmetric);

(10) disorder in stacks;

(11) steric interaction between stacks;

(12) Peierls and other lattice instabilities;

(13) coupling of electrons to intermolecular and intramolecular vibrations (vibronic effects); and

(14) coupling to librational modes (librons).

Torrance compiled values of p for the four classes of cations, where p is considered to be equal to $1/n$ where n is the number of anions in molecules when D (donor) is unity. For example, D–TCNQ$_n$, $p = 1/n$. Salts with 1:1 stoichiometry show $p = 1$; 1:2 stoichiometry, $p = \frac{1}{2}$, etc. If $p = 1$ (class I) conduction can occur down the stack at a cost of the large Coulombic energy U_0 causing these salts to be insulators (charge transfer is complete), e.g., KTCNQ. If the charge transfer is incomplete, and, as determined from x-ray and neutron data, $p = <1$, as for TTF–TCNQ, a highly conducting solid may be formed. For 1:2 complexes, $p = \frac{1}{2}$ or $<\frac{1}{2}$ (classes III and IV) and conductivities are not as high as for class II. Table IV summarizes this discussion. The insulator (class I) salts have complete charge transfer from the donor to TCNQ, and $p = 1$ electron per TCNQ. Conduction can occur down the stack at a cost of a large Coulombic energy (V) making these materials insulators. Class II salts have incomplete charge transfer of electrons from D $\rightarrow$ TCNQ so that $p < 1$ electrons per TCNQ molecule on the average. The unpaired electron can move along the stack without having to overcome large Coulombic interactions and these systems are conductors. Class III salts ($p = \frac{1}{2}$) are considered to have complete charge transfer and the limiting factor to conduction is the weaker interaction (V_1), and thus, conduction is intermediate. Class IV salts are considered to have incomplete charge-transfer and are not limited by high Coulombic energies. Class II and IV systems have a common incomplete charge transfer, and as a result are conductors.

For incomplete charge-transfer molecules, mixed-valence stacks are found. For example, for TCNQ the reaction with a donor might occur as

$$D^0 + TCNQ^0 \underset{I-A}{\overset{E_m}{\rightleftarrows}} D^+ + TCNQ^-$$

and both $TCNQ^0$ and $TCNQ^-$ occur, as well as D^0 and D^+: E_m is the binding energy of solid formed, I the ionization potential, and A the electron affinity; When I is low, E_m is large (e.g., alkali metal), and the equilibrium is displaced to the right in the direction of an ionic ground state. This favors the insulator state. When I is higher, E_m is lower (e.g., TTF), and the equilibrium is shifted toward the nonionic ground state, favoring

TABLE IV

Relationship of 1:1 versus 1:2 Salts of TCNQ with Respect to CT (Charge Transfer), ρ ($1/n$ When n Is the Number of Anions), and σ (Conductivity)

	Simple 1:1		Complex 1:2	
Class	I	II	III	IV
CT complete	Yes	No	Yes	No
ρ	1	<1	1/2	$<1/2$
σ	Low	High	Intermediate	High
Example	KTCNQ	TTF–TCNQ	TEA–$(TCNQ)_2$	$Q_n(TCNQ)_2{}^a$

^{a} Q is quinolinium.

the conductive state. Table V compares results for alkali–metal TCNQ versus TTF–TCNQ salts.

Figure 9 shows the oxidation potentials for a large number of donor molecules and reduction potentials for a variety of acceptors. This figure is useful in matching the donors with the acceptors to predict compound formation (Torrance, 1985). This can be quite helpful in designing a synthesis program for new charge-transfer electrical conductors.

The term characterized as the ionization potential (I) or ($I - A$) is not directly measurable experimentally. However, by measuring and comparing the electrochemical reduction potential (E_p) in solution, one can obtain an estimate of (I). Figure 10 plots $E_p(\alpha I)$ versus σ (conductivity) for a series of conductors, E_p being equivalent to E_{REDOX}. Torrance (1985) demonstrated that many of the differences in behavior of charge-transfer solids can be described and comprehended in terms of ΔE_{REDOX} ($\sim E_p$), which can be defined as the difference between the electrochemical oxidation potential of the donor and the reduction potential of the acceptor. The plot shows that the highly conducting compounds occur over a narrow range of ΔE_{REDOX} and correspond to materials showing mixed valences with $p < 1$. (Note: we are using p instead of ρ, which is the nomenclature Torrance uses, to prevent confusion with ρ = resistance.) The degree of charge transfer occurring in the donor–acceptor reaction is equivalent to ρ.

In 1979, using some of the variants previously discussed, it was discovered that when the TMTSF–DMTCNQ salt was subjected to a pressure of 10 kbar and then cooled, very high σ (conductivity) resulted. This process stabilized the conductivity of 2×10^5 (ohm cm)$^{-1}$ at low temperatures (Andrieux *et al.*, 1979). However, superconductivity was not observed. This result stimulated new interest in the field—the application of pressure at low temperature in search of superconductivity by suppression of metal–insulator transitions. In an attempt to explain these findings it was theorized that pressure might suppress the metal–insulator transition

TABLE V

Comparison of Alkali Metal TCNQ versus TTF–TCNQ Salts

Alkali-metal TCNQ	TTF–TCNQ
$\rho = 1$	$\rho < 1$
No mixed valence states	Mixed valence states
Ionic ground state	Nonionic state
σ = insulator	σ = conductor

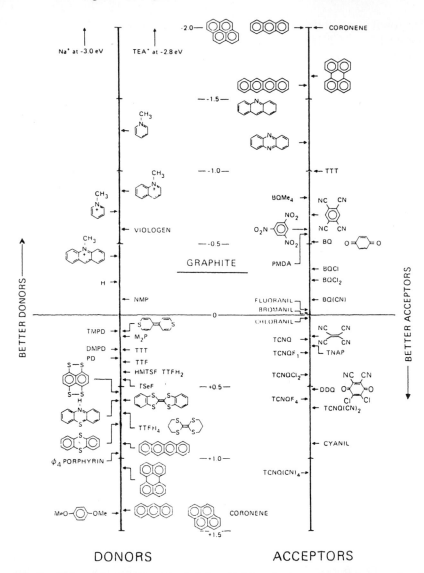

Fig. 9.   Values of oxidation potential (in volts) for a large number of donor molecules (left) and the reduction potential for a variety of acceptors (right). [From Torrance (1985).]

by suppressing a Peierls transition. HMTSF–TCNQ, HMTTF–TCNQ, HMTSF–TNAP, and (TSeT)$_2$Cl (tetraselenatetracene) subjected to pressure also retained their conductivities to ∼1 K (Andrieux *et al.,* 1979). The high conductivites obtained at low temperature were eventually attributed to a crossover from a one-dimensional metallic state to a three-dimensional semimetal state at lower temperature and high pressure.

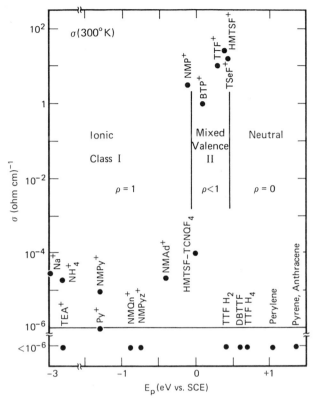

Fig. 10. Plot of electrochemical reduction potential (*Ep*) in solution versus conductivity. [From Torrance (1979).]

III. BECHGAARD SALTS—(TMTSF)₂X CONDUCTORS

In 1979 Bechgaard *et al.* (1980) began to study other salts of TMTSF involving the stoichiometry $(TMTSF)_2X$, where X is a monovalent inorganic anion, such as PF_6^-, AsF_6^-, TaF_6^-, NbF_6^-, ClO_4^-, ReO_4^-, BF_4^-, BrO_4^-, IO_4^-, NO_3^-, FSO_3^-, $CF_3SO_3^-$, and TeF_5^-. These studies were initiated to remove the properties of the molecular acceptor species from the solid in order to focus on TMTSF itself. These materials have since become known as the Bechgaard salts. They are the first organic molecular compounds to become superconductive, albeit at low temperature (~1 K) and high pressure (8–12 kbar), and have attracted considerable interest. The $(TMTSF)_2ClO_4$ salt was the first superconductive material ($T_c \simeq 1.3$ K) at *ambient pressure* and only when *slow cooled* (vide infra) (Bechgaard, 1982; Bechgaard *et al.*, 1981). Recently, a new class of superconductive organic materials with much higher transition temperatures (~5 K) at ambient pressure have been synthesized and are discussed in Section V.

A. Synthesis of TMTSF

The original synthesis of TMTSF required CSe_2 as a starting material (Bechgaard *et al.,* 1974); Anderson and Bechgaard, 1975; Bechgaard *et al.,* 1975) and that route has been patented (Cowan *et al.,* 1981). Since CSe_2 is difficult to handle and extremely malodorous (rotten radishes!), syntheses in which gaseous H_2Se replaced CSe_2 in the synthesis of TMTSF were subsequently reported using selenoureas (Shu *et al.,* 1977) or N,N-di-methylphosgeneiminium chloride (Wudl and Nalewajek, 1980; Wudl *et al.,* 1981). A synthesis based on the use of H_2Se, and which can easily be accomplished by students, has very recently become available (Braam *et al.,* 1986). However, gaseous H_2Se is extremely toxic, with approximately LC_{50} (30 min) in guinea pigs: 6 ppm, and must be handled with great care (Spector, 1956). Therefore, it is not surprising that a TMTSF synthesis has been developed that does not require either CSe_2 or H_2Se, but rather uses elemental selenium as shown in Fig. 11 (Moradpour *et al.,* 1982).

Although these improvements in the synthesis of TMTSF were welcome, it appears that the presence of a minor sulfur ($<1–2\%$) impurity is observed in any of the procedures, other than that using CSe_2, in which N,N-dimethylphosgeneiminium chloride is used as an intermediate. This sulfur impurity apparently suppresses the superconducting transition temperatures of $(TMTSF)_2X$ derivatives and when $X = ClO_4^-$ the T_c is reduced

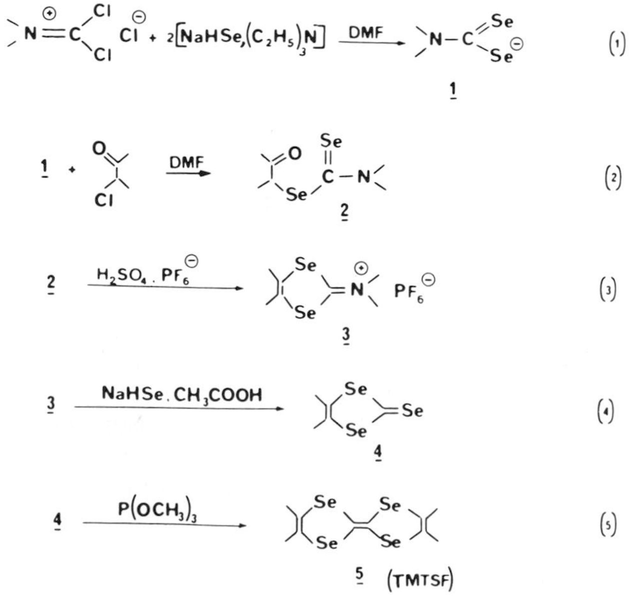

Fig. 11.   Synthetic scheme for TMTSF. [From Moradpour *et al.* (1982).]

by ~0.1–0.3 K from 1.3 K. Careful vacuum gradient sublimation of the N,N-dimethylphosgenciminium chloride apparently results in the removal of the sulfur contaminant (Moradpour *et al.*, 1985). In passing, it should be noted that the preparation of perdeuterio-TMTSF (Wudl *et al.*, 1981), tetraselenafulvalene-TCNQ (Engler and Patel, 1974), and dibenzotetra-selenafulvalene charge-transfer salts have also been reported (Lerstrup *et al.*, 1983; Johannsen *et al.*, 1983).

B. Electrocrystallization of 2:1 Derivatives of TMTSF

Only (2:1) (radical-cation: monovalent anion) derivatives of organic synmetals appear to form salts that exhibit superconductivity. Crystals of these conductors are produced using simple electrochemical oxidation techniques in an H-shaped cell, or a variation of this type cell, as shown in Fig. 12.

Solutions of the organic donor (TMTSF) and a salt of the desired anion, such as a tetraalkylammonium derivative to increase solubility in the

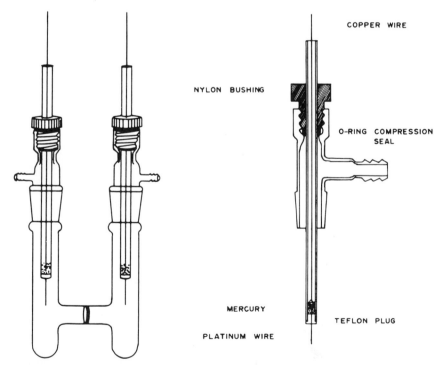

Fig. 12.   Cell and electrodes typically used in the electrocrystallization of organic conductors. [From Lee *et al.* (1982).]

organic medium, are prepared using redistilled, dried, and deoxygenated organic solvents such as tetrahydrofuran or 1,2,2-trichloroethane. Tetra-butylammonium (n-Bu_4N^+) salts are the most commonly used because high-quality crystals of ~0.5 mm in width and thickness, and with lengths of ~10 mm of TMTSF derivatives, are frequently grown. Some typical crystals are shown in Fig. 13.

Before electrocrystallization is initiated the solvent and anionic deriv-ative are placed in the cathode compartment of the cell while the solvent, anionic derivative, and organic donor are loaded in the anode (oxidizing) compartment. Platinum electrodes are then inserted in both compartments and oxidation, with concomitant crystal growth at the anode, is accom-plished using either constant voltage or constant current techniques. In the case of $(TMTSF)_2X$, crystals are grown on the anode according to the reaction

$$2n \cdot (TMTSF) + nX^- \xrightarrow{-n.e} [(TMTSF)_2X]_n$$

Generally speaking, only one crystalline phase grows when employing TMTSF and an octahedral or tetrahedral anion. Using the constant current technique (current = 1–3 μA) long shiny-black needle-shaped crystals form when TMTSF is used. In all cases the crystals formed have a metallic luster but may have electrical properties that vary from insulating to semiconducting to metallic in nature. A detailed description of the elec-

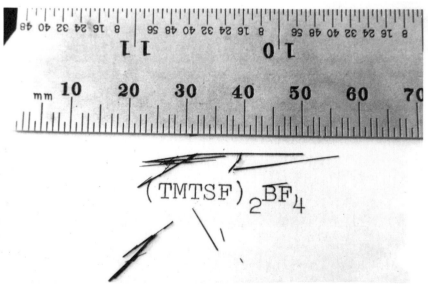

Fig. 13. Photograph of freshly harvested $(TMTSF)_2BF_4$ crystals which were grown using cell shown in Fig. 12. [From Williams and Carneiro (1985).]

trocrystallization of (TMTSF)$_2$ClO$_4$, including donor concentration, solvents used, current densities, etc. has been given (Bechgaard *et al.*, 1981), as well as for other TMTSF salts (Braam *et al.*, 1986). A brief review of the preparation and electrical properties of TMTSF derivatives, some with stoichiometries other than the common 2:1 phases, has been published (Bechgaard, 1982).

C. Crystal Structures of (TMTSF)$_2$X Conductors

Understanding the crystal structures of the known organic superconductors is essential because they provide some of the vital insight needed to unravel the variety of physical properties exhibited by these materials. Their understanding also provides a means for varying the electrical properties and for engineering new organic metals. For example, although *all* (TMTSF)$_2$X compounds possess the same triclinic (space group $P\bar{1}$) crystal structure at room temperature, which is an unusual characteristic of these materials, they may have vastly different properties at low temperature or high pressures. This is demonstrated by the fact that a pressure of ~9 kbar must be applied to (TMTSF)$_2$PF$_6$ before it will become superconducting at 0.9 K, (TMTSF)$_2$ReO$_4$ has an anion-assisted disorder–order transition at 180 K where it loses its metallic properties (unless pressure is applied which suppresses the transition), and slow-cooled (TMTSF)$_2$ClO$_4$ is the only known ambient pressure Se-based superconductor with $T_c = 1.3$ K (Bechgaard, 1982). The differences in physical behavior can, in part, be associated with very minute differences in the crystallographic structure. Thus, although the anion plays no obvious role in the actual conduction process, which occurs through a network of Se$\cdots$Se interactions (vide infra), it does cause pronounced changes in electrical properties which we shall discuss in a later section.

The crystal structure of (TMTSF)$_2$BrO$_4$ is shown in Fig. 14 as representative of the series (Williams *et al.*, 1982a). The basic architectural feature of the isostructural (TMTSF)$_2$X salts is the zigzag columnar stacking of nearly planar TMTSF molecules parallel to the high conductivity *a* axis (Bechgaard, 1982; Bechgaard *et al.*, 1981; Williams *et al.*, 1982a; Thorup *et al.*, 1981; Soling *et al.*, 1982; Tanaka *et al.*, 1983; Soling *et al.*, 1983; Rindorf *et al.*, 1982; Guy *et al.*, 1983; Williams *et al.*, 1983a; Williams *et al.*, 1983b). The structures of the salts are different from that of neutral TMTSF itself (Kistenmacher *et al.*, 1979). The TMTSF molecules in (TMTSF)$_2$X salts form stacks along the *a* crystallographic axis, which is also the direction of highest electrical conductivity. Quite by accident, these stacks also result in the formation of infinite two-dimensional *molecular sheets*, which extend in the *a–b* plane, with the TMTSF molecules

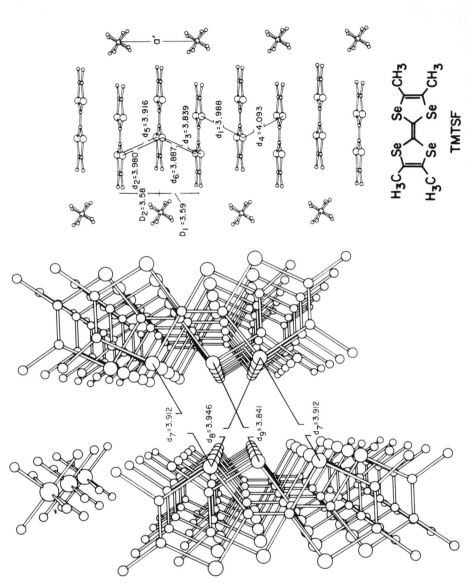

Fig. 14. Perspective views of the crystal structure of $(TMTSF)_2BrO_4$, looking down the stacks along the a axis (left), and perpendicular to them approximately along the b axis (right). [From Williams and Carneiro (1985).]

connected through interstack Se···Se interactions. These interactions result in added structural "dimensionality" in these systems beyond that provided solely by the one-dimensional stacking of TMTSF molecules. However, the TMTSF molecules themselves do not form a three-dimensional network because the sheets are separated along c by the anions (X). Possibly the most important structural feature that has been revealed from crystallographic studies performed at two temperatures (298 and 125 K) is the existence of an "infinite sheet network" (Williams et al., 1982a) of Se···Se interactions as shown in Fig. 15.

At room temperature the intermolecular intra- and interstack Se···Se distances are all similar and have values of 3.9–4.9 Å compared to the selenium atom van der Waals radius sum (Pauling, 1960) of ~4.0 Å. However, as the temperature is lowered (298 to 125 K) rather unusual changes occur, namely, the ratio of the decrease in the interstack:intrastack Se···Se distances is not unity but is approximately 2:1 (Williams et al., 1982a, 1983b). Thus, the distances between the "chains" shown in Fig. 15 decrease, on the average, by twice as much as the distances between TMTSF molecules within each stack. This most certainly leads to increased interchain bonding and electronic delocalization through the selenium atom network as the temperature is decreased (Whangbo et al., 1983).

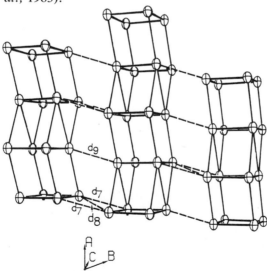

Fig. 15. Representation of the selenium atom "infinite sheet network" in (TMTSF)₂X salts. This network is the main pathway for electrical conduction. Lines connect selenium atoms that are closer than the van der Waals' radius sum (Se···Se) of 4.0 Å. Full lines indicate intrastack and broken lines indicate interstack distances d_7, d_8, and d_9. [From Williams and Carneiro (1985).]

Given the importance of the specific anions for the physical behavior of the $(TMTSF)_2X$ salts it seems worthwhile to dwell on the influence of their sizes and shapes on the crystallographic properties. Contrary to previous reports, even the centrosymmetric anions AsF_6^- and PF_6^- are in crystallographic disorder with their central atom most likely always residing at the inversion center ($\bar{1}$ site) in the triclinic unit cell (Williams *et al.*, 1983a). When the anion, X^-, is highly symmetric (octahedral or tetrahedral), the network of interstack $Se \cdots Se$ distances expands and contracts in a surprisingly predictable fashion as the size of the anion is varied. This network expansion, not surprisingly, is accompanied by systematic changes in the crystallographic unit cell volume. These features are demonstrated in Figs. 16 and 17. We define the anion volume V_A by using effective ionic radii (Huheey, 1978). Adopting the method of Shannon and Prewitt (1969) for deriving effective multiatomic radii we define V_A by the volume of the sphere that has the correct multiatomic radius. Hence, $V_A = \frac{4}{3}\pi(r_i + 2r_o)^3$, where r_i is the radius of the inner ion (e.g., Cl in ClO_4^-) and r_o is that associated with the outer ion.

By plotting known V_c's (where V_c is the unit cell volume) for six TMTSF salts ($X = PF_6^-$, ReO_4^-, BrO_4^-, ClO_4^-, BF_4^-, and FSO_3^-) versus V_A, a linear relation between V_c and V_A is observed, both at $T = 295$ K as shown in Fig. 16 and at $T = 125$ K in Fig. 17. This suggests that V_c may be accurately estimated, by using a linear least-squares fit, yielding a calculated volume, $V_{cp} = 0.65\ V_A + 645$ ($T = 298$ K) or $V_{cp} = 0.42\ V_A + 642$

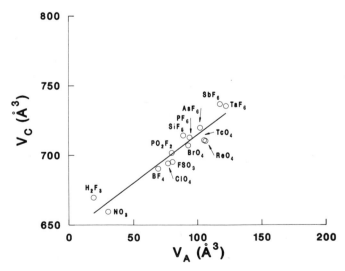

Fig. 16. Unit cell volume V_c of $(TMTSF)_2X$ salts at room temperature versus anion volume V_A. [From Williams and Carneiro (1985).]

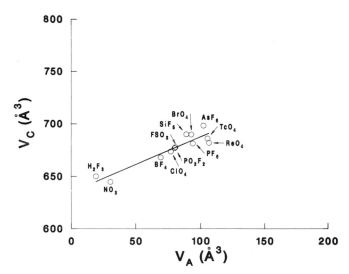

Fig. 17.   Unit cell volume V_c of (TMTSF)$_2$X salts at 120–125 K versus anion volume V_A. [From Williams and Carneiro (1985).]

($T = 125$ K) (Williams *et al.*, 1983b). For example, for ClO$_4^-$, $r_{Cl}^{7+} = 0.22$ Å and $r_O^{2-} = 1.21$ Å, the calculated V_c is 675 Å^3 and the observed V_c is 673.7 Å^3 ($T = 125$ K). Concerning bond lengths within an anion, the use of effective radii appears well justified because the results are identical to those obtained from molecular orbital calculations (Teramae *et al.*, 1982, 1983). But the empirical effective anion volume, i.e., its contribution to V_c is only about 50% [65% (298 K) and 42% (125 K)] of the calculated V_A. Values of these volumes are given in Table VI and it is concluded that this approach provides a very good measure of the anion volumes, in particular their relative sizes.

From this discussion it appears that the search for new superconducting (TMTSF)$_2$X derivatives, especially at ambient pressure, should center around those for which the unit cell volume is close to that of (TMTSF)$_2$ClO$_4$ [$V_c = 694.3$ Å (298 K) and 673.7 Å^3 (125 K)]. Thus, the equations and methodology given here are of practical use because, for any imaginable tetrahedral or octahedral anion, the unit cell volume can be predicted before the salt is prepared. For example, based on the equations above candidates for superconductivity might be (TMTSF)$_2$PO$_2$F$_2$ [$V_{cp} = 675.6$ Å^3 (125 K)], (TMTSF)$_2$CrO$_3$F [$V_{cp} = 679.1$ Å^3 (125 K)], and (TMTSF)$_2$WF$_6$ [$V_{cp} = 692.4$ Å^3 (125 K)]. While (TMTSF)$_2$PO$_2$F$_2$ has been prepared (Cox *et al.*, 1984; Eriks *et al.*, 1984), it undergoes a metal–insulator transition at 137 K and pressures up to 15 kbar do not induce superconductivity, most likely because of an unusual anion disorder that

TABLE VI

Internuclear Distances (r) and Anionic Volumes (V_A) From Molecular Orbital Calculations (MO) and Effective Ionic Radii (EIR), Compared to Experimental Values (EXP)

Anion		Distance (Å)			V_A (Å^3)	
		MO[a]	EIR	EXP[a]	EIR	EXP[b]
H_2F_3					19	25
NO_3	r_{N-O}	1.277	1.28	1.22–1.27	30	20
BF_4	r_{B-F}	1.380	1.42	1.40–1.43	70	45
ClO_4	r_{Cl-O}	1.542	1.44	1.41–1.43	77	49
FSO_3	r_{S-F}	1.601	1.42	1.55–1.555	81	50
	r_{S-O}	1.459	1.48	1.424–1.455		
BrO_4					93	62
PF_6	r_{P-F}	1.612	1.68	1.60–1.63	94	59
AsF_6	r_{As-F}	1.744	1.76	1.59–1.96	102	74
TcO_4					105	66
ReO_4					106	65
SbF_6	r_{Sb-F}	1.895	1.90	1.53–1.99	118	72
TaF_6					122	91
CF_3SO_3					153	95

[a] From Teramae *et al.* (1982, 1983).
[b] Room temperature value $= V_C - 645$ Å^3.

occurs below 137 K (Eriks *et al.*, 1984); the CrO_3F^- anion oxidizes TMTSF and a derivative cannot be prepared (Williams, 1985); and $(TMTSF)_2WF_6$ has not yet been prepared but it is expected that pressure would be required to induce superconductivity because of the larger (predicted) unit cell volume compared to that of $(TMTSF)_2ClO_4$. Another possible candidate is $(TMTSF)_2CF_3O$ because the CF_3O^- anion has a V_C similar to ClO_4^-.

Important further illumination of the relationships between the structural and transport properties will be provided in the future by crystallographic studies performed under pressure. As yet, very limited information is available. However, from studies of $(TMTSF)_2PF_6$, the volume compressibility is estimated to be 0.5%/kbar (Morosin *et al.*, 1982). It, therefore, requires ~6 kbar to compress $(TMTSF)_2PF_6$ into the same volume as $(TMTSF)_2ClO_4$ which is in rough agreement with the critical pressure for superconductivity in the PF_6^- compound.

As indicated above, the major structural changes upon cooling involve the interstack Se $\cdots$ Se distances (d_7, d_8, and d_9 in Fig. 15), which are given in Table VII. Inspection of the interstack Se $\cdots$ Se contact distances in Table VII reveals that upon cooling large decreases in Se $\cdots$ Se distances occur, which are as much as 0.30 Å less than the van der Waals radius

TABLE VII

Interstack Se $\cdots$ Se Contact Distances (Å) and Unit Cell Volume (Å³)
for (TMTSF)₂X Salts $\left(T = \dfrac{298 \text{ K}}{125 \text{ K}}\right)$

Anion (X^-)	d_7	d_8	d_9	V_c (Å³)
AsF_6^-	3.9449 (9)	3.9627 (11)	3.9053 (13)	719.9
	3.8159 (5)	3.8861 (7)	3.7894 (7)	695.9
PF_6^-	3.9342 (20)	3.9586 (27)	3.8786 (28)	714.3
	3.7847 (14)	3.8706 (22)	3.7413 (22)	681.3
ReO_4^-	3.902 (2)	3.933 (2)	3.827 (2)	710.5
	3.794 (4)	3.845 (5)	3.699 (4)	681.9
BrO_4^-	3.9118 (9)	3.9457 (14)	3.8411 (13)	707.2
	3.8149 (8)	3.8618 (12)	3.7216 (11)	689.8
FSO_3^-	3.8676 (7)	3.9516 (11)	3.7815 (9)	695.3
	3.7565 (5)	3.8382 (8)	3.6601 (7)	677.7
ClO_4^-	3.8653 (14)	3.9553 (24)	3.7783 (20)	694.3
	3.7596 (5)	3.8485 (9)	3.6549 (8)	673.7
BF_4^- [a]	3.850	3.978	3.743	690.4
	3.7526 (11)	3.8792 (17)	3.6394 (15)	668.1

[a] Values in the numerator were given without estimated standard deviations [Kobayashi et al. (1982)].

sum for Se, suggesting considerable bonding interaction (Whangbo et al., 1983). It has also been observed (Williams et al., 1983b) that the Se $\cdots$ Se distances are anion dependent, and vary systematically depending on the anion size, suggesting that *correlations* between crystallographic unit cell volumes, which reflect anion size, and interstack Se $\cdots$ Se distances might exist. As shown in Fig. 18 there is a striking correlation between the anion volume V_A (and hence, unit cell volume V_c) and the average interstack Se $\cdots$ Se distance [$d_{avg} = (2d_7 + d_9)/3$] in (TMTSF)₂X metals and these structural features also correlate well with the observation of pressure induced superconductivity in the majority of these systems.

Within the series of octahedral and tetrahedral anions the ClO_4^- anion has a small volume V_A, and also a very small d_{avg} and V_c. It is, therefore, tempting to make (TMTSF)₂X salts with even smaller anions than ClO_4^-, and examples of such ions are the triangular (pancake-shaped) NO_3^- and (banana-shaped) $H_2F_3^-$. It is now interesting to compare the unit cell volumes V_c and the interstack distances d_{avg} of these compounds to what is expected from the larger and more symmetric anions treated above. From Figs. 16 and 17 it appears that V_c behaves regularly, i.e., it follows

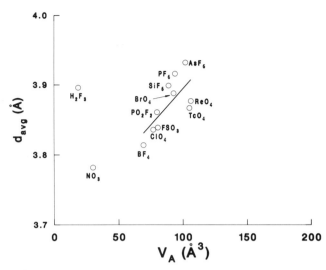

Fig. 18.   Average interstack distance, d_{avg}, (see text) in $(TMTSF)_2X$ salts at room temperature versus anion volume V_A. [From Williams and Carneiro (1985).]

the usual linear behavior versus V_A. However, the d_{avg} values depart appreciably from the expected dependence as is shown in Fig. 18. Hence, although the very small anions NO_3^- and $H_2F_3^-$ do make the unit cell volume smaller than that for $X = ClO_4^-$, the interstack selenium atom network does not contract accordingly. It is tempting to relate this structural observation with the fact that superconductivity is not observed in $(TMTSF)_2NO_3$ and $(TMTSF)_2H_2F_3$ (Williams *et al.*, 1983c). Parkin (1984) has given another opinion in attempting to explain the behavior of these salts. He plotted the unit cell volume of the $(TMTSF)_2X$ salt versus that of the same TMTTF salt. A straight line was obtained as the anion was varied indicating that the unit cell volume is determined by the molecular volume plus a contribution from the anion volume.

Another possibly important empirical correlation between crystal structure and the occurrence of superconductivity in $(TMTSF)_2X$ materials may be related to short cation–anion contact distances which occur through Se–F or Se–O interactions (Thorup *et al.*, 1983). In the $(TMTSF)_2X$ series superconductivity has definitely been observed for $X = PF_6^-$, AsF_6^-, TaF_6^-, SbF_6^-, ReO_4^-, and ClO_4^-; in the latter case superconductivity is observed at ambient pressure. The van der Waals contact distances for $Se \cdots Se$, $Se \cdots F$, and $Se \cdots O$ are approximately 4.00, 3.35, and 3.40 Å, respectively (Pauling, 1960). The cation–anion contacts through Se–(F, O) are given in Table VIII and they reveal very short contact distances except for $(TMTSF)_2NO_3$ which does not undergo a superconducting transition.

TABLE VIII

Short Cation–Anion Se–(O, F) Distances in (TMTSF)₂X Materials
at Room Temperature

	Anion (X⁻)					
	NO_3^-	ClO_4^-	ReO_4^-	PF_6^-	TaF_6^-	$PO_2F_2^-$
Se-(O, F) Å	3.94	3.34	3.16	3.23	3.09	2.91[a]

[a] This short contact, which is the shortest Se · · · O distance observed in any (TMTSF)₂X salt, arises because below the metal–insulator transition at 137 K (Cox *et al.*, 1984), the anion is shifted off the center of symmetry resulting in a crystallographic disorder (Eriks *et al.*, 1984).

Since the short contacts are approximately in the c direction in the crystal, Parkin *et al.* (1982) used these observations to suggest that it is the length of the c axis that determines the critical pressure for superconductivity. However, compressibility studies have firmly established that it would require a much higher pressure to compress the c axis length of (TMTSF)₂PF₆ into that of (TMTSF)₂ClO₄ than the pressure needed for superconductivity (Morosin *et al.*, 1982). Hence, the c-axis criterion does not appear to play a determining role for the occurrence of superconductivity in (TMTSF)₂X salts.

Another interesting aspect concerning the detailed structures of (TMTSF)₂X materials and the role the anion plays, in addition to the correlations previously described and the finding of short Se–(F, O) anion distances discussed above, is the very recent observation that the peripheral atoms of the anions, namely, AsF_6^-, PF_6^-, and ClO_4^-, are involved in weak van der Waals interactions with the hydrogen atoms of the methyl groups indicative of weak hydrogen-bond-like interactions (Williams *et al.*, 1983b; Beno *et al.*, 1983). For example, the immediate nearest neighbor environment about the disordered (Williams *et al.*, 1983b) octahedral AsF_6^- anion in (TMTSF)₂AsF₆ reveals a nearly isotropic (symmetric) sea of nearby ($d < 2.6$ Å) hydrogen atoms (see Fig. 19) arising from the fact that in these materials the anion resides in a "methyl-group H atom cavity" (Beno *et al.*, 1983). By contrast, the tetrahedral ClO_4^- anion in (TMTSF)₂ClO₄ possesses a very asymmetric methyl-group H atom environment as shown in Fig. 20. This asymmetric distribution of oxygen atom to methyl-group H atom [H_2C—H · · · $OClO_3^-$] bonding interactions results in a "pinning" of the anion which may be associated with the critical anion-ordering phase transition, a necessary prerequisite to superconductivity in (TMTSF)₂ClO₄, observed at 24 K using x-ray diffraction techniques (Pouget *et al.*, 1983a). As illustrated in Fig. 20, the lower relative thermal motion, as observed in their thermal vibration ellipsoids, of O(1)

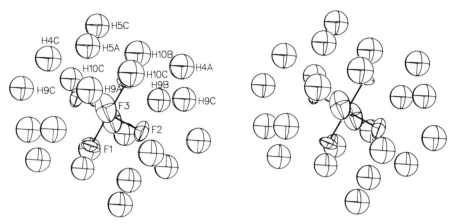

Fig. 19.   The H bonding environment (stereoview) about the AsF_6^- anion, derived from the low-temperature (125 K) x-ray crystal structure of $(TMTSF)_2AsF_6$, is far more symmetrical than that of the ClO_4^- anion. For clarity the AsF_6^- anion has been drawn in an *ordered* configuration with six F atom positions. A disordered model with twelve partially occupied fluorine positions gives similar results. [From Williams *et al.* (1983a).]

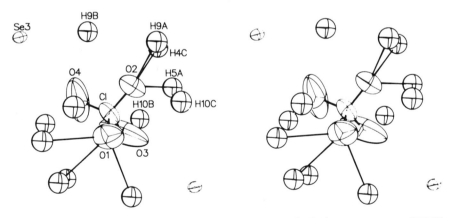

Fig. 20.   A stereoview of the ClO_4^- anion environment in the low-temperature (125 K) x-ray-determined crystal structure of $(TMTSF)_2ClO_4$. Short H_2C—$H \cdots O$—ClO_3^- hydrogen bonding intersections (drawn as faint lines for $O \cdots H < 3.0$ Å) exist for $0(1)$ and $0(2)$ which limit the thermal motion of these atoms and may be responsible for "pinning" the ClO_4^- anion in the lattice. [From Beno *et al.* (1983) and Williams and Carneiro (1985). Reprinted with permission of *Solid State Commun.*, **48**, 99, M. A. Beno, G. S. Blackman, P. C. W. Leung, and J. M. Williams, Hydrogen bond formation and anion ordering in superconducting $(TMTSF)_2ClO_4$, $(TMTSF)_2AsF_6$, Copyright 1983, Pergamon Press, Ltd.]

and 0(2) compared to 0(3) and 0(4), undoubtedly results from the greater involvement of 0(1) and 0(2) in what could be termed "weak H bond formation." It has also been proposed that the anion-ordering phenomena observed in many (TMTSF)$_2$X compounds may be associated with a precursor methyl-group ordering which occurs at low temperature in these materials (Beno *et al.*, 1983). If this is the case, then the synthesis of new superconducting materials requires anions that interact with the methyl groups in such a fashion that they produce anion-ordered derivatives. It is, perhaps, pertinent to point out that controlling the formation C—H$\cdots$X interactions in a crystal is an extremely difficult task.

As a final point regarding the structure of (TMTSF)$_2$X conductors, it seems worth mentioning that contrary to a previous report (Emery *et al.*, 1982), there is no significant dimerization in the (TMTSF) stacks down to $T = 125$ K, as indicated by the interplanar distances D_1 and D_2 given in Table IX (Williams *et al.*, 1983c). Such a dimerization could influence the electronic properties, since it would create a gap in the electronic spectrum at the wave vector $2k_F (= \pi/a)$, where k_F is the Fermi vector (see Chapter 7). Usually only a gap at k_F will affect transport properties of a metal, but

TABLE IX

Interplanar Distances in Å for (TMTSF)$_2$X Salts $\left(T = \dfrac{298 \text{ K}}{125 \text{ K}} \right)$

Anion[a] (X$^-$)	D_1[b]	D_2[b]	$\Delta(D_1 - D_2)$
AsF$_6^-$	3.65	3.62	0.03
	3.57	3.57	0.00
PF$_6^-$	3.66	3.63	0.03
	3.59	3.59	0.00
ReO$_4^-$	3.64	3.64	0.00
	3.59	3.56	0.03
BrO$_4^-$	3.63	3.65	-0.02
	3.59	3.58	0.01
FSO$_3^-$	3.62	3.63	-0.01
	3.58	3.57	0.01
ClO$_4^-$	3.63	3.63	0.00
	3.58	3.57	0.01
BF$_4^-$	3.63	3.63	0.00
	3.57	3.55	0.02

[a] From Williams and Carneiro (1985), except for (TMTSF)$_2$ReO$_4$ (Rindorf *et al.*, 1984).
[b] The interplanar distances D_1 and D_2 are the distances between the best plane for the Se atoms of a TMTSF molecule.

if Coulomb repulsion between electrons is of importance, the "dimerization gap" is effective and decreases conductivity. With this in mind, it should be noted that both the dimerization (observed from crystallographic studies) and Coulomb repulsion effects (derived from transport measurements) are small in $(TMTSF)_2X$ salts but appear appreciable in the sulphur family of $(TMTTF)_2X$ materials.

D. Discussion of $(TMTSF)_2X$ Conductors

The first Bechgaard salt synthesized was $(TMTSF)_2PF_6$ (Bechgaard *et al.*, 1980), which was found to be superconductive at 0.9 K and ~6.5 kbar pressure (Parkin *et al.*, 1981). The compound was also found to exhibit a Meissner effect (Andres *et al.*, 1980), which is a true test of volume superconductivity. These results were so promising that a whole family of charge-transfer organic cation-radical salts was synthesized (Bechgaard *et al.*, 1980, vide supra). In the past five years the Bechgaard salts have garnered a major portion of the research interest in organic metals because of their promise as new synmetal superconductors (see Greene and Street, 1984).

Several significant results have been forthcoming from this research:

(1) high low-temperature conductivities, becoming superconductive with application of modest levels of pressure; $(TMTSF)_2ClO_4$ being the only selenium based superconductor found at ambient pressure (1.3 K) (Bechgaard *et al.*, 1981, 1982);

(2) low $T_{M \to I}$ transitions at <20 K; exceptions being $(TMTSF)_2BF_4$ (41 K) and $(TMTSF)_2ReO_4$ (180 K) (Bechgaard *et al.*, 1981, 1982); it has been suggested that the $T_{M \to I}$ transition is controlled by the electronegativity of the anion in the $(TMTSF)_2X$ family [according to Wudl (1984)]; in later discussion it will be seen that numerous other factors are involved;

(3) insulating ground states created by spin-density wave (SDW) formation and by anion-ordering mechanism rather than by a Peierls charge-density wave (CDW) interaction; and

(4) unique phase transitions upon application of a magnetic field.

Table X summarizes the results for the Bechgaard salts, tabulating $T_{M \to I}$, origin of insulating state, T_C, and P_{SC} (Bechgaard, 1982). It may be observed that the octahedral anions possessing a center of symmetry show $T_{M \to I}$ at <20 K at ambient pressures. Anions with less symmetry (e.g., ReO_4^- and BF_4^-) show $T_{M \to I}$ at higher temperature, with $(TMTSF)_2ClO_4$ being exceptional. The salts with non-(2:1)-stoichiometries have metal–insulator transitions, and show no superconductivity.

TABLE X

Tabulation of (TMTSF)$_2$X Conductors with Respect to $T_{M \to I}$ and T_c[a]

Anion	$T_{M \to I}$ (K)	Origin[b]	T_c (K)	P_{SC} (kbar)
Octahedral				
PF$_6$	12–17	SDW	0.9	~6.5
AsF$_6$	12–16	SDW	1.1	12
SbF$_6$	17	—	0.4	11
TaF$_6$	—	—	1.4	12
NbF$_6$	12	—	Absent	12
Tetrahedral				
ClO$_4$	Absent	—	1.4	1 (bar)
ReO$_4$	180	Disorder–order	1.3	9.5
BF$_4$	40	—	—	
BrO$_4$	Metallic	—		
IO$_4$	Semiconductor	—		
Planar				
NO$_3$	12	SDW	Absent	12
[Perdeuterio-(TMTSF)$_2$ClO$_4$ behaves similarly]				
Nonsymmetrical anions				
SO$_3$F	Soft transitions	—		
CF$_3$SO$_3$	Soft transitions	—		
TeF$_5$	Conductor at 5 K	—		
Other stoichiometries				
Br	(1:0.8) semiconductor			
SCN	(1:0.5) semiconductor			
Br$_3$	(1:0.8) semiconductor			
NO$_3$	(1:1) insulator			
Δ_{R-SC_3}[c]	(1:1) insulator			
ϕ_{SiF_6}[d]	(2:1) insulator			
TeF$_5$	(3:2) semiconductor			

[a] Bechgaard (1982).
[b] Origin of metal–insulator transition.
[c] Δ_R = Phenyl or p-tolyl.
[d] ϕ = divalent anion.

The origin of the $T_{M \to I}$ transition in the octahedral anion systems appears to be associated with an antiferromagnetic transition (SDW) rather than a Peierls instability (Parkin *et al.*, 1982). In the case of (TMTSF)$_2$ReO$_4$ the transition is associated with an anion ordering. Figure 21 shows the various interactions possible and responsible for the metal–insulator transition. Why the magnetic (SDW) state is favored over the Peierls state in the Bechgaard salts is not well understood at present.

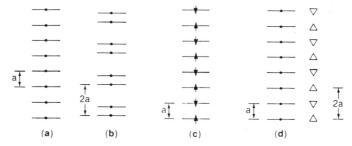

Fig. 21.   Several possible configurations of a linear chain of molecules with one electron in the highest orbital: (a) metal with uniform spacing of lattice constant a; (b) insulator with dimerization caused by Peierls transition; (c) SDW insulator with spin (↑) periodicity caused by Coulomb interactions; (d) insulator with periodicity caused by ordering of nonsymmetric anions (Δ). [From Greene and Street (1984). Copyright 1984 by the AAAS.]

E.   X-Ray Diffuse Scattering Studies of Anion Ordering in Bechgaard Salts

In the mid 1960s theoreticians predicted that in a quasi-one-dimensional metal, characterized as having a one-dimensional gas of weakly interacting electrons, instabilities could arise leading to transitions with various ground states, such as charge-density (CDW), spin-density wave (SDW) or superconducting (Bychkov *et al.*, 1966). Previously, it had been predicted that electron–phonon coupling in a one-dimensional metal would lead to a CDW state [the Peierls instability (Peierls, 1955)], and in this case a structural distortion accompanies the formation of an insulating state. In a simple case such as that which occurs in one-dimensional platinum chain systems (Williams *et al.*, 1982b), the lattice instability is produced by the softening of a phonon branch at wave vectors of component $2\mathbf{k}_F$ ($\mathbf{k}_F$ is the Fermi wave vector) in the direction of the chains therby forming a Kohn anomaly in the phonon spectrum (Kohn, 1959).

Because the amplitude of the lattice distortion in the Peierls insulator is very small, the x-ray scattering associated with it is very weak and yet this technique offers a unique probe of anion-ordering phenomena. Furthermore, the one-dimensional nature of this distortion gives rise to diffuse Bragg planes instead of the usual well-defined Bragg reflections. These two facts have led to the development of a special diffuse x-ray photographic technique often referred to as the "monochromatic Laue technique" or XDS (for x-ray diffuse scattering) (Cómes, 1975; Renker *et al.*, 1975; Renker and Cómes, 1975; Megtert *et al.*, 1978).

For (TMTSF)$_2$X salts the one-dimensional lattice distortion seems to play a less important role than in other quasi-one-dimensional conductors, although faint diffuse $2\mathbf{k}_F$ scattering has been observed in a few systems (Cómes, 1975; Renker *et al.*, 1975; Renker and Cómes, 1975; Megtert *et al.*, 1978; Mortensen *et al.*, 1983; Pouget *et al.*, 1982). Therefore, the usual Peierls instability is not (or only weakly) active in (TMTSF)$_2$X materials. This is easy to understand in the case of centrosymmetric anions (TMTSF)$_2$X (X = PF$_6^-$, AsF$_6^-$, SbF$_6^-$, TaF$_6^-$) since they undergo phase transitions to an antiferromagnetic SDW ground state (Pouget *et al.*, 1982; Torrance *et al.*, 1982), a transition which according to theory does not involve electron–phonon coupling. Under applied pressure the SDW state is suppressed resulting in a superconducting ground state (see Section VIC and Chapter 8). Hence, for centrosymmetric anions no new structural features appear at low temperatures.

However, the (TMTSF)$_2$X derivatives containing noncentrosymmetric anions (X = ClO$_4^-$, ReO$_4^-$, FSO$_3^-$, H$_2$F$_3^-$, BrO$_4^-$, NO$_3^-$) show structural phase transitions at relatively high temperatures compared to the temperatures associated with magnetic (SDW) or superconducting transitions (Moret *et al.*, 1983). These phase transitions are generally associated with anion-ordering phenomena. In this regard it should be remembered that for the noncentrosymmetric anion cases, the anion is located at an inversion center (space group P$\bar{1}$), resulting in orientational disorder of the anion at ambient temperature, and XDS studies show that as the temperature is reduced, anion-ordering phase transitions occur (Moret *et al.*, 1983). The transition temperatures, associated wave vectors and superstructure unit cells are given in Table XI (Mortensen *et al.*, 1983; Pouget

TABLE XI

Superstructures in (TMTSF)$_2$X and (BEDT–TTF)$_2$X Salts[a]

Salt	Anion symmetry	Transition temperature T_0 (K)	Wave vector[b]	Unit cell[b]
(TMTSF)$_2$H$_2$F$_3$		63	$\frac{1}{2}, \frac{1}{2}, \frac{1}{2}$	$2a, 2a, 2c$
(TMTSF)$_2$NO$_3$	Triangular	41	$\frac{1}{2}, 0, 0$	$2a, b, c$
(TMTSF)$_2$ClO$_4$	Tetrahedral	24	$0, \frac{1}{2}, 0$	$a, 2b, c$
(TMTSF)$_2$BrO$_4$	Tetrahedral	~250	$\frac{1}{2}, ?, ?$	$2a, ?, ?$
(TMTSF)$_2$ReO$_4$	Tetrahedral	177	$\frac{1}{2}, \frac{1}{2}, \frac{1}{2}$	$2a, 2b, 2c$
(TMTSF)$_2$FSO$_3$		88	$\frac{1}{2}, \frac{1}{2}, \frac{1}{2}$	$2a, 2b, 2c$
(ET)$_2$ReO$_4$	Tetrahedral	<300	$0, \frac{1}{2}, 0$	$a, 2b, c$
(ET)$_2$ClO$_4 \cdot$(TCE)$_{0.5}$	Tetrahedral	≈200	$\frac{1}{2}, \frac{1}{2}, \frac{1}{2}$	$2a, 2b, 2c$

[a] Taken in part from Williams and Carneiro (1985).
[b] Referred to the room temperature (TMTSF)$_2$X unit cell. Also for (ET)$_2$ClO$_4$ and (ET)$_2$ReO$_4$.

et al., 1983a,b; Tomíc *et al.*, 1983; Pouget *et al.*, 1981; Jacobsen *et al.*, 1982; Moret *et al.*, 1982; Wudl *et al.*, 1982; Takahashi *et al.*, 1982).

With the exception of $X = NO_3^-$ and ClO_4^-, these compounds exhibit a doubling of their crystallographic axes at the phase transition corresponding to a superstructure with wave vector $(\frac{1}{2}, \frac{1}{2}, \frac{1}{2}) \cong (2\mathbf{k}_F, \frac{1}{2}, \frac{1}{2})$. This has the usual symmetry of the Peierls distortion. A metal–insulator transition occurs simultaneously with the lattice ordering and as such, the ground state is indistinguishable from a CDW. But the lack of one-dimensional XDS above the transition suggests that the anions play a direct role in establishing the ground state. The two salts with anomalous superstructures also have peculiar electronic properties. Thus, $(TMTSF)_2NO_3$ with its $(\frac{1}{2}, 0, 0)$ superstructure has a SDW ground state and $(TMTSF)_2ClO_4$ with its $(0, \frac{1}{2}, 0)$ superstructure becomes superconducting only when *slowly cooled* through the anion-ordering transition at 24 K (Gubser *et al.*, 1982; Pouget *et al.*, 1983a,b).

Except in the case of $(TMTSF)_2ClO_4$, the only ambient pressure organic superconductor based on TMTSF, the period of the *a* axis (organic molecule chain axis) doubles at the structural transition. For $(TMTSF)_2ClO_4$ this is an important observation considering the band structure of these materials since $0.5a^*$ (a^* is the reciprocal lattice vector) corresponds to $2\mathbf{k}_F$ for the one-dimensional electron system, i.e., one electron is shared by two molecules in the TMTSF molecular chain. This could lead to the possible opening of a gap at the Fermi level thereby leading to an insulating state. Again, $\mathbf{k}_F$ is the in-chain Fermi wave vector for, or weakly interacting, independent electrons. However, $(TMTSF)_2ClO_4$ undergoes a transition which may be characterized as being of the $4\mathbf{k}_F$ wave vector type which could result from another type of instability of the electron gas arising from Coulomb interactions between electrons. This $4\mathbf{k}_F$ instability can be coupled to the lattice and induce the softening of the $4\mathbf{k}_F$ phonon. However, the lack of other evidence of strong Coulomb interactions (not the least that superconductivity occurs) suggests a nonelectronic origin of the superstructure. As discussed earlier, in slowly cooled $(TMTSF)_2ClO_4$ (R = relaxed state) samples only, anion ordering is a precursor to superconductivity and if samples are cooled suddenly (Q = quenched state) superconductivity does not develop (Fournel *et al.*, 1981). Thus, fast cooling of the sample suppresses superconductivity and stabilizes the SDW ground state and associated antiferromagnetism. The situation is not yet well understood in terms of the detailed structural changes in $(TMTSF)_2ClO_4$ and this is a topic of intense investigation. In the only other detailed XDS study, of $(TMTSF)_2ReO_4$, the situation is complicated because the ReO_4^- tetrahedra are both ordered below the 180 K phase transition and displaced from their centrosymmetric high-temperature position and this is accompanied by a $2\mathbf{k}_F$ distortion of the TMTSF stack (Moret *et* *al.,* 1982; Fournel *et al.,* 1981). It should be noted that from a crystallographic point of view, it is often technically difficult to arrive at a refined crystal structure based on both the usual strong reflections, and the very weak superstructure reflections, unless nontrivial structural changes have occurred.

The similarity in ordering transitions between some (TMTSF)$_2$X salts and (ET)$_2$ReO$_4$ makes it interesting to look into the structure of the latter salt using as a reference point the room temperature structure of a (TMTSF)$_2$X salt. Bis(ethylenedithio)tetrathiafulvalene (ET) is discussed in Section V. As indicated in Table XI the room temperature structure of (ET)$_2$ReO$_4$ resembles the $(0, \frac{1}{2}, 0)$ superstructure, i.e., that of (TMTSF)$_2$ClO$_4$ below 24 K. However, transport measurements suggest a new ordering in (ET)$_2$ReO$_4$ at 81 K, so the analogy is not entirely clear between the structural characteristics of the two superconductors. However, (ET)$_2$ReO$_4$ requires pressure to suppress the transition at 81 K while (TMTSF)$_2$ClO$_4$ requires no pressure to achieve the superconducting state.

Diffuse x-ray scattering experiments reveal a superstructure in (ET)$_2$ClO$_4$(TCE)$_{0.5}$ below 200 K of wave vector $(\frac{1}{2}, 0, \frac{1}{2})$ which is associated with anion ordering (Kagoshima *et al.,* 1983). When compared to the (TMTSF)$_2$X structure, it corresponds to a transition from $(0, \frac{1}{2}, 0)$ at high temperatures to $(\frac{1}{2}, \frac{1}{2}, \frac{1}{2})$ at low temperature without any noticeable changes in conductivity or susceptibility. Hence, in this case, where the two basic structures of (ET)$_2$ClO$_4$(TCE)$_{0.5}$ and of (TMTSF)$_2$X are very different; there is no analogy between the effects of their superstructures.

In conclusion, XDS studies of numerous TMTSF systems have provided a great deal of information on the nature of the anion-ordering transitions that occur at various temperatures. However, detailed single-crystal structural analyses are still required in order to determine the precise structural changes associated with these transitions.

IV.   (TMTTF)$_2$X CONDUCTORS

Parkin *et al.* (1982) have reported a second family of organic charge-transfer salts of the form, (TMTTF)$_2$X, which are isostructural to the (TMTSF)$_2$X materials. The only difference involves sulfur atoms replacing the selenium atoms. The charge transfer in the salts is determined by the valence stacks of the anion X, which occur in a 2:1 ratio for singly charged anions, and the conduction band is half-filled for both types of materials.

The similarity in chemical and crystallographic structures of TMTSF and TMTTF families suggest similar electrical properties, but at ambient pressure they are quite different; (TMTSF)$_2$X salts have higher room temperature conductivities [~300–500 versus ~50–300 (ohm cm)$^{-1}$]. The (TMTTF)$_2$X salts demonstrate higher $T_{M\rightarrow I}$ transition temperatures, especially where no anion ordering is in effect.

The electrical properties of (TMTTF)$_2$X salts are very sensitive to applied pressure, and at sufficiently high pressures, they show similar behavior to those demonstrated by (TMTSF)$_2$X salts. However, superconductivity has not yet been observed in (TMTTF)$_2$X systems. Both (TMTTF)$_2$PF$_6$ and (TMTTF)$_2$ClO$_4$ show room temperature conductivity maxima at ~230 K and 1 kbar. With elevated pressure (~30 kbar) the $T_{M\rightarrow I}$ transition is lowered to 10 and 30 K, respectively (see Table XII). The order–disorder transition in (TMTTF)$_2$ClO$_4$ observed at 75 K without pressure is frozen out with pressure. Maximum conductivity for (TMTTF)$_2$Br occurs at 100 K at ambient pressure and it undergoes a transition to a metallic state at 22 kbar and ~10 K, with a high conductivity state at 25 kbar and 4 K [$\sigma = 5 \times 10^5$–10^6 (ohm cm)$^{-1}$]. Figure 22a,b illustrates the normalized resistence plot for (TMTTF)$_2$Br at 25 kbar (Jérome and Schulz, 1982). Thus, to date, superconductivity has not been observed in any TMTTF derivative.

V. (BEDT–TTF)$_2$X OR (ET)$_2$X CONDUCTORS

Another modification of the TTF molecule occurred in 1978 when BEDT–TTF, or ET [bis(ethylenedithio)tetrathiafulvalene], was first reported by Mizuno *et al.* (1978) (see Fig. 1 for structure). The molecule contains eight sulfur atoms, twice that of TMTTF. Thus, ET is a significant electron donor, and a complete new family of electrical conductors has since been synthesized having the stoichiometry of (ET)$_2$X, where again X is the charge-compensating monovalent anion (vide infra). The discovery of metallic conductivity down to $T = 1.4$ K in (ET)$_2$ClO$_4$(TCE)$_{0.5}$ provided the impetus that caused numerous studies of the ET:X system (Saito *et al.*, 1982).

TABLE XII

(TMTTF)$_2$X Materials Under Pressure

Anion	$T_{M\rightarrow I}$ (K)		P (kbar)
	Ambient pressure	With pressure	
PF$_6^-$	230	10	30
ClO$_4^-$	230	30	30
Br$^-$	100	10	22

V.   (BEDT–TTF) or (ET)$_2$X Conductors   **45**

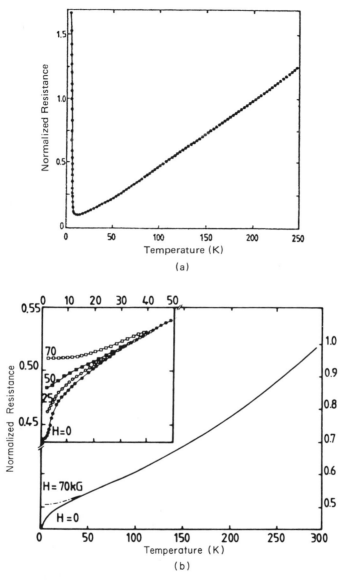

Fig. 22.   (a) Normalized resistance versus temperature for (TMTTF)$_2$Br at 22 kbar. [From Parkin *et al.* (1982).] (b) Normalized resistance versus temperature for (TMTTF)$_2$Br at 25 kbar. Inset shows the result of the application of a magnetic field at 25, 50, and 70 kG along some arbitrary direction perpendicular to the *a* axis. [From Parkin *et al.* (1982).]

A.   The Synthesis of BEDT–TTF (ET) and Chalcogenide Derivatives

Sulfur based ET and alkyl derivatives were prepared (Mizuno 1978; Krug 1977) in the late 1970s using CS_2 as starting material. A preparative procedure suitable for student use that involves the reduction of CS_2 with metallic sodium has also been developed (Reed et al., 1985) as indicated in Fig. 23. Preparative schemes for the systematic modification of the ET framework through the insertion of selenium, and sulfur-selenium combinations, have been advanced quite recently and some derivatives have been prepared (Schumaker et al., 1983). At this time the derivatives of ET that form superconductors are for β-$(ET)_2X$, X is trihalide (I_3^-) ($T_c \simeq$ 1.5 K) (Yagubskii et al., 1984b; Crabtree et al., 1984; Williams et al., 1984a; Carlson et al., 1985a), IBr_2^- ($T_c = 2.8$ K) (Carlson, 1985b; Williams et al., 1984b), AuI_2^- ($T_c = 5$ K) (Williams et al., 1985; Wang et al., 1985), and ReO_4^- ($T_c \simeq 2$ K) (Parkin et al., 1983b). Somewhat surprisingly, all of these systems are ambient pressure organic superconductors except when X is ReO_4^-.

Electrocrystallization procedures similar to those used for $(TMTSF)_2X$ salts are used in the preparation of the $(ET)_2X$ salts. However, in the case of ET, as many as four or more different crystallographic phases, with differing electrical properties, may form in one growth cycle. As might be

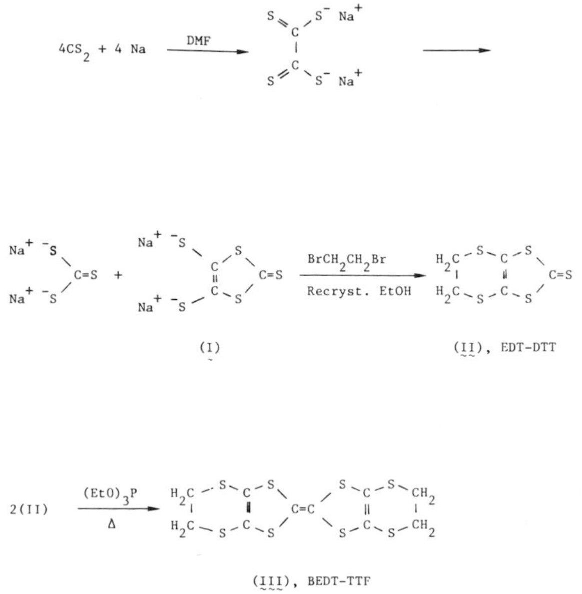

Fig. 23.   Synthetic scheme for BEDT–TTF. [From Williams and Carneiro (1985).]

expected, sorting out the different ET:X phases is a very complicated task but one that can be accomplished by using ESR techniques (Leung *et al.*, 1985a). Using the constant current technique (current = 1–5 μA), metallic black crystals of differing morphologies are found when using ET (Parkin *et al.*, 1983a; Saito *et al.*, 1982; Williams *et al.*, 1984a). The detailed electrocrystallization procedures for (ET)$_2$X, when X is ReO$_4^-$ and FSO$_3^-$, have also been described (Reed *et al.*, 1985). Parkin *et al.* (1985) have attempted to prepare, by electrochemical procedures, new modifications of ET, such as BEDSe–TSeF (ES) (Lee *et al.*, 1983; Schumaker *et al.*, 1983) and BVDT–TTF (VT). Here the extreme C—C bonds at either end of the ET molecule are left unsaturated. These systems have thus far proven to have disappointing electrical properties. The 1:1 BEDSe–TSeF salts with ReO$_4^-$ and ClO$_4^-$ are semiconductors. The (VT)$_3$(FSO$_3$)$_2$ salt has been synthesized and appears to have properties similar to (ET)$_3$(FSO$_3$)$_2$.

B. Crystal Structures of (ET)$_2$X Conductors

Turning now to a discussion of the crystal structure of (ET)$_2$X conductors, we start by noting that these do not always crystallize with one single type of structure. Therefore, at this time, it is not feasible to carry the analysis of their structure–property relationships to nearly the same degree of detail as was done for the (TMTSF)$_2$X series.

As an illustration of the structures adopted by the tetrahedral anion materials, the crystal structure of (ET)$_2$BrO$_4$ is shown in Fig. 24 (Williams *et al.*, 1984c). At room temperature it crystallizes with a triclinic unit cell (space group $P\bar{1}$) and this compound is isostructural (Williams *et al.*, 1984c) with the pressure-induced superconductor (ET)$_2$ReO$_4$ (Parkin *et al.*, 1983a,b). At first sight, this structure somewhat resembles that of the (TMTSF)$_2$X materials for the following reason: loosely packed molecular stacks are formed along *a* with the molecular planes placed approximately perpendicular to the stacking axis. The stacks are in close contact along *b* and are separated along *c* by the anions, so that the *ab* planes contain sheets of interacting sulphur atoms. However, significant differences occur between the (TMTSF)$_2$X and (ET)$_2$X structures which are important to point out. Compared to the TMTSF salts the ET donor molecules are far from planar and do not stack in the same zigzag array as shown in Fig. 14, but clearly in a more complicated array. And a striking difference between the sulphur atom "corrugated sheet network" (Williams *et al.*, 1984c) of (ET)$_2$X (X = ReO$_4^-$ or BrO$_4^-$) shown in Fig. 25, is revealed when compared to the selenium atom infinite sheet network in (TMTSF)$_2$X of Fig. 15. It appears that there are no intrastack S$\cdots$S contact distances (along the stacks) shorter than the S$\cdots$S van der Waals radius sum of 3.6 Å, either

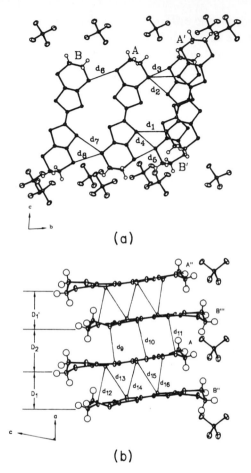

(a)

(b)

Fig. 24.   View of the intermolecular S···S interactions in $(ET)_2BrO_4$. (a) Indicates the interstack S···S contact distances less than the van der Waals sum of 3.60 Å (298/125 K): $d_1 = 3.581(2)/3.505(2)$, $d_2 = 3.499(2)/3.448(2)$, $d_3 = 3.583(2)/3.483(2)$, $d_4 = 3.628(2)/3.550(2)$, $d_5 = 3.466(2)/3.402(2)$, $d_6 = 3.497(2)/3.450(2)$, $d_7 = 3.516(2)/3.434(2)$, and $d_8 = 3.475(2)/3.427(2)$ Å. (b) The S···S contact distances, d_9–d_{16} are, by contrast, all longer than 3.60 Å even at 125 K. In addition the loose zigzag molecular packing of ET molecules is such that they are not equally spaced, $D_1 = 4.01/3.95$ Å and $D_2 = 3.69/3.60$ Å. As a result of the (apparently) weak intrastack and strong interstack interactions, $(ET)_2X$ molecular metals are structurally different from the previously discovered $(TMTSF)_2X$ based organic superconductors. Almost identical S···S distances and interplanar spacings are observed in $(ET)_2ReO_4$ at both 298 K and 125 K. [From Williams *et al.* (1984c).]

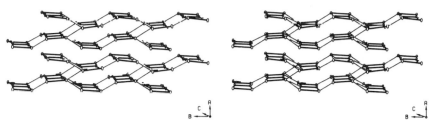

Fig. 25.   A stereoview of the short (<3.60 Å) intermolecular interstack S···S interactions in $(ET)_2ReO_4$ and $(ET)_2BrO_4$ which form a two-dimensional "corrugated sheet"
network. This network, which is the principal pathway for electrical conduction, is much
different from that observed in $(TMTSF)_2X$ salts, but similar to the network of interstack
S···S interactions observed in $ET_2(ClO_4)(TCE)_{0.5}$. [From Kobayashi *et al.* (1983a) and
Williams *et al.* (1984c).]

at room temperature or at 120 K. Only between chains are short interstack
contacts formed in $(ET)_2X$ (X = BrO_4^-, ReO_4^-). Hence, from a structural
point of view these two ET salts are not quasi-one-dimensional, a fact,
however, which does not preclude that the electronic structure may well
be so. The recently prepared ambient pressure superconductor β-$(ET)_2I_3$
has a crystal structure which is markedly similar to the previously mentioned $(ET)_2X$ conductors (Kaminskii, 1984; Mori *et al.*, 1984; Williams *et
al.*, 1984a) in terms of the long intrastack and short interstack S···S
contacts. It is also noteworthy that compared to the $(ET)_2X$ (X = BrO_4^-
and ReO_4^- conductors), the molecular stacks in β-$(ET)_2I_3$ are not perpendicular to any specific crystallographic axis, but rather to the [110] direction (see Fig. 26). At first sight, β-$(ET)_2I_3$ bears remarkable resemblance

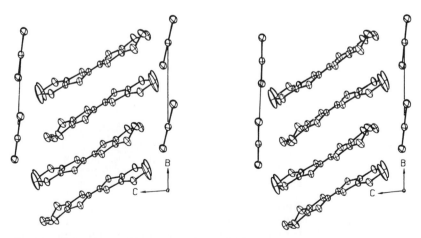

Fig. 26.   Molecular packing (stereoview) of ET molecules and linear (centrosymmetric) I_3^- anions in β-$(ET)_2I_3$. Note that the loose molecular stacks occur in the [110]
direction rather than along a crystallographic axis. [From Williams and Carneiro (1985).]

to the 1:2 M(TCNQ)$_2$ salts where the conducting molecules form stacks of dimers and where the molecules are tilted with respect to the conducting axis. But as previously pointed out, one observes mainly short interstack distances compared to the intrastack S$\cdots$S separations, and this superconductor appears to be rather two dimensional in structure (see Fig. 27).

Finally, it should be noted that exactly the same 2:1, ET:I$_3$ stoichiometry results, during electrocrystallization, in the simultaneous formation of two different forms of (ET)$_2$I$_3$, i.e., α- (room temperature ESR linewidth $\Delta H_{pp} \simeq 95G$) and β- (ΔH_{pp}, 298 K, $\simeq$ 20G), respectively. The crystal structure of α-(ET)$_2$I$_3$ (Mori *et al.*, 1984; Bender *et al.*, 1984) differs markedly from that of β-(ET)$_2$I$_3$ (Kaminskii *et al.*, 1984; Mori *et al.*, 1984; Williams *et al.*, 1984a; Shibaeva *et al.*, 1985). Not surprisingly, the electrical properties are also very different, with the α-form undergoing a metal–insulator transition at 135 K (Bender *et al.*, 1984) while the β-form is the first sulfur based ambient pressure organic superconductor (Yagubskii *et al.*, 1984b; Crabtree *et al.*, 1984; Williams *et al.*, 1984a; Carlson *et al.*, 1985a). A most surprising structural feature of β-(ET)$_2$I$_3$ is the development, at 200 K and down to at least 9 K, of a novel incommensurate "modulated" structure (Emge *et al.*, 1984; Leung *et al.*, 1984b, 1985a,b) observed for the first time in an organic superconductor. The main features of the modulated structure involve displacements from the "average" crystal structure positions of the I$_3^-$ anion and ET molecules, which have *different* direction and magnitude, namely, 0.281(1) Å for I$_3^-$ and 0.124(3) Å for the ET molecules, respectively (Emge *et al.*, 1984; Leung *et al.*, 1984a, 1985a,b). An interesting feature of this system is that although there is a structural modulation, the overall pattern of S$\cdots$S overlaps is preserved because while the S$\cdots$S contacts expand in one unit cell they contract in the adjacent cell. The modulated structure is shown in Fig. 28 and it must be noted that the resulting local fluctuations of the interatomic

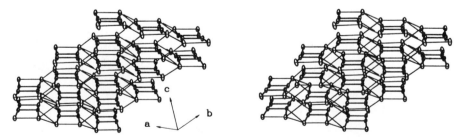

Fig. 27.   Stereoview of the "corrugated sheet network" of short interstack S$\cdots$S interactions (faint lines) between nonplanar and nonparallel ET molecules in β-(ET)$_2$I$_3$ at 125 K (90% ellipsoids). Only the S atoms of ET are shown for clarity. [From Williams *et al.* (1984a).]

S···S distances due to the displacive modulation are very significant and will have to be taken into account in any future theoretical studies of this material.

The same type of corrugated sheet network (see Fig. 29) of short ($d <$ 3.6 Å) S···S contacts found in β-(ET)$_2$I$_3$ is also observed in β-(ET)$_2$IBr$_2$, with the latter having a much higher superconducting transition temperature of ∼2.4–2.8 K (Carlson *et al.*, 1985b; Williams *et al.*,1984b). More important, because the IBr$_2^-$ anion is ∼7% shorter than the I$_3^-$ anion, the unit cell volume of the former is less (828.7 Å^3 versus 855.9 Å^3) with the result that the average interstack S···S contact distances in β-(ET)$_2$IBr$_2$

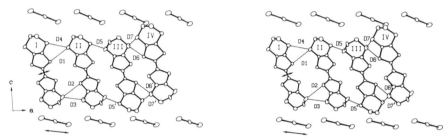

Fig. 28. Stereoview of the molecular packing in β-(ET)$_2$I$_3$ on the *a–c* plane showing the observed structural modulations of the ET molecules and the I$_3^-$ anions. The allowed displacement vectors of an ET molecule (0.124 Å) and an I$_3^-$ anion (0.281 Å) are indicated by a pair of arrows whose length is approximately 5 times the magnitude of the observed displacements. [From Leung *et al.* (1985a).]

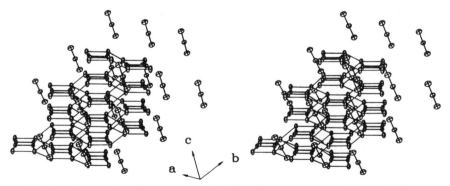

Fig. 29. Stereoview of the novel sandwich or *layered* structure of β-(ET)$_2$IBr$_2$ composed of alternating two-dimensional sheets of linear (Br–I–Br)$^-$ anions between which a "corrugated sheet" network of short interstack S···S interactions are inserted. Only the sulfur atoms of the ET molecules are shown in the network, and light lines indicate the interstack ($d_{S···S} < 3.60$ Å) interactions. The —CH$_2$ groups of the ET molecule protrude from both ends of the molecule (directly out of the plane of the page) and grasp the X$_3^-$ anions in a pincer hold. Therefore, by varying the length of the X$_3^-$ anion, the interstack and intrastack S···S distances can be directly altered. [From Williams *et al.* (1984b).]

are markedly shorter [0.02 Å] than in β-(ET)$_2$I$_3$. Thus, for the first time in any organic system, ambient pressure superconductivity has been maintained in two materials having the same donor molecule. The synthesis of β-(ET)$_2$IBr$_2$ resulted from a rational and systematic approach involving polyhalide anion displacement with the aim of modifying only slightly the corrugated sheet network in β-(ET)$_2$X materials (Williams *et al.*, 1984b). This design strategy results in systematic variation of the interstack S⋯S distances with anion replacement (see Fig. 27 caption). This occurs because the trihalide anion resides in a cavity formed by CH$_2$ group hydrogen atoms (Williams *et al.*, 1984b), similar to the methyl-group hydrogen atom cavity in (TMTSF)$_2$X systems, and by varying the anion length the S⋯S distances can be changed. Newly synthesized ET:I salts of various compositions have been reported to be ambient pressure superconductors with T_c's of ~2.5 K (Yagubskii *et al.*, 1984a,b; Shivaeva *et al.*, 1985). Table XIII summarizes crystal data and other characteristics for the organic metals of the ET:I system.

For (ET)$_2$ClO$_4$(TCE)$_{0.5}$ (TCE is trichloroethane), solvent is incorporated into the structure at positions similar to those occupied by the anions, i.e., separating the sheets of ET molecules (Kobayashi *et al.*, 1983a). For this structure definite stacks cannot be identified at all, and as a consequence, the electronic structure is also very two dimensional. It is not unlikely that

TABLE XIII

Crystal Data and Some Other Characteristics for Organic Metals of the (BEDT–TTF)–I System[a,b]

Crystal phase	α	β	γ	δ	ε
a (Å)	10.785	6.609	13.76	10.728	13.974
b (Å)	9.172	9.083	14.73	34.14	18.77
c (Å)	17.39	15.267	33.61	34.92	17.40
Alpha angle (°)	82.08	85.63	90	95.00	67.3
Beta angle (°)	96.92	95.62	90	90	90
Gamma angle (°)	89.13	70.22	90	90	90
V (Å^3)	1690.3	852.2	6812	12736	4211
(BEDT–TTF):I	1:1.5	1:1.5	1:2.5	1:3	1:3.5
Space group	$P\bar{1}$	$P\bar{1}$	*Pbnm*	*C2/c*	*P2$_1$/c*
Z	2	1	4	24	4
d_{calc} (g/cm^3)	2.26	2.24	2.06	2.41	2.27
σ_{RT} (ohm^{-1} cm^{-1})	~20–30	~20–30	~20	~10–20	~20
T (K)	137(M → I)	1.5(M → SC)	2.5(M → SC)	130(M → I)	2.5(M → SC)

[a] Taken from Shibaeva *et al.* (1985).

[b] α = α-(BEDT–TTF)$_2$I$_3$; β = β-(BEDT–TTF)$_2$I$_3$; γ = γ-(BEDT–TTF)$_3$(I$_3$)$_{2.5}$; δ = δ-(BEDT–TTF)I$_3$; ε = ε-(BEDT–TTF)$_2$I$_3$(I$_8$)$_{0.5}$.

this structure (triclinic with space group $P\bar{1}$) is also adopted by (ET)$_2$ReO$_4$(THF)$_{0.5}$ with the only difference being that the solvent here is tetrahydrofuran (Parkin *et al.*, 1983b).

As an example of an (ET) conductor which has clearly separated stacks, we mention β-(ET)$_2$PF$_6$ (Kobayashi *et al.*, 1983b) and (ET)$_2$AsF$_6$ (Leung *et al.*, 1984). They have structures which clearly resemble those typical of the earlier mentioned M(TCNQ)$_2$ conductors and while β-(ET)$_2$PF$_6$ undergoes a metal–insulator transition at 297 K (Kobayashi *et al.*, 1983b), the same type of transition occurs above 125 K in (ET)$_2$AsF$_6$ (Leung *et al.*, 1984b).

Although the space group symmetries and the stoichiometries of these ET salts appear to vary widely, the modes of the crystal packing or the ET molecules appear limited (vide infra). In general, there are two different types of packing along the interstack directions (see Fig. 30) designated as L- and Z-modes for linear and zigzag, respectively (Leung *et al.*, 1985a). Along the loosely packed ($d_{S\cdots S} > 3.60$ Å) intrastack direction, there appear to be three different packing modes exhibited by the ET molecules in the ET:X systems. For convenience, these intrastack-packing modes are designated as a-, b-, and c- modes (see Fig. 31). In the a-mode, the ET

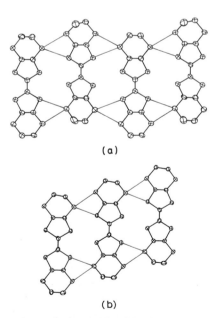

(a)

(b)

Fig. 30.   Illustration of two distinct types of interstack packing of the ET molecules in ET:X salts: (a) zigzag or Z-mode and (b) linear or L-mode. The hydrogen atoms are omitted for clarity. [From Leung *et al.* (1985a).]

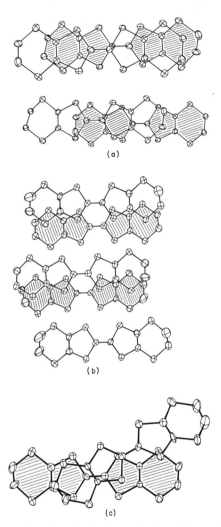

Fig. 31.   Illustration of the three distinct intrastack packing motifs of the ET molecules currently observed in ET:X systems: (a) a-mode, (b) b-mode, and (c) c-mode. The ET molecules in the bottom layer are shaded. [From Leung *et al.* (1985a).]

molecules pack in a face-to-face manner. As illustrated in Fig. 31, quantitative differences in the intermolecular overlaps can occur due to the relative molecular displacements along the long in-plane axes of the organic molecules. This displacement, designated "D" by Mori *et al.* (1984), is a very important factor in determining the overlap integrals between adjacent ET molecules. Interestingly enough, the known superconducting phases of the ET salts adopt the a-type packing mode. For the b-mode, there is no distinct stacking mode. Each ET molecule overlaps two others along a direction perpendicular to the ET sheets. All known ET salts which adopt this mode are metallic, but superconductivity has never been observed in these systems. For example, (ET)$_2$ClO$_4$(TCE)$_{0.5}$ remains metallic to 1.4 K, but is never superconducting (Saito *et al.*, 1982; Kagoshima *et al.*, 1983). To date, ET salts that stack in the c-mode are semiconductors.

C. Discussion of (ET)$_2$X Conductors

Since the initial synthesis of ET a number of charge-transfer salts that are electrical conductors have been synthesized which have the stoichiometries (ET)$_2$X, where X is the monovalent anion. Anions include I$_3^-$, IBr$_2^-$, AuI$_2^-$, ReO$_4^-$, BrO$_4^-$, InBr$_4^-$, ClO$_4^-$, AsF$_6^-$, and PF$_6^-$. The first such salt found to be superconducting was (ET)$_2$ReO$_4$, at a temperature of ~2 K and at a pressure of ~4–6 kbar (Parkin *et al.*, 1983a). Inasmuch as the T_c occurred at higher temperatures in these salts than the (TMTSF)$_2$X salts, much interest in these systems was generated. Subsequently, superconductivity was found at ambient pressure in β-(ET)$_2$I$_3$ (T_c = 1.4 K) (Yagubskii *et al.*, 1984b; Williams *et al.*, 1984a; Crabtree *et al.*, 1984), β-(ET)$_2$IBr$_2$ (T_c = 2.8 K) (Williams *et al.*, 1984a; Carlson *et al.*, 1985b), and β-(ET)$_2$AuI$_2$ (T_c = 5 K) (Wang *et al.*, 1985). These were the first charge-transfer compounds in this family that were found to be superconductive without requiring applied pressure.

It has been found that under electrocrystallization reactions several different crystal phases may grow with ET and X. For example, in the system ET:ReO$_4$, at least three phases may be found: (ET)$_2$ReO$_4$, (ET)$_3$(ReO$_4$)$_2$, and (ET)$_2$(ReO$_4$)(THF)$_{0.5}$. As previously mentioned, (ET)$_2$I$_3$ crystallizes in several different forms. As expected, there can be considerable variation in the electrical properties associated with the various phases. Parkin *et al.* (1985) have suggested that these different phases are due to the extreme and unexpected flexibility of the outer rings of the ET molecules, which can make an angle of ~45° with the plane of the inner portion of the molecule.

As we have pointed out for some (TMTSF)$_2$X and (ET)$_2$X salts it is necessary to both lower the temperature and increase pressure to obtain superconducting phases. It has been also observed that a further increase in pressure suppresses the superconducting transition ($dT_c/dp < 0$).

Laukhin et al. (1985) and Shchegolev et al. (1985) found that β-(ET)$_2$I$_3$ is converted into a superconductor with pressure ($\sim$1 kbar) with T_c being raised to 7.0–8.0 K. This was later confirmed by Murata et al. (1985). These results are very interesting, since the normal effect of increased pressure is to suppress the superconductivity. Studies have shown that pressure-induced superconductivity in β-(ET)$_2$I$_3$ depends on the pressure-application method (Schirber et al., 1986). When a hydrostatic (isotropic) pressure is applied, T_c is suppressed below 1.5 K in the manner normally expected while an anisotropic pressure results in production of the high T_c ($\sim$8 K) state. It has been suggested that this latter case of "shear-induced" superconductivity is derived from a change in the structure of β-(ET)$_2$I$_3$ and likely involves a suppression of the modulated structure (vide supra) and possibly complete CH$_2$ group ordering which would favor increased T_c (Schirber et al., 1986). The role of the intermolecular S$\cdots$S interactions in the ambient pressure superconductors β-(ET)$_2$X (X = I$_3^-$, IBr$_2^-$) have been discussed (Whangbo et al., 1985a) as well as the band structures (Whangbo et al., 1985b).

The Russian scientists (Merkanov et al., 1985) have reported that crystals of β-(ET)$_2$I$_3$ prepared by thermal vacuum processing of iodine-rich crystals of the ζ phase [(ET)$_4$(I$_3$)$_2$I$_8$] show an elevation of T_c from 1.5 to 6–7 K. This elevation of T_c has been accomplished without pressure and is most unusual, but has not been confirmed. The authors attribute this increase in T_c to a consequence of a small deviation of iodine content, change in number of carriers, neither of which appear to lead to changes in crystal structure.

A new ET salt not containing a trihalide anion has been prepared, which is also superconducting at ambient pressure. This is β-(ET)$_2$AuI$_2$ (Williams et al., 1985a; Wang et al., 1985), which is the first linear symmetric (IAuI)$^-$ triatomic metal-containing anion derivative, and is isostructural with β-(ET)$_2$X (X = I$_3^-$, IBr$_2^-$). The structure consists of discrete layers of AuI$_2^-$ anions between which is sandwiched a corrugated sheet network of ET molecules with short ($d_{S\cdots S} \leq 3.60$ Å, the van der Waals radii sum for sulfur) interstack S$\cdots$S distances. The loosely packed stacking of ET entities is characterized by intrastack S$\cdots$S distances greater than 3.60 Å. The material is triclinic with a space group $P\bar{1}$ and $Z = 1$. The T_c for this salt is 4.98 $\pm$ 0.08 K. Figure 32 shows the RF field penetration depth results, as a function of T, showing the onset of bulk superconductivity at $\sim$5 K (Wang et al., 1985). As is the case for the β-(ET)$_2$X (X = I$_3^-$ and IBr$_2^-$ systems), different crystals from the same preparation exhibit different T_c's varying by as much as 1 K and this result is due to impurities.

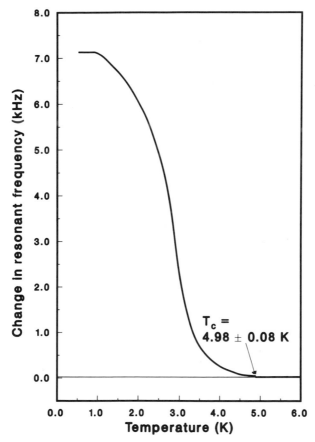

Fig. 32. Radiofrequency field penetration depth measurements of β-(ET)$_2$AuI$_2$ indicating the onset temperature for bulk superconductivity to be ~5 K at ambient pressure. [From Wang *et al.* (1985).]

VI. ELECTRICAL CONDUCTION

Despite the great similarity in crystallographic structures of (TMTSF)$_2$X compounds, they have very varied electronic properties. This is clear from Fig. 33, where the electrical resistivities of some salts are shown as a function of temperature. One finds that the TMTSF molecule is able to reproduce almost all previously studied molecules in giving highly conducting salts with a great variation in metal–insulator transition temperatures ($T_{M \to I}$) and even superconductivity (at T_c). This conclusion contra-

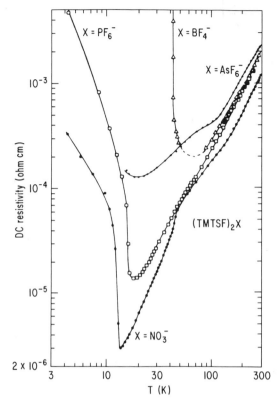

Fig. 33. The electrical resistivities of selected (TMTSF)$_2$X salts having anions of various geometries, at ambient pressure. [Redrawn from Bechgaard *et al.* (1980) and Williams and Carneiro (1985).]

dicts former suggestions that mostly molecular features are responsible for the variation of solid state properties, because in (TMTSF)$_2$X compounds identical molecules in very similar crystallographic environments behave quite differently.

The situation in ET:X salts is different, since they form very different crystallographic structures. Even the same anion may give rise to a multiplicity of crystallographic phases as demonstrated by X = ReO$_4^-$ where four phases have been found. It is, therefore, not astonishing that ET:X salts have different conduction properties as shown in Fig. 34.

The fact that β-(ET)$_2$I$_3$ and β-(ET)$_2$IBr$_2$ are the second and third ambient pressure organic superconductors, respectively, and that one phase of (ET)$_2$ReO$_4$ becomes superconducting under pressure makes the other phases extremely interesting from the point of view of addressing the

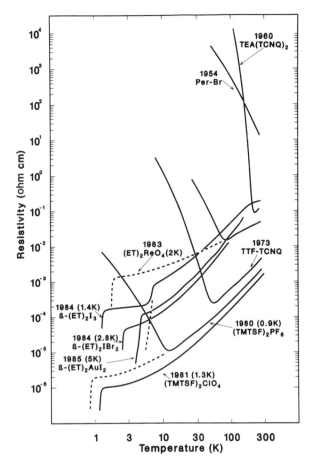

Fig. 34.   The electrical resistivities of selected $(ET)_2X$ salts at ambient pressure (solid curves) and nonambient pressure (dashed curves) are compared with TMTSF salts, TTF–TCNQ, TEA(TCNQ)$_2$, and Per–Br. Superconducting temperatures and pressure are indicated in parenthesis.

following question: Which features of the crystallographic structure and anion environment provide the grounds for expecting superconductivity to occur in an organic conductor? We now turn to this question.

A.   Resistivity along the Chains

1.   (TMTSF)$_2$X Salts

The usual features of the electrical resistivities of (TMTSF)$_2$X compounds shown in Fig. 33 are as follows. The room temperature value along the needle axis is typically $\rho_a = 1.5 \times 10^{-3}$ ohm cm (Bechgaard et al., 1980; Jacobsen et al., 1982). This is very similar to the value for TTF–TCNQ and somewhat higher than the best conducting organic salt HMTSF–TNAP (0.5×10^{-3} ohm cm) (Bechgaard et al., 1978). The temperature dependence of the resistivity follows a behavior typical of organic conductors when the temperature is above the electronic transition temperatures, $T_{M \to I}$ or T_c. In the cases where the anion-ordering transition [T_O] described above does not coalesce with the electronic transition, it results in a small deviation from the T^2 dependence of the resistivity.

Below the electronic transition temperature $T_{M \to I}$ the properties change from metallic to either semiconducting (X = PF_6^-, SbF_6^-, BF_4^-, FSO_3^-, ReO_4^-, $H_2F_3^-$) (Bechgaard et al., 1980; Jacobsen et al., 1982; Mortensen et al., 1983; Wudl et al., 1982; Maaroufi et al., 1983) or superconducting (X = ClO_4^-) (Mortensen et al., 1984), where, in both cases, the electronic density of states is characterized by an energy gap 2Δ. Although it cannot be concluded from the electrical resistivity, we note that the semiconducting salts fall into one of two classes. One contains the usual dielectric semiconductors (X = ReO_4^-, FSO_3^-, BF_4^-, $H_2F_3^-$) (Bechgaard et al., 1980; Jacobsen et al., 1982; Mortensen et al., 1983; Wudl et al., 1982; Maaroufi et al., 1983) and they may be characterized as anion-assisted Peierls insulators, since their properties in many respects are like those of the Peierls charge-density wave (CDW) insulator; but some aspects of their behavior can only be understood as stemming from a direct influence of the anions. The other class of semiconductors has an antiferromagnetic ground state often referred to as the spin-density wave (SDW) ground state; it contains salts of anions (X = SbF_6^-, AsF_6^-, PF_6^-, and possibly also NO_3^-) (Bechgaard, 1982). Published values of the activation energies Δ are given in Table XIV.

This classification is documented to varying degrees for the different compounds. Fully compelling evidence exists for $(TMTSF)_2ClO_4$ (superconductor), $(TMTSF)_2ReO_4$ (dielectric semiconductor), and $(TMTSF)_2PF_6$ (magnetic semiconductor), so that these salts may be considered as prototypes for the three kinds of low-temperature behavior of the regularly behaving $(TMTSF)_2X$ salts.

Some salts show behaviors which are different from those previously described; but most of these have not been well characterized. For example, for X = BrO_4^-, which has only recently been studied, anion disorder seems to lead to localized electronic states (Mortensen et al., 1984).

2. $(ET)_2X$ Salts

The electrical resistivities in the molecular stacking direction of some ET:X salts are shown in Fig. 34. The first material studied $(ET)_2ClO_4(TCE)_{0.5}$ (TCE is trichloroethane) has a room temperature value

TABLE XIV

Conduction Characteristics of $(TMTSF)_2X$ and $(ET)_2X$ Salts[a,b]

X^-	V_A (Å³)	V_c (Å³) (at 298 K)	T_0 (K)	$T_{M \to I}$ (K)	Δ (meV)[c]	P_{SC} (kbar)	T_c (K)
$(TMTSF)_2X$							
H_2F_3	10	669.6	63	63[d]	—	—	—
NO_3	30	659.5	41	12[e]	—	—	—
BF_4	70	690.4	40	40[d]	—	—	—
ClO_4	77	694.3	24	—	—	0	1.4
FSO_3	81	695.8	87.5	86[d]	47	~7	1.4
BrO_4	93	707.2	≈250	≈220[f]	—	—	—
PF_6	94	712.9	—	12[e]	2.0	~9	0.9
AsF_6	102	719.9	—	12[e]	2.0	12	0.9
TcO_4	105	711.0	—	—	—	—	—
ReO_4	106	710.5	177	182[d]	83	9.5	1.0
SbF_6	118	737.0	—	17[e]		11	0.8
TaF_6	122	735.6	—	11[e]	1.8	12	0.8
CF_3SO_3	153	739.5	≈280	280[d]	—	—	
$(ET)_2X$							
$ClO_4(TCE)_{0.5}$	—	1684 ($Z = 2$)	—	—	—	—	—
$PF_6(\alpha)$	94	794.3	—	54	—	—	—
$PF_6(\beta)$	94	3256.4 ($Z = 4$)	297	276	—	—	—
ReO_4	106	1565 ($Z = 2$)	81	45	4–6	1.3	
$I_3(\beta)$		855.9	—	—	0	1.4	
$I_3(\gamma)$[g]						2.5	
$I_3(\epsilon)$[h]						2.5	
$IBr_2(\beta)$		828.7	—	—	0	2.7	
$AuI_2(\beta)$		845.2				4.98	
AsF_6	102	3274	125	—	—	—	

[a] Taken from Williams and Carniero (1985).
[b] V_A and V_c are the calculated anion volume and measured unit cell volume; T_0 is the anion ordering temperature, and $T_{M \to I}$ the metal–insulator transition temperature.
[c] Semiconductor gap parameter.
[d] Ground state; anion assisted charge-density wave.
[e] Ground state; spin-density wave.
[f] Ground state; localized states.
[g] $(ET)_3(I_2)_{2.5}$.
[h] $(ET)_2I_3(I_8)_{0.5}$.

of 4×10^{-2} ohm cm decreasing to about 1×10^{-3} ohm cm at $T = 16$ K (Saito *et al.*, 1982). At this temperature the resistivity goes through a broad minimum rising to 1.5×10^{-3} ohm cm at the lowest temperatures. This situation is similar to what has been observed in HMTSF–TCNQ under pressure (Friend *et al.*, 1978) and in HMTSF–TNAP (Bechgaard *et al.*, 1978).

In the ET:ReO_4 salts the conductivity varies (Parkin *et al.*, 1983a,b). The 2:1 salt, which becomes superconducting under pressure, has a sharp metal insulator transition at 81 K whereas the 3:2 salt has a broader transition at a similar temperature. In the 2:1:$\frac{1}{2}$ salt (where $\frac{1}{2}$ indicates incorporation of $\frac{1}{2}$ solvent molecule THF (tetrahydrofuran) per formula unit) there is no indication of a transition from the conductivity data.

A third 2:1 salt, β-$(ET)_2PF_6$, shows a usual Peierls instability as judged from the resistance data of Fig. 34 with an activation energy of 230 meV (Kobayashi *et al.*, 1983b). In fact, it looks much like the classical platinum chain conductor $K_2[Pt(CN)_4]Br_{0.3}\cdot3.0H_2O$ (KCP) (Williams *et al.*, 1982b) as do some of the structurally similar $M(TNCQ)_2$ conductors (Acker *et al.*, 1960). It appears that α-$(ET)_2PF_6$ has an activated conductivity over the entire temperature region with a low activation energy of 50 meV (Kobayashi *et al.*, 1983b). In terms of the optical conductivity derived from polarized specular reflectance data, β-$(ET)_2I_3$ exhibits, for the first time in an organic conductor, a "crossover" from one-dimensional behavior at 298 K to three-dimensional behavior at 40 K (Jacobsen *et al.*, 1985).

Finally, the resistivity characteristics of the ambient pressure superconductor β-$(ET)_2I_3$ resemble those of $(TMTSF)_2ClO_4$ a great deal except for the higher T_c of the triiodide salt (its room temperature value is ~0.03 ohm cm) (Yagubskii *et al.*, 1984b; Crabtree *et al.*, 1984; Williams *et al.*, 1984a; Carlson *et al.*, 1985a).

3. Discussion

The ratio $k_B T_{M\rightarrow I}/\Delta$ can be used to examine the anisotropy of the electronic structure in a quasi-one-dimensional material. The ratio is 0.567 in the mean field (or molecular field) approximation which holds fairly well for three-dimensional instabilities (e.g., BCS superconductivity); but it approaches zero as the dimensionality goes to unity. With $k_B = 0.086$ meV/K the values of Table XIV yield results for $k_B T_{M\rightarrow I}/\Delta$ close to 0.567 for $(TMTSF)_2X$ suggesting that these salts are not very one dimensional. However, the lack of a transition in $(ET)_2ClO_4(TCE)_{0.5}$ should not be confused with extreme one dimensionality. Rather, this compound would appear to be so slightly anisotropic that the one-dimensional instabilities which give rise to the transitions are not active. The extraordinary variation of conductivity behavior of $(TMTSF)_2X$ and ET:X salts at ambient pressure make the two donor molecules involved unique. In the metallic phase at high temperatures they behave much like previously studied organic conductors apart from the possible features at the anion-ordering temperature T_O. But their low-temperature behavior has provided novel superconductivity, spin-density waves and anion assisted Peierls insulators to condensed matter physics. Measurements of the resistivity itself contributes to the characterization of these phases by giving information about the metal to insulator transition temperatures $T_{M \to I}$, T_c, and T_O, respectively, and the semiconducting gap parameter Δ below $T_{M \to I}$. Values for the different salts are given in Table XIV.

B. Conduction Anisotropy

1. (TMTSF)$_2$X and (ET)$_2$X

The electrical resistivity in directions other than the molecular stacking axis has been measured in a few cases. In (TMTSF)$_2$PF$_6$ one finds $\rho_a : \rho_{b'} : \rho_{c*} \simeq 1 : 200(3000) : 3 \times 10^4 (10^6)$, where numbers in parentheses refer to $T = 20$ K, i.e., just above the transition (Jacobsen et al., 1981). Here b' is a vector in the ab plane perpendicular to a, and $c*$ is the reciprocal lattice vector orthogonal to the same plane. The situation in (TMTSF)$_2$PF$_6$ is probably typical for the series, whereas in (ET)$_2$ClO$_4$(TCE)$_{0.5}$ one finds $\rho_\perp / \rho_\parallel \simeq 1$–2 (Friend et al., 1978), which suggests a very two-dimensional electronic structure in agreement with other experimental results, as well as band structure calculations (Mori et al., 1982).

2. Discussion

The anistropy of a quasi-one-dimensional conductor provides considerable insight into the directional dependence of the electronic structure of the compound. It is, however, difficult to measure in (TMTSF)$_2$X salts and in some of the ET:X salts. Because of the low crystallographic symmetry, crystal faces are not perpendicular to high symmetry directions. Nevertheless, adopting a simple tight-binding form for the electronic energy band:

$$\epsilon(k) = 2t_a \cos(k_a a) + 2t_{b'} \cos(k_{b'} b') + 2t_{c*} \cos(k_{c*} 2\pi/c*)$$

with

$$k = k_a \hat{a} + k_b \hat{b}' + k_{c*} \hat{c}*$$

(where the caret denotes a unit vector) one may estimate the ratios between the transfer integrals t_a, $t_{b'}$, t_{c*} from the conductivity results. For a quasi-one-dimensional conductor a rough estimate (Soda et al., 1977) $p_\parallel/p_\perp = (t_\perp/t_\parallel)^2$ of the anisotropy given above for (TMTSF)$_2$PF$_6$ yields $t_a : t_{b'} : t_{c*} \approx 1 : \frac{1}{10} : \frac{1}{200}$, which is in rough agreement with more accurate determinations. In (ET)$_2$ClO$_4$(TCE)$_{0.5}$ the low anisotropy of the conductivity suggests a rather two-dimensional electronic structure in accordance with the crystallographic structure and the associated "corrugated sheet network" of interstack S$\cdots$S interactions.

C. Pressure Studies

Because of the "softness" of organic metals one expects them to show interesting behavior under applied pressures. This had been demonstrated earlier by Jérome and co-workers on several compounds and in the case of TMTSF–DMTCNQ a pressure of 10 kbar transforms it abruptly from a Peierls semiconductor at $T = 50$ K to a metal at all temperatures (Andrieux et al., 1979). When the temperature-dependent resistance of the $(TMTSF)_2X$ family became known, the very low transition temperatures in some of the compounds suggested that these salts would easily become metallic under pressure and maybe even superconducting.

1. $(TMTSF)_2X$ Salts

Indeed, the pressure dependence of the resistivity of $(TMTSF)_2X$ salts shows a very rich behavior. $(TMTSF)_2PF_6$ was the first organic metal to show superconductivity at a pressure of ~6.5 kbar at 0.9 K (Jérome et al., 1980). Up to this pressure the metal–insulator transition temperature gradually decreases, possibly with a strong dependence close to P_{SC}. The superconducting transition temperature also decreases, but slowly, with increasing pressure. Several of the octahedral anion salts show the same behavior with somewhat different P_{SC}'s, as one would expect from the anion volumes discussed above (P_{SC}'s are listed in Table XIV). The octahedral anion salts ($X = PF_6^-$, AsF_6^-) show a greater variation in behavior than their tetrahedral counterparts, namely, $(TMTSF)_2ReO_4$ with $T_{M \to I} = 182$ K at ambient pressure and which becomes superconducting at ~1.0 K above $P_{SC} \simeq 9.5$ kbar, whereas $(TMTSF)_2ClO_4$ becomes superconducting at ~1.4 K at ambient pressure (Gubser et al., 1982). Both of these two results are exciting and astonishing. First, the success with the ClO_4^- salt may be considered the victory of more than two decades of struggle to find an organic superconductor; but, secondly, the ReO_4^- salt demonstrated that superconductivity can be achieved with modest pressures even if the metal–insulator transition temperature is quite high.

The common pressure dependence of the $(TMTSF)_2X$ salts is illustrated in Fig. 35. Below the critical pressure P_{SC} the electronic ground state is an insulator of either the spin-density wave type or the anion assisted Peierls type. Above P_{SC} the ground state is superconducting. P_{SC} varies monotonically with the anion volume previously discussed; but the transition temperature at ambient pressure and the type of low-pressure ground state depend specifically on the symmetry of the anion. The latter observation demonstrates a correspondence between the lattice superstructure and the nature of the ground state, and this is substantiated by studies of $(TMTSF)_2ClO_4$ where the anion order can be obstructed by rapid cooling

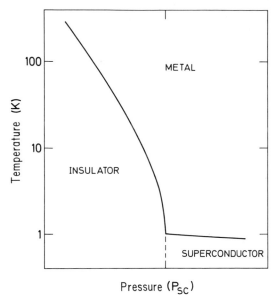

Fig. 35. Illustration of the temperature–pressure phase diagram for (TMTSF)$_2$X salts. The pressure P_{SC} varies between ambient pressure and ~12 kbar for different compounds. [From Williams and Carneiro (1985).]

through T_O = 24 K (Takahashi *et al.*, 1982; Garoche *et al.*, 1982). In this quenched state the salt has a spin-density wave ground state associated with the freezing in of a ClO$_4^-$ anion disorder.

2. (ET)$_2$X Salts

Only in the case of the 2:1 ReO$_4^-$ salt, and of course X = I$_3^-$, IBr$_2^-$, and AuI$_2^-$, do the properties of the (ET)$_2$X salts resemble the (TMTSF)$_2$X series (Williams and Carneiro, 1985). With a critical pressure of 4–6 kbar, the ReO$_4^-$ salt fits the general phase diagram given in Fig. 35. The great similarity in properties of salts based on *different* molecules suggests that the solid state environment is as crucial to the physical behavior as are the molecular characteristics.

VII. MAGNETIC PROPERTIES

The magnetic susceptibility $\chi(T)$ of organic conductors is of interest because, in several ways, it reflects other properties than just the conductivity. In particular, it provides a clear means of discriminating between the charge-density wave and the spin-density wave states which conductivity measurements cannot distinguish. The value of χ may be measured by either static or resonance techniques. Static measurements contain in general contributions from (i) molecular core diamagnetism, (ii) Curie-like susceptibility stemming from localized spin-carrying defects, and (iii) spin susceptibility from the conduction electrons. Electron spin resonance (ESR) measures only the two latter contributions, which contain the interesting information about the electronic properties of the conductor. In order to compare results of static and ESR measurements the temperature independent contribution (i) must, therefore, be subtracted from the former. Furthermore, the ESR signal has a width that measures the interactions of the spins.

A. Magnetic Susceptibility

1. (TMTSF)$_2$X

The room temperature electronic susceptibility is typically 3×10^{-4} emu cm^3/mole for (TMTSF)$_2$X compounds and decreases approximately linearly with decreasing temperature until the transitions occur (Scott 1982; Pederson *et al.*, 1982). Below the transition there is a clear distinction between the three possible ground states previously discussed (Bychkov *et al.*, 1966; Scott *et al.*, 1981). This is illustrated in Fig. 36.

In the case of an anion assisted charge-density wave system with X = ReO$_4^-$, χ becomes activated with an activation energy similar to Δ from conductivity, both when measured statically and by ESR (Pedersen *et al.*, 1982). For spin-density wave systems, e.g., X = PF$_6^-$, the situation is much more complicated (Scott *et al.*, 1981; Mortensen *et al.*, 1982). Whereas the ESR signal vanishes both from decreasing χ as well as from line broadening, the static susceptibility vanishes only when the magnetic field is in the direction of the spins. Consequently χ has a strong directional dependence and in a powder, which is most often used in static measurements, only a very small anomaly is seen at $T_{M \rightarrow I}$. For superconducting (TMTSF)$_2$ClO$_4$ the two susceptibilities give evidence for an incomplete spin-density wave transition at $T \simeq 5$ K at intermediate cooling rates reminiscent of the features in some conductivity measurements (Scott, 1982). At the superconducting transition the susceptibility vanishes, but the transition temperature is lower than that deduced from conductivity (Carneiro *et al.*, 1984).

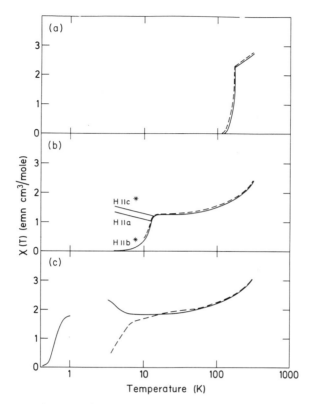

Fig. 36. Magnetic susceptibility of $(TMTSF)_2X$ salts with X = (a) ReO_4^-, (b) PF_6^-, and (c) ClO_4^-. Full lines indicate static measurements on powders (ReO_4^- and ClO_4^-) and on single crystals (PF_6^-). Dashed lines show ESR single crystal results. The three cases illustrate anion assisted charge-density wave (ReO_4^-), spin-density wave (PF_6^-), and possible spin-density wave precursors of superconductivity (ClO_4^-). A separate result showing the superconducting transition is also shown for X = ClO_4^-. [From Williams and Carneiro (1985).]

2. ET:ReO₄

ESR measurements have been performed on three ET:ReO₄ phases (Carneiro *et al.*, 1984), but $\chi(T)$ is not known on an absolute scale. The results are shown in Fig. 37.

The 2:1 salt has features similar to $(TMTSF)_2ReO_4$; it has a constant susceptibility in the metallic regime and an abrupt transition to a very low value below $T_{M\rightarrow I}$. The 3:2 salt has a $\chi(T)$ similar to what one expects from a Peierls transition without influence of anion ordering, i.e., some-

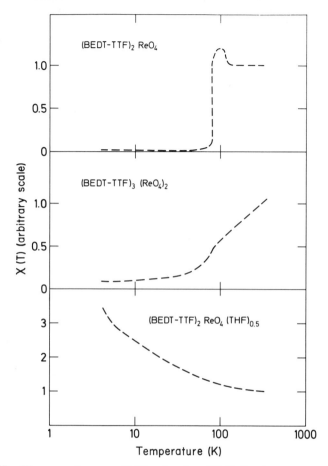

Fig. 37. The magnetic susceptibility of various ET : ReO$_4^-$ phases on an arbitrary scale. [From Carneiro *et al.* (1984).]

what similar to TTF–TCNQ if scaled to the present $T_{M \to I}$. Finally, the 2:1:0.5 salt has a rather featureless susceptibility with a Curie-like $(1/T)$ tail at low temperatures signifying a high concentration of defects. An ESR study of both α-(ET)$_2$I$_3$ and β-(ET)$_2$I$_3$ has been reported (Venturini *et al.*, 1985).

3.   Discussion

The high-temperature susceptibility may be understood as originating from two contributions. Firstly, the so-called Pauli susceptibility stems from the band nature of the electronic states. It may be calculated from the transfer integral $t_{\parallel}$, which from optical measurements and calculations is 260 meV for $(TMTSF)_2X$ salts and corresponds to a bandwidth of 1.1 eV. The contribution to $\chi(T)$ is a temperature-independent susceptibility of 0.6×10^{-4} emu cm^3/mole, approximately half the measured χ at temperatures just above $T_{M \to I}$. This excess susceptibility is attributed to a "correlation enhancement" owing to the Coulombic repulsion between electrons. This effect is often parameterized into an on-site Coulombic energy, the "Hubbard U." An enhancement factor of 2 as observed for $(TMTSF)_2X$ gives $U = 4t_{\parallel}$ according to the zero-temperature theory of Takahashi (1970), but later analysis with the temperature dependence of χ taken into account yields $U/4t_{\parallel} = 0.4$ (Tanaka and Tanaka, 1983). These values for $U/4t_{\parallel}$ are typical of good organic conductors but the importance of correlations is difficult to assess from such estimates of $U/4t_{\parallel}$. However, it should be noted that there is no evidence for strong effects on the conductivity originating from the dimerization gap in the electronic band structure, so in this respect U is not large in $(TMTSF)_2X$ salts.

The low value of U in $(TMTSF)_2X$ salts, in view of their three-quarters filled electron band (or one-quarter hole band), is in striking contrast to the high U's in one-quarter filled $M(TCNQ)_2$ conductors which often have a susceptibility enhancement over the Pauli susceptibility of factors of 10–30, suggesting that $U/4t_{\parallel} \gg 1$. As pointed out by Mazumdar and Bloch (1983), U is an effective parameter which is magnified at the band filling of one-quarter. This makes it much easier to understand why $M(TCNQ)_2$ salts show strong correlation effects and why in $(TMTSF)_2X$ salts U is so low.

With respect to the $ET:ReO_4$ salts, a similar analysis cannot be carried out since the absolute value of the χ is not known. However, the fact that $\chi(T)$ is constant above $T_{M \to I}$ in the 2:1 salts suggests a low value of $U/4t_{\parallel}$.

The low-temperature behavior of $\chi(T)$ can be understood from the usual concepts of a charge-density wave, where the abruptness of the phase transition seems to be the major signature of the effects of anion ordering on the magnetic properties. In the case of the spin-density wave, Overhausers' theory for itinerant antiferromagnetism gives a satisfactory description of the observed phenomena (Overhauser, 1962). Regarding the superconducting transition temperature in $(TMTSF)_2ClO_4$, which is lower when measured by susceptibility than by conductivity, a possible explanation is as follows. At some temperature superconducting lamellae develop, presumably parallel to the ab planes, giving a zero resistance; but in between the lamellae, the normal metal still contributes to $\chi(T)$. Only when the entire crystal at some lower temperature is in its zero resistance state does one observe the magnetic transition.

B. ESR Linewidths

In Fig. 38 we show the ESR linewidths ΔH_{pp} for the six compounds previously discussed. The large widths of $(TMTSF)_2X$ salts, compared to the $ET:ReO_4$ derivatives, is due to the larger spin-orbit coupling in selenium than in sulfur. Apart from this, the linewidths show the following features:

(1)   In the case of an anion assisted Peierls transition, the linewidth changes abruptly, e.g., $(TMTSF)_2ReO_4$ and $(ET)_2ReO_4$, whereas the usual Peierls transition in $(ET)_3(ReO_4)_2$ is more gradual.

(2)   The linewidth increases below a spin-density wave transition.

(3)   Even in the absence of a transition, the linewidth does change appreciably with temperature. Hence, the ESR linewidth, ΔH_{pp}, provides a good means of discriminating between the different phases of the salt (Williams *et al.*, 1985b; Carneiro *et al.*, 1984). Table XV summarizes crystallographic, ESR, and conductivity data on ET salts.

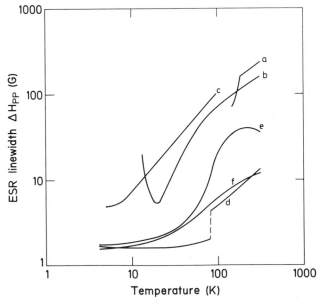

Fig. 38.   A comparison of the ESR linewidths of various $(TMTSF)_2X$ [X = monovalent anion, ReO_4 (a), PF_6^- (b), and ClO_4^- (c)] and $ET:ReO_4^-$ [2:1 (d), 3:2 (e), 2:1:0.5 (f)] salts. [From Carneiro *et al.* (1984).]

TABLE XV

Summary of Crystallographic, ESR, and Conductivity Data on ET Salts[a]

Compound	Space group	V_c (Å³) at 298 K	Z	Interstack mode	Intrastack mode	ESR[b] linewidth (G) (Avg. ± esd)	Comment	$T_{M \to I}$ (K)
β-(ET)₂I₃	$P\bar{1}$	856	1	L	a	21 ± 4	T_c = 1.4 K[c], h	—
α-(ET)₂I₃	$P\bar{1}$	1717	2	—	—	94 ± 15	Metallic	135
γ-(ET)₃(I₃)₂.₅	$Pbnm$	6812	4	—	—	—	T_c = 2.5 K	—
ε-(ET)(I₃)(I₈)₀.₅	$P2_1/c$	4211	—	—	—	—	T_c = 2.5 K	—
β-(ET)₂IBr₂	$P\bar{1}$	828	1	L	a	21 ± 2	T_c = 2.7 K	22
β-(ET)₂ICl₂	$P\bar{1}$	814	1	L	"b"[d]	9.8	Metallic	—
β-(ET)₂BrICl	$P\bar{1}$	821	1	L	"b"[d]	10	Metallic	—
β-(ET)₂I₂Br	$P\bar{1}$	842	1	L	a	—	Metallic	—
β-(ET)₂AuI₂	$P\bar{1}$	845	1	L	a	—	T_c = 5 K	81
(ET)₂ReO₄	$P\bar{1}$	1565	2	L	a	16 ± 3	T_c = 2 K (P > 4 kbar)	—
(ET)₂ReO₄(THF)₀.₅	$P2_1/c$	1948	—	—	—	15 ± 3	Semiconductor(?)	—
α-(ET)₃(ReO₄)₂	$P2_1/n$	2418	2	Z	b	35 ± 4	Metallic(?)	50
α-(ET)₃(BrO₄)₂	$P\bar{1}$	1213	1	Z	b	—	Metallic	—
(ET)₂(BrO₄)(TCE)₀.₅	$P\bar{1}$	1668	2	Z	b	31 ± 6	—	—
(ET)₂BrO₄	$P\bar{1}$	1589	2	L	a	—	Metallic	—
(ET)₂InBr₄	$P\bar{1}$	1820	2	L	"c"[e]	9 ± 1	Semiconductor	—
(ET)₂(ClO₄)(TCE)₀.₅	$P\bar{1}$	1684	2	Z	b	30 ± 3	Metallic	None
(ET)₃(ClO₄)₂	$P\bar{1}$	1182	1	Z	b	—	Metallic	170
(ET)₂(ClO₄)(C₄H₈O₂)	$P2/c$	1814	2	L	c	—	Semiconductor	—
(ET)₂AsF₆	$A2/a$	3274	4	L	c	13 ± 5	Semiconductor	—
α-(ET)₂PF₆	$P\bar{1}$	794	1	L	a	—	Semiconductor	—
β-(ET)₂PF₆	$Pnna$	3255	4	L	c	—	Semiconductor	—
α'-(ET)₂Ag(CN)₂	$P2/n$	1646	4			38	Pseudo-1D	—
α'-(ET)₂Au(CN)₂	$P2/n$	1658	4			34	Pseudo-1D	—
α'-(ET)₂AuBr₂	$P2/n$	1622	4			54	Pseudo-1D	—
(ET)Ag₄(CN)₂[f]	$Fddd$		8	3D network		13	Metallic	120
(ET)₃Ag_xI₈[g], x = 6.4	$P\bar{1}$	1566	2				Metallic	298

[a] Taken in part from Leung et al. (1985a).

[b] Average ± estimated value.

[c] T_c = 8 K at 1.3 kbar (Murata et al., 1985). T_c = 7.5 K at 1 kbar (Laukhin et al., 1985; Shchegolev et al., 1985).

[d] "b" is a combination of a + b modes.

[e] "c" is a combination of a + c modes.

[f] See Geiser et al. (1985).

[g] See Geiser et al. (1986).

[h] The actual "critical pressure" for assessing the high-T_c β* state (8 K) is ≥ 0.5 kbar (see Schirber et al., 1986).

VIII. NEW DEVELOPMENTS (ET–X CONDUCTORS)

A. Raising T_c in ET–X Conductors

Higher T_c's may be obtained if molecular and anionic disorder and structural modulation can be avoided in β-$(ET)_2X$ conductors containing monovalent triatomic anions (Williams et al., 1986). The anion appears to shorten or lengthen cation network $S \cdots S$ distances through variations in anion length. The approximate anion lengths are I_3^- (10.1 Å) $> I_2Br^-$ (9.7 Å) $>$ AuI_2^- (9.4 Å) $>$ IBr_2^- (9.3 Å). In addition, for superconductivity to occur the linear anion must be symmetric, as I–I–Br$^-$ is disordered in the crystal, and, therefore, generates a random electrical potential which is sensed by the conducting $S \cdots S$ network and superconductivity is not observed (Emge et al., 1985). Most importantly, the hydrogen–anion interactions $(CH_2 \cdots X^-)$ in turn control the $S \cdots S$ contact distances in these systems. Large anions such as I_3^-, I_2Br^-, AuI_2^-, and IBr_2^- in β-$(ET)_2X$ compounds provide ET networks that cause nearly isotropic two-dimensional metallic properties (Williams, 1985c; Williams, et al., 1984a; Crabtree et al., 1984; Carlson et al., 1985c; Whangbo et al., 1985a,b; Jacobsen et al., 1985; Mori et al., 1984). However, when the anion is smaller, as in the case of ICl_2^-, a more compact packing of ET molecules is possible, and stronger $CH_2 \cdots X^-$ interactions are expected (from the hydrogen bonding trend $Cl^- > Br^- > I^-$). This leads to a drastically altered $S \cdots S$ network as observed in β-$(ET)_2ICl_2$ (Emge et al., 1985), which causes it to be a one-dimensional metal, and which undergoes a $T_{M \to I}$ at 22 K (Emge et al., 1985).

Figure 39 shows the plot of the unit cell volume V_c (which reflects the anion lengths) versus T_c for β-$(ET)_2X$ (X = I_3^-, IBr_2^-, AuI_2^-). A nearly linear relationship results between V_c and T_c when the "high T_c β* state" (0.5 kbar pressure) (Schirber et al., 1986) of the I_3^- salt is added to the plot. This suggests that the T_c associated with the 1.5 K state is artifically low (Williams et al., 1986).

B. New ET Salts

Several new and unusual ET salts have recently been synthesized. One of these is $(ET)Ag_4(CN)_5$ (Geiser et al., 1985). In the course of a synthesis program to obtain new superconducting β-$(ET)_2X$ derivatives containing linear anions X^- of varying lengths, the synthesis of the ET salt of the $Ag(CN)_2^-$ anion was attempted. A 2:1 salt α-$(ET)_2Ag(CN)_2$ (Beno et al., 1986) was prepared and it was found to be a one-dimensional metal like β-$(ET)_2ICl_2$ (Emge et al., 1986), as well as an unexpected salt $(ET)Ag_4(CN)_5$, which contains a three-dimensional network of ET^+ cations. Every ET^+

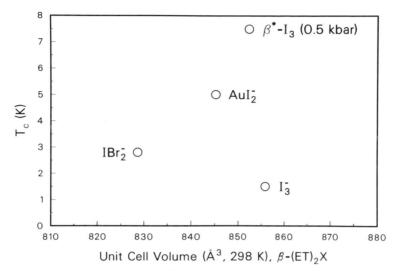

Fig. 39. A plot of superconducting transition temperatures (T_c's) versus unit cell volume (V_c) for β-(ET)$_2$X when X = I$_3^-$, IBr$_2^-$, AuI$_2^-$ (ambient pressure) and X = I$_3^-$ (P = 0.5 kbar) indicating the "high T_c β^* state" (8 K) in I$_3^-$ when under applied pressure. [From Williams *et al.* (1986) and Schirber *et al.* (1986).]

cation is surrounded by four nearest-neighbor ET$^+$ cations in a pseudo-tetrahedral manner. This three-dimensional donor network is interwoven with honeycomblike networks of silver cations and cyanide anions. (ET)Ag$_4$(CN)$_5$ is the *first* example of a compound containing a three-dimensional donor network based on a single radical-cation donor species and is also the first to contain a polymeric anion. Band electronic structure calculations predict the salt to be a three-dimensional metal. The salt has an *Fddd* space group and $Z = 8$. Figure 40 shows a projection view of a layer of ET$^+$ cations surrounded by a "necklace" of silver cyanide anions consisting of ten Ag$^+$ cations and ten CN$^-$ anions.

Another new ET salt synthesized is (ET)$_3$Ag$_x$I$_8$ ($x \simeq 6.4$) (Geiser *et al.*, 1986). The material is prepared by electrocrystallization, using a new type of supporting electrolyte (K[18-crown-6]$_2$Ag$_4$I$_6$) and ET in methylene chloride solvent. The crystal structure is triclinic, space group $P\bar{1}$ ($Z = 2$). The crystal structure consists of alternating layers of ET organic-donor networks and silver iodide polyanions. The layer of the organic donor exhibits a number of S$\cdots$S contacts between 3.35–3.65 Å, thus forming a two-dimensional network. The polyanion consists of a double layer of iodide ions, forming channels parallel to $3a + b$, where the Ag$^+$ ions are distributed over at least 20 interstitial sites. The Ag$^+$ sites are partially occupied in a highly nonrandom way, typical of an inorganic solid

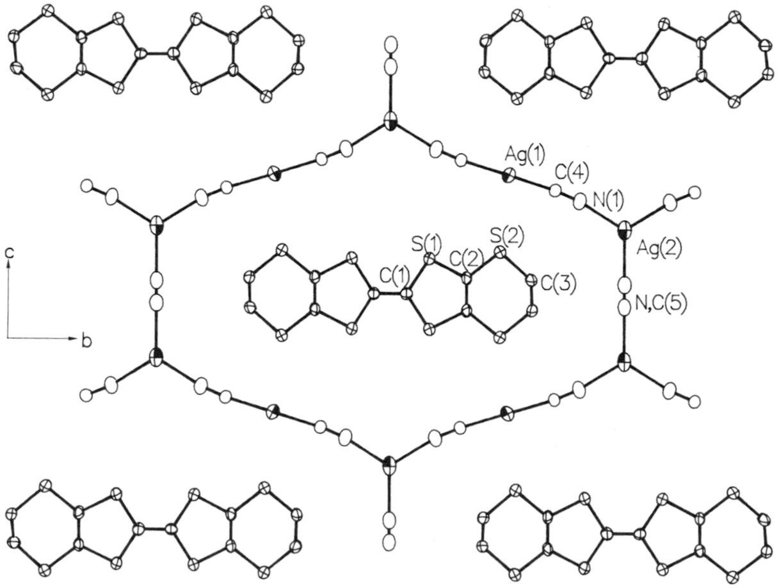

Fig. 40. A projection view of a layer of ET^+ cations interwoven with a honeycomb-like network of polymeric silver cations and cyanide anions in $(ET)Ag_4(CN)_5$. This salt contains the first three-dimensional donor molecule network observed in an organic conductor and a unique polymeric anion. [From Geiser *et al.* (1985).]

electrolyte. Figure 41 shows a perspective view of the unit cell. The anion layer is a two-dimensional analog of the solid electrolytes MAg_4I_5 [$M = K^+$, Rb^+, NH_4^+) (Geller, 1967), $(C_5H_5NH)Ag_5I_6$ (Geller and Owens, 1972), $[(CH_3)_4N]_2Ag_{13}I_{15}$ (Geller and Lind, 1970), and α-AgI (Tubandt and Lorenz, 1914)]. $(ET)_3Ag_xI_8$ is the first synthetic metal to contain a truly two-dimensional polymeric anion, and one which combines the structural features of both an organic electronic conductor and an inorganic solid electrolyte (ionic conductor)! Preliminary conductivity studies indicate the material is metallic at room temperature with conductivity $\sigma = 50$ (ohm cm)$^{-1}$. Conductivity drops by a factor of three at 180 K and by three orders of magnitude between 70 and 30 K. Heating one sample of the solid above 300 K produced an increase in resistivity, contrary to the expected decrease for ionic conductivity. Thus, the conductivity appears to be dominated by electronic, rather than ionic contributions. However, this salt may be the first in a possible series of organic conductor-solid electrolyte hybrids.

Another class of new organic conductors are the α'-$(ET)_2X$ salts with $X = Ag(CN)_2^-$, $Au(CN)_2^-$, and $AuBr_2^-$ (Beno *et al.*, 1986). These salts possess a two-dimensional network, drastically different from the α- and

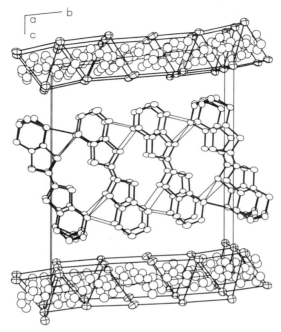

Fig. 41. The unit cell of $(ET)_3Ag_xI_8$ ($x \sim 6.4$) along the a axis. The figure shows one layer of the two-dimensional ET donor-molecule network (center) with $S \cdots S$ contacts (3.35–3.65 Å) indicated by thin lines and two silver iodide layers (top and bottom). This conductor is the first of its type being an electronic solid electrolyte hybrid. [From Geiser *et al.* (1986).]

β-$(ET)_2X$ ($X = I_3^-$, IBr_2^-, and AuI_2^-). Band-electronic structures calculated for the ET network of the α'-$(ET)_2X$ salts show that they are pseudo-one-dimensional metals and are likely susceptible to metal–insulator transitions. The salts are monoclinic space group $P2/n$ with $Z = 4$ and a unit cell volume of ~ 1600 Å^3.

IX. CONCLUDING REMARKS

From structural and transport studies of the organic conductors based on $(TMTSF)_2X$ and $(ET)_2X$, their varied properties appear exceedingly rich and complex. Although these systems are quite complicated, we are able to point out several cases of correspondence between their crystallographic and physical properties, which may be summarized as follows:

(1) As synthetic metals they are by no means unique in terms of high metallic conductivity since several salts of other systems have higher

conductivity. This is in agreement with both experimental and theoretical results based on their intrachain electronic bandwidth $4t_\parallel$ which is not particularly high and certainly very small compared to ordinary metals.

(2)   A unique feature is, however, the very small role that Coulomb interactions appear to play, in particular when considering their stoichiometry. Their low values of U are, at least in part, a molecular property but studies of different phases of $ET:ReO_4$ indicate that U depends specifically on the crystallographic environment. That the electrons are not strongly influenced by Coulomb repulsions appears to be a necessary condition for superconductivity.

(3)   As one-dimensional conductors, these systems are often not very one-dimensional. From a crystallographic and electronic point of view these compounds exhibit rather strong coupling between the stacks. There is variation in this coupling within both the $(TMTSF)_2X$ and the $ET:X$ series, although the variation is much greater in the latter case. It is likely that the richness in behavior is related to a crossover in effective dimensionality, from compound to compound, or as a function of pressure in one compound, so that both one-dimensional instabilities (spin-density and charge-density waves) as well as three-dimensional superconductivity may occur.

For the isostructural series $(TMTSF)_2X$ the differences in physical behavior for different anions X are associated with very minute changes in the crystallographic structure. The anion volume determines uniquely the unit cell volume, but in particular, the average interstack $Se\cdots Se$ distance, which is directly related to the dimensionality, and this seems to correlate with the occurrence of superconductivity. The symmetry of the anion also plays a determining role as it is responsible for structural order–disorder transitions of varying symmetries. In the case of the Peierls transition there is a direct relation between the symmetry (or periodicity) of this superlattice and the symmetry of the Peierls instability, but the role of symmetry is not understood in detail. The same is true for the role of disorder since superconductivity is not destroyed by disorder in ordinary superconductors, whereas it is in $(TMTSF)_2X$ and β-$(ET)_2X$ organic superconductors. A possible explanation for this could be that ordering influences the interchain distances enough to change the effective dimensionality.

Tellurium analogues have been synthesized, and it has been of interest to compare these compounds with the $(TMTSF)_2X$ and $(ET)_2X$ families (Wudl and Aharon-Shalom, 1982; Lerstrup et al., 1982). At this writing the results do not appear promising largely due to the considerable insolubility of the Te-based systems in most organic solvents.

It is clear that the field of organic conductors is continually providing new and unusual materials for study by scientists of widely varying disciplines. Clearly, the potential for future surprises in this field of study is enormous!

X. SUMMARY

The following general statements may be made with regard to charge-transfer conductors discussed in this chapter:

(a) AB type (e.g., TTF–TCNQ)

(1) In the case of charge-transfer molecules (e.g., AB, where A = TTF or variant and B = TCNQ or variant, improved conductivity is provided by the stacking of both molecules. Segregated stacks provide the highest conductivities.

(2) Unsaturation is necessary to provide π-overlap that occurs between molecules in stacks.

(3) Incomplete charge transfer is preferred, which provides the following properties:

   (i)   nonionic ground state,
   (ii) mixed valence stacks,
   (iii) lower energy in systems (E_{ACT}),
   (iv) I (ionization potential $- A$ (electron affinity) $> E_m$ (binding energy)—favors nonionic ground state and conductive state,
   (v) $E_M > I - A$—favors ionic ground state and provides the insulator state.

(4) Selenium compounds provide larger orbitals (compared to S), greater overlap, higher room temperature conductivity and lower metal–insulator transition temperatures.

(b) (TMTSF)$_2$X conductors

(1) Anion size is important—unit cell volume (V_c) controls average Se$\cdots$Se distances; produces pronounced changes in electrical properties.

(2) Exhibit metallic behavior and the conductivity increases at lower temperature, until one reaches the metal–insulator transition.

(3) Metal–insulator state involves SDW or anion ordering and not a Peierls lattice distortion.

(4) Pressure can suppress the SDW ground state resulting in a superconducting ground state.

(5) Pressure also lowers the $T_{M \to I}$ transition temperature.

(6) Below the critical pressure, P_{SC}, the electronic ground state is an insulator of either the SDW type or the anion assisted Peierls type. Above P_{SC} the ground state is often superconducting.

(7) P_{SC} varies monotonically with the anion volume.

(8) T_c at ambient pressure, and the type of low-pressure ground state, depends specifically on the anion symmetries.

(9) Anions are often disordered at room temperature. Lower temperature tends to promote anion ordering. However, rapid cooling "freezes in" anion disorder in the ClO_4^- system.

(10) Pressure increase narrows bandgap.

(11) Molecules not often one-dimensional because of rather strong coupling between stacks especially at lower T and higher P. This *interstack* coupling appears greater in $(ET)_2X$ conductors compared to $(TMTSF)_2X$ conductors.

(12) For $(TMTSF)_2X$ compounds intra- and interstack $Se \cdots Se$ distances are similar at room temperature (~ 3.9–4.0 Å compared to 4.0 Å for van der Waals radius sum). As temperature is lowered from 298 K to 125 K the ratio of the decrease in the interstack–intrastack $Se \cdots Se$ distances in ~ 2. This leads to increased interchain bonding and electronic delocalization through the selenium atom network as T is lowered.

(13) In $(TMTSF)_2X$ conductors when X is AsF_6^-, PF_6^-, or ClO_4^-, weak van der Waal interactions between hydrogen atoms of methyl groups and flourine or oxygen atoms may exist. Anion ordering may be driven by methyl group ordering and attempts to control this behavior, which may be important to the onset of superconductivity, will be extremely difficult.

(b) $(ET)_2X$ conductors

(1) $(ET)_2X$ conductors favor short interstack versus intrastack $S \cdots S$ packing distances; i.e., different from $(TMTSF)_2X$. Whereas planarity is favored in TMTSF molecules, ET molecules deviate considerably from planarity. Two types of interstack packing are found and designated as the L-mode (linear) and Z-mode (zigzag).

(2) Intrastack packing motifs appear to be of three types and are called a, b, and c.

a-stacking produces superconductive phases.

b-stacking is found to produce metallic, but not superconducting materials.

c-stacking produces semiconductors.

(3) Intrastack $S \cdots S$ distances are longer than the interstack separations and generally exceed the sulfur atom van der Waals sum of 3.6 Å at

room temperature. Theoretical studies reveal that both intra- and inter-stack distances contribute to the electrical properties and band structure.

(4)   Interstack distances are generally shorter than 3.6 Å therby producing two-dimensionality in the structure which tends to suppress one-dimensional transitions.

(5)   There is a crossover from one-dimensional behavior at 298 K to three-dimensional behavior at 40 K in β-(ET)$_2$I$_3$ (Jacobsen, et al., 1985).

(6)   Higher T_c's ($\sim$8 K) obtained for β^*-(ET)$_2$I$_3$ with the application of pressure appear to be due to the shear effects whenever nonhydrostatic pressures are applied. When hydrostatic pressures are applied T_c remains suppressed ($<$1.5 K).

(7)   Higher T_c's in ET–X conductors may be obtained if molecular and anionic disorder and structural modulation are avoided.

LIST OF SYMBOLS AND ABBREVIATIONS

TCNQ	7,7,8,8-Tetracyano-p-quinodimethane
TTF	Tetrathiafulvalene
TMTSF	Tetramethytetraselenafulvalene
TMTTF	Tetramethyltetrathiafulvalene
HMTSF	Hexamethylenetetraselenafulvalene
HMTTF	Hexamethylenetetrathiafulvalene
BEDT-TTF (or ET)	Bis(ethylenedithio)tetrathiafulvalene
TNAP	11,11,12,12-Tetracyanonaphtho-2,6-quinodimethane
TSeT	Tetraselenatetracene
$T_{M \rightarrow I}$	Temperature for metal–insulator phase transition
ρ	Resistivity in ohm cm units
V_A	Anion volume (Å^3)
V_{cp}	Calculated volume for unit cell (Å^3)
V_c	Unit cell volume (Å^3)
$\mathbf{k}_F$	In-chain Fermi wave vector; gap in electronic spectrum occurs at wave vector 2 $\mathbf{k}_F$ ($= \pi/a$)
SDW	Spin-density wave
CDW	Charge-density wave
T_c	Transition temperature to superconducting phase
T_O	Anion-ordering temperature
P_{SC}	Pressure at onset of superconducting phase
ΔH_{pp}	ESR peak-to-peak linewidths
p	Number of unpaired electrons per TCNQ; equivalent to degree of charge transfer occurring in the donor–acceptor reaction
U_0	Coulombic repulsion energy between two electrons on same molecule in the nomenclature of Torrance (1979)
V_1	Coulombic repulsion between electrons on neighboring molecules in the nomenclature of Torrance (1979)
χ	Magnetic susceptibility

$k_B T_{MI}/\Delta$ Anisotropy of the electronic structure as a quasi-one-dimensional
            material
k_B 0.086 meV/K
Δ Semiconducting band gap parameter
Hubbard U Difference of on-site Coulomb repulsion energy before and after
            electron transfer in the nomenclature of Tanaka and Tanaka
            (1983) and Hubbard (1978)

REFERENCES

Acker, D. S., Harder, R. J., Hertler, W. R., Mahler, W., Melby, L. R., Benson, R. E.,
    and Mochel, W. E. (1960), *J. Am. Chem. Soc.* **82**, 6408.

Akamatu, H., Inokuchi, H., and Matsunaga, Y. (1954), *Nature* **173**, 168.

Anderson, J. R., and Bechgaard, K. (1975), *J. Org. Chem.* **40**, 2016.

Andres, K., Wudl, F., McWhan, D. B., Thomas, G. A., Nalewajek, D., and Stevens,
    A. L. (1980), *Phys. Rev. Lett.* **45**, 1449.

Andrieux, A., Duroure, C., Jérome, D., and Bechgaard, K. (1979), *J. Phys. Lett.* **40**, 381.

Bardeen, J., Cooper, L. N., and Schrieffer, J. R. (1957) *Phys. Rev.* **108**, 1175.

Bechgaard, K. (1982), *Mol. Cryst. Liq. Cryst.* **79**, 1.

Bechgaard, K., Cowan, D. O., and Bloch, A. N. (1974), *J. Chem. Soc. Chem. Commun.*,
    p. 937.

Bechgaard, K., Cowan, D. O., and Henriksen, L. (1975), *J. Org. Chem.* **40**, 746.

Bechgaard, K., Cowan, D. O., and Bloch, A. N. (1976), *Mol. Cryst. Liq. Cryst.* **32**, 227.

Bechgaard, K., Jacobsen, C. S., and Anderson, N. H. (1978), *Solid State Commun.* **25**,
    875.

Bechgaard, K., Jacobsen, C. S., Mortensen, K., Pedersen, H. J., and Thorup, N. (1980),
    Solid State Commun. **33**, 1119.

Bechgaard, K., Carneiro, K., Rasmussen, F. B., Olsen, H., Rindorf, G., Jacobsen,
    C. S., Pedersen, H., and Scott, J. E. (1981), *J. Am. Chem. Soc.* **103**, 2440.

Bechgaard, K., Carneiro, K., Eg, O., Olsen, M., Rasmussen, F. B., Jacobsen, C., and
    Rindorf, G. (1982), *Mol. Cryst. Liq. Cryst.* **79**, 271.

Bender, K., Dietz, K., Endres, H., Helberg, K. W., Hennig, I., Keller, H. J., Schafer,
    H. W., and Schweitzer, D. (1984), *Mol. Cryst. Liq. Cryst.* **107**, 451.

Beno, M. A., Blackman, G. S., Leung, P. C. W., and Williams, J. M. (1983), *Solid State
    Commun.* **48**, 99.

Beno, M. A., Firestone, M. A., Leung, P. C. W., Sowa, L. M., Wang, H. H., Williams,
    J. M., and Whangbo, M.-H. (1986), *Solid State Commun.* **57**, 735.

Braam, J. M., Carlson, K. D., Stephens, D. A., Rehan, A. E., Compton, S. J., and
    Williams, J. M. (1986), *Inorg. Synth.* **24**, 132.

Bychkov, Y. A., Gor'kov, L. P., and Dzyaloshinskii, I. E. (1966), *JETP Lett. (Engl.
    Transl.)* **23**, 489.

Carlson, K. D., Crabtree, G. W., Hall, L. N., Copps, P. T., Wang, H. H., Emge, T. J.,
    Beno, M. A., and Williams, J. M. (1985a), *Mol. Cryst. Liq. Cryst.* **119**, 357.

Carlson, K. D., Crabtree, G. W., Hall, L. N., Behroozi, F., Copps, P. T., Sowa, L. M.,
    Nuñez, L., Firestone, M. A., Wang, H. H., Beno, M. A., Emge. T. J., and Williams,
    J. M. (1985b), *Mol. Cryst. Liq. Cryst.* **125**, 159.

Carlson, K. D., Crabtree, G. W., Choi, M., Hall, L. N., Copps, P. T., Wang, H. H.,
    Emge, T. J., Beno, M. A., and Williams, J. M. (1985c), *Mol. Cryst. Liq. Cryst.* **125**,
    145.

Carneiro, K., Scott, J. C., and Engler, E. M. (1984), *Solid State Commun.* **50**, 477.

Cómes, R. (1975), "One Dimensional Conductors" (H. G. Schuster, ed.), p. 32, Springer-Verlag, Berlin and New York.

Cowan, D. O., Bloch, A. N., and Bechgaard, K. (1981), U. S. Patent No. 4246173.

Cox, S., Boysel, R. M., Moses, D., Wudl, F., Chen, J., Ochsenbein, S., Heeger, A. J., Walsh, W. M., and Rupp, L. N. (1984), *Solid State Commun.* **49**, 259.

Crabtree, G. W., Carlson, K. D., Hall, L. N., Copps, P. T., Wang, H. H., Emge, T. J., Beno, M. A., and Williams, J. M. (1984), *Phys. Rev. B* **30**, 2958.

Emery, V. J., Bruinsma, R., and Barisíc, S. (1982), *Phys. Rev. Lett.* **48**, 1039.

Emge, T. J., Leung, P. C. W., Beno, M. A., Schultz, A. J., Wang, H. H., Sowa, L. M., and Williams, J. M. (1984), *Phys. Rev. B* **30**, 6780.

Emge, T. J., Wang, H. H., Beno, M. A., Leung, P. C. W., Firestone, M. A., Jenkins, H. C., Cook, J. D., Carlson, K. D., Williams, J. M., Venturini, E. L., Azevedo, L. J., and Schirber, J. E. (1985), *Inorg. Chem.* **24**, 1736.

Emge, T. J., Wang, H. H., Leung, P. C. W., Rust, P. R., Jackson, P. L., Carlson, K. D., Williams, J. M., Whangbo, M.-H., Venturini, E. L., Schirber, J. E., Azevedo, L. J., and Ferraro, J. R. (1986), *J. Am. Chem. Soc.* **108**, 695.

Engler, W., and Patel, V. V. (1974), *J. Am. Chem. Soc.* **96**, 7376.

Eriks, K., Wang, H. H., Reed, P. E., Beno, M. A., Appelman, E. H., and Williams, J. M. (1984), *Acta Crystallogr. Sect. C* **41**, 257.

Fournel, A., More, C., Roger, G., Sorbier, J. P., Delrieu, J. M., Jérome, D., Ribault, M., and Bechgaard, K. (1981), *J. Phys. Lett.* **42**, 445.

Friend, R. H., Jérome, D., Fabre, J. M., Giral, L., and Bechgaard, K. (1978), *J. Phys. C* **11**, 263.

Garoche, P., Brussetti, R., and Bechgaard, K. (1982), *Phys. Rev. Lett.* **49**, 1346.

Geiser, U., Wang, H. H., Gerdom, L. E., Firestone, M. A., Sowa, L. M., Williams, J. M., and Whangbo, M.-H. (1985), *J. Am. Chem. Soc.* **107**. 8305.

Geiser, U., Wang, H. H., Donega, K. M., Anderson, B. A., Williams, J. M., and Kwak, J. F. (1986), *Inorg. Chem.* **25**. 402.

Geller, S. (1967), *Science* **157**, 310.

Geller, S., and Lind, M. D. (1970), *J. Chem. Phys.* **52**, 5854.

Geller, S., and Owens, B. B. (1972), *J. Phys. Chem. Solids* **33**, 5854.

Greene, R. E., and Street, G. B. (1984), *Science (Washington, D. C.)* **226**, 651.

Gubser, D. U., Fuller, W. W., Poehler, T. O., Stokes, J., Cowan, D. O., Lee, M., and Bloch, A. N. (1982), *Mol. Cryst. Liq. Cryst.* **79**, 225.

Guy, D. R. P., Boebinger, G. S., Marseglia, E. A., and Friend, R. H. (1983), *J. Phys. C* **16**, 691.

Hubbard, J. (1978), *Phys. Rev. B*. **17**, 494.

Huheey, J. E. (1978), "Inorganic Chemistry—Principles of Structure and Reactivity," 2nd ed., p. 71, Harper, New York.

Isett, L. C., and Perez-Albuerne, E. A. (1978), *in* "Synthesis and Properties of Low-Dimensional Materials" (J. S. Miller and A. J. Epstein, eds.) *Ann. N. Y. Acad. Sci.* **313**, 395.

Jacobsen, C. S., Mortensen, K., Thorup, N., Tanner, D. B., Weger, M., and Bechgaard, K. (1981), *Chem. Scr.* **17**, 103.

Jacobsen, C. S., Pedersen, H. J., Mortensen, K., Rindorf, G., Thorup, N., Torrance, J. B., and Bechgaard, K. (1982), *J. Phys. C* **15**, 2651.

Jacobsen, C. S., Williams, J. M., and Wang, H. H. (1985), *Solid State Commun.* **54**, 937.

Jérome, D., Mazaud, A., Ribault, M., and Bechgaard, K. (1980), *J. Phys. Lett.* **41**, 95.

Jérome, D., and Schulz, H. J. (1982), *Adv. Phys.* **31**, 299.

Johannsen, I., Bechgaard, K., Mortensen, K., and Jacobsen, C. S. (1983), *J. Chem. Soc. Chem. Commun.*, p. 295.

Kagoshima, S., Pouget, J. P., Saito, G., and Inokuchi, H. (1983), *Solid State Commun.* **45**, 1001.

Kaminskii, V. F., Prokhorova, T. G., Shibaeva, R. P., and Yagubskii, E. B. (1984), *JETP Lett. (Engl. Transl.)* **39**, 17.

Kistenmacher, T. J., Emge, T. J., Shu, P., and Cowan, D. O. (1979), *Acta Crystallogr. Sect. B* **35**, 772.

Kobayashi, H., Kobayashi, A., Saito, G., and Inokuchi, H. (1982), *Chem. Lett. J. Chem. Soc. Japan,* p. 245.

Kobayashi, H., Kobayashi, A., Sasaki, Y., Saito, G., Enoki, T., and Inokuchi, H. (1983a), *J. Am. Chem. Soc.* **105**, 297.

Kobayashi, H., Mori, T., Kato, R., Kobayashi, A., Sasaki, Y., Saito, G., and Inokuchi, H. (1983b), *Chem. Lett. J. Chem. Soc. Japan,* p. 681.

Kohn, W. (1959), *Phys. Rev. Lett.* **2**, 393.

Krug, W. P. (1977), Ph.D. dissertation, Johns Hopkins Univ., Baltimore, Maryland.

Laukhin, V. N., Kostyochenko, E. E., Sushko, Yu. B., Shchegolev, I. F., and Yugubskii, E. B. (1985), *Zh. Eksp. Teor. Fiz. Pis'ma Red.* **41**, 68.

Lee, M. M., Stokes, J. P., Wiygul, F. M., Kistenmacher, T. J., Poehler, T. J., Bloch, A. N., Fuller, W. W., and Gubser, D. U. (1982), *Mol. Cryst. Liq. Cryst.* **79**, 501.

Lee, V. Y., Engler, E. M., Schumacker, R. R., and Parkin, S. S. P. (1983), *J. Chem. Soc. Chem. Commun.,* p. 235.

Lerstrup, K., Talham, D., Bloch, A., Poehler, T., and Cowan, D. (1982), *J. Chem. Soc. Chem. Commun.,* p. 336.

Lerstrup, K., Lee, M., Wiygul, F. M., Kistenmacher, T. J., and Cowan, D. O. (1983), *J. Chem. Soc. Chem. Commun.,* p. 294.

Leung, P. C. W., Beno, M. A., Blackman, G. S., Coughlin, B. R., Miderski, C. A., Joss, W., Crabtree, G. W., and Williams, J. M. (1984a), *Acta Crystallogr. Sect. C* **40**, 1331.

Leung, P. C. W., Emge, T. J., Beno, M. A., Wang, H. H., Williams, J. M., Petricek, V., and Coppens, P. (1984b), *J. Am. Chem. Soc.* **106**, 7644.

Leung, P. C. W., Beno, M. A., Emge. T. J., Wang. H. H., Bowman, M. K., Firestone, M. A., Sowa, L. M., and Williams, J. M. (1985a), *Mol. Cryst. Liq. Cryst.* **125**, 113.

Leung, P. C. W., Emge, T. J., Beno, M. A., Wang, H. H., Williams, J. M., Petricek, V., and Coppens, P. (1985b), *J. Am. Chem. Soc.* **107**, 6184.

Little, W. A. (1964), *Phys. Rev. A* **134**, 1416.

Maaroufi, A., Coulon, C., Flandrois, S., Delhaes, P., Mortensen, K., and Bechgaard, K. (1983), *Solid State Commun.* **48**, 555.

Mazumdar, S., and Bloch, A. N. (1983), *Phys. Rev. Lett.* **50**, 207.

McCoy, H. H., and Moore, W. C. (1911), *J. Am. Chem. Soc.* **33**, 1273; Kraus, H. (1913), *J. Am. Chem. Soc.* **34**, 1732.

Megtert, S., Pouget, J. P., and Cómes, R. (1978), *Ann. N. Y. Acad. Sci.* **313**, 234.

Melby, L. R., Harder, R. J., Hertler, W. R., Mahler, W., Benson, R. E., and Mochel, W. E. (1962), *J. Am. Chem. Soc.* **84**, 3374.

Merkhanov, V. A., Kostyuchenko, E. E., Lauklin, V. N., Lobkovskaya, R. M., Makova, M. K., Shibaeva, R. P., Shchegolev, I. F., and Yagubskii, E. B. (1985), *Zh. Eksp. Teor. Fiz. Pis'ma Red.* **41**, 146.

Miller, J. S. (1978), *Ann. N. Y. Acad. Sci.* **313**, 25–60.

Mizuno, M., Garito, A. F., and Cava, M. P. (1978), *J. Chem. Soc. Chem. Commun.,* p. 18.

Moradpour, A., Peyrussan, V., Johansen, I., Bechgaard, K. (1982), *J. Org. Chem.* **48**, 388.

Moradpour, A., Bechgaard, K., Barrie, M., Lenoir, C., Murata, K., Lacoe, R. C., Ribault, M., and Jérome, D. (1985), *Mol. Cryst. Liq. Cryst.* **119**, 69.

Moret, R., Pouget, J. P., Cómes, R., and Bechgaard, K. (1982), *Phys. Rev. Lett.* **49**, 1008.

Moret, R., Pouget, J. P., Cómes, R., and Bechgaard, K. (1983), *J. Phys. (Paris) Colloq.* **44**, C3–957.

Mori, T., Kobayashi, A., Sasaki, Y., Kobayashi, H., Saito, G., and Inokuchi, H. (1982), *Chem. Lett., J. Chem. Soc. Japan*, p. 1963.

Mori, T., Kobayashi, A., Sasaki, Y., Kobayashi, H., Saito, G., and Inokuchi, H. (1984), *Chem. Lett., J. Chem. Soc. Japan*, p. 957.

Morosin, B., Schirber, J. E., Greene, R. L., and Engler, E. M. (1982), *Phys. Rev. B* **26**, 2660.

Mortensen, K., Tomkiewicz, Y., and Bechgaard, K. (1982), *Phys. Rev. B* **25**, 1982.

Mortensen, K., Jacobsen, C. S., Lindegaard-Andersen, A., and Bechgaard, K. (1983), *J. Phys. (Paris) Colloq.* **44**, C3–963.

Mortensen, K., Jacobsen, C. S. Bechgaard, K., and Williams, J. M. (1984), *Synth. Met.* **9**, 63.

Murata, K., Tokumoto, M., Anzai, H., Bando, H., Saito, G., Kajimura, K., and Ishiguro, T. (1985), *J. Phys. Soc. Jpn.* **54**, 1.

Overhauser, A. W. (1962), *Phys. Rev.* **128**, 1437.

Parkin. S. S. P. (1984), "Physics and Chemistry of Electrons and Ions in Condensed Matter" (J. V. Acrivos *et al.*, eds.), pp. 655–666. Reidel Publ., Boston, Massachusetts.

Parkin, S. S. P., Ribault, M., Jérome, D., and Bechgaard, K. (1981), *J. Phys. C* **14**, 5305.

Parkin, S. S. P., Creuzet, F., Ribault, M., Jérome, D., Bechgaard, K., and Fabre, J. M. (1982), *Mol. Cryst. Liq. Cryst.* **79**, 249.

Parkin, S. S. P., Engler, E. M., Schumaker, R. R., Lagier, R., Lee, V. Y., Scott, J. C., and Greene, R. L. (1983a), *Phys. Rev. Lett.* **50**, 270.

Parkin, S. S. P., Engler, E. M., Schumaker, R. R., Lagier, R., Lee, V. Y., Voiron, J., Carneiro, K., Scott, J. C., and Greene, R. L. (1983b), *J. Phys. (Paris) Colloq.* **44**, C3-791.

Parkin, S. S. P., Engler, E. M., Lee, V. Y., and Schumaker, R. R. (1985), *Mol. Cryst. Liq. Cryst.* **119**, 375.

Pauling, L. (1960), "The Nature of the Chemical Bond," 3rd ed., pp. 537–540, Cornell Univ. Press, Ithaca, New York.

Pedersen, H. J., Scott, J. C., and Bechgaard, K. (1982), *Phys. Scr.* **25**, 849.

Peierls, R. E. (1955), "Quantum Theory of Solids," p. 107, Oxford Univ. Press, London and New York.

Pouget, J. P., Moret, R., Cómes, R., and Bechgaard, K. (1981), *J. Phys. Lett. (Orsay, Fr.)* **42**, 543.

Pouget, J. P., Moret, R., Cómes, R., Bechgaard, K., Fabre, J. M., and Giral, L. (1982), *Mol. Cryst. Liq. Cryst.* **79**, 129.

Pouget, J. P., Shirane, G., Bechgaard, K., and Fabre, J., (1983a), *Phys. Rev. B* **27**, 5203.

Pouget, J. P., Moret, R., Cómes, R., Shirane, G., Bechgaard, K., and Fabre, J. M. (1983b), *J. Phys. (Paris) Colloq.* **44**, C3-969.

Reed, P. E. Braam, J. M., Sowa, L. M., Barkhau, R. A., Blackman, G. S., Cox, D. D., Ball, G. A., Wang. H. H., and Williams, J. M. (1985), *Inorg. Synth.* (in press).

Renker, B., and Cómes, R. (1975) *in* "Low Dimensional Cooperative Phenomena" (J. H. Keller, ed.), p. 235, Plenum, New York.

Renker, B., Pintshovius, L., Glaser, W., Rietschel, K., and Cómes, R. (1975) *in* "One Dimensional Conductors" (H. G. Schuster, ed.), p. 32, Springer-Verlag, Berlin and New York.

Rindorf, G., Soling, H., and Thorup, N. (1982), *Acta Crystallogr. Sect. B* **38**, 2805.

Rindorf, G., Soling, H., and Thorup. N. (1984), *Acta Crystallogr. Sect. C* **40**, 1137.

Saito, G., Enoki, T., Toriumi, T., and Inokuchi, H. (1982), *Solid State Commun.* **42**, 557.

Schirber, J. E., Azevedo. L. J., Kwak, J. F., Venturini, E. L., Leung. P. C. W., Beno, M. A., Wang. H. H., and Williams, J. M. (1986), *Phys. Rev. B* **33**, 1987.

Schirber, J. E., Azevedo, L. J., Kwak, J. F., Venturini, E. L., Beno, M. A., Wang, H. H., and Williams, J. M. (1986). *Solid State Commun.* **59**, 525.

Schumaker, R. R., Lee, V. Y., and Engler, E. M. (1983), *J. Phys. (Paris) Colloq.* **44**, C3-1139.

Scott, J. C. (1982), *Mol. Cryst. Liq. Cryst.* **79**, 49.

Scott, B. A., LaPlaca, S. J., Torrance, J. B., Silverman, B. B., and Welber, B. (1978), *Ann. N. Y. Acad. Sci.* **313**, 369.

Scott, J. C., Pedersen, H. J., and Bechgaard, K. (1981), *Phys. Rev. B* **24**, 475.

Shannon, R. D., and Prewitt, C. T. (1969), *Acta Crystallogr. Sect. B* **25**, 925; Shannon, R. D. (1969), *Acta. Crystallogr. Sect. A* **32**, 751.

Shchegolev, I. F., Yagubskii, E. B., and Laukhin, V. N. (1985), *Mol. Cryst. Liq. Cryst.* **126**, 365.

Shibaeva, R. P., Kaminskii, V. F., and Yagubskii, E. B. (1985), *Mol. Cryst. Liq. Cryst.* **119**, 361.

Shu, P., Bloch, A. N., Carruthers, T. F., and Cowan, D. O. (1977), *J. Chem. Soc. Chem. Commun.,* p. 505.

Siemons, L. J., Bierstedt, P. E., and Kepler, R. G. (1963), *J. Chem. Phys.* **39**, 3523.

Soda, G., Jérome, D., Weger, M., Alizon, J., Gallice, J., Robert, H., Fabre, J. M., and Giral, L. (1977), *J. Phys. (Paris)* **38**, 931.

Soling, G., Rindorf, G., and Thorup, N. (1982), *Cryst. Struct. Commun.* **11**, 1980.

Soling, H., Rindorf, G., and Thorup, N. (1983), *Acta Crystallogr. Sect. B* **38**, 2805.

Spector, W. S., ed. (1956), "Handbook of Toxicology," Vol. I., p. 340, Saunders, Philadelphia.

Takahashi, T. (1970), *Prog. Theor. Phys.* **40**, 348.

Takahashi, T., Jerome, D. D., and Bechgaard, K. (1982), *J. Phys. (Paris) Lett.* **43**, 565.

Tanaka, J., and Tanaka, C. (1983), *J. Phys. (Paris) Colloq.* **44**, C3-997.

Tanaka, C., Tanaka, J., Dietz, K., Katayama, C., and Tanaka, M. (1983), *Bull. Chem. Soc. Jpn.* **56**, 405.

Teramae, H., Tanaka, K., and Yanabe, T. (1982), *Solid State Commun.* **44**, 431.

Teramae, H., Tanaka, K., Shiotani, K., and Yamabe, T. (1983), *Solid State Commun.* **46**, 633.

Thorup, N., Rindorf, G., Soling, H., and Bechgaard, K. (1981), *Acta Crystallogr. Sect. B* **37**, 1236.

Thorup, N., Rindorf, G., Soling, H., Johannsen, I., Mortensen, K., and Bechgaard, K. (1983), *J. Phys. (Paris) Colloq.* **44**, C3-1017.

Tomíc, S., Pouget, J. P., Jérome, D., Bechgaard, K., and Williams, J. M. (1983), *J. Phys. (Paris)* **44**, 375.

Torrance, J. B. (1978), *in* "Synthesis and Properties of Low-Dimensional Materials" (J. S. Miller and A. J. Epstein, eds.), *Ann. N. Y. Acad. Sci.* **313**, 210.

Torrance, J. B. (1985), *Mol. Cryst. Liq. Cryst.* **126**, 55.

Torrance, J. B. (1979) *in* "Molecular Metals" (W. E. Hatfield, ed.), pp. 7–14, Plenum, New York.

Torrance, J. B., Pedersen, H. J., and Bechgaard, K. (1982), *Phys. Rev. Lett.* **49**, 881.

Tubandt, C., and Lorenz, E. (1914), *Z. Phys. Chem.* **87**, 513.

Venturini, E. L., Azevedo, L. J., Schirber, J. E., Williams, J. M., and Wang. H. H. (1985), *Phys. Rev. B: Condens. Matter* **32**, 2819.

Wang, H. H., Beno, M. A., Geiser, U., Firestone, M. A., Webb, K. S., Nuñez, L., Crabtree, G. W., Carlson, K. D., Williams, J. M., Azevedo, L. J., Kwak, J. F., and Schirber, J. E. (1985), *Inorg. Chem.* **24**, 2465.

Whangbo, M.-H., Williams, J. M., Beno, M. A., and Dorfman, J. R. (1983), *J. Am. Chem. Soc.* **105**, 1645.

Whangbo, M.-H., Williams, J. M., Leung, P. C. W., Beno, M. A., Emge, T. J., and Wang, H. H. (1985a), *Inorg. Chem.* **24**, 3500.

Whangbo, M.-H., Leung, P. C. W., Beno, M. A., Enge, T. J., Wang, H. H., Carlson, K. D., and Crabtree, G. W. (1985b), *J. Am. Chem. Soc.* **107**, 5815.

Williams, J. M. (1985)—work in progress.

Williams, J. M., and Carneiro, K. (1985), *Adv. Inorg. Chem. Radiochem.* **29**, 249–296.

Williams, J. M., Beno, M. A., Appelman, E. K., Capriotti, J. M., Wudl, F., Aharon-Shalom, E., and Nalewajek, D. (1982a), *Mol. Cryst. Liq. Cryst.* **79**, 319.

Williams, J. M., Schultz, A. J., Underhill, A. E., and Carneiro, K. (1982b), *in* "Extended Linear Chain Compounds" (J. S. Miller, ed.), Vol. I, p. 73, Plenum, New York.

Williams, J. M., Beno, M. A., Sullivan, J. C., Banovetz, L. M., Braam, J. M., Blackman, G. S., Carlson, C. D., Greer, D. L., and Loesing, D. M. (1983a), *J. Am. Chem. Soc.* **105**, 643.

Williams, J. M., Beno, M. A., Banovetz, L. M., Braam, J. M., Blackman, G. S., Carlson, C. D., Greer, D. L., Loesing, D. M., and Carneiro, K. (1983b), *J. Phys. (Paris) Colloq.* **44**, C3-941.

Williams, J. M., Beno, M. A., Sullivan, J. C., Banovetz, L. M., Braam, J. M., Blackman, G. S., Carlson, C. D., Greer, D. L., Loesing, D. M., and Carneiro, K. (1983c), *Phys. Rev. B: Condens. Matter* **28**, 2873.

Williams, J. M., Emge, T. J., Wang, H. H., Beno, M. A., Copps, P. T., Hall, L. N., Carlson, K. D., and Crabtree, G. W. (1984a), *Inorg. Chem.* **23**, 2558.

Williams, J. M., Wang, H. H., Beno, M. A., Emge, T. J., Sowa, L. M., Copps, P. T., Behroozi, F., Hall, L. N., Carlson, K. D., and Crabtree, G. W. (1984b), *Inorg. Chem.* **23**, 3839.

Williams, J. M., Beno, M. A., Wang, H. H., and Reed, P. E. (1984c), *Inorg. Chem.* **23**, 1790.

Williams, J. M., Beno, M. A., Geiser, U., Firestone, M. A., Webb, K. S., Nuñez, L., Crabtree, G. W., Carlson, K. D., Azevedo, L. J., Kwak, J. F., and Schirber, J. E. (1985a), paper presented at International Conference "Materials and Mechanisms of Superconductivity," Ames, Iowa, May 29–31, *Physica B* **135**, 520.

Williams, J. M., Bowman, M. K., Wang, H. H., and Firestone, M. A. (1985b)—work in progress.

Williams, J. M., Beno, M. A., Wang, H. H., Geiser, U. W., Emge, T. J., Leung, P. C. W., Crabtree, G. W., Carlson, K. D., Azevedo, L. J., Venturini, E. L., Schirber, J. E., Kwak, J. F., and Whangbo, M.-H. (1986), paper presented at International Conference on Neutron Scattering, Aug. 19–23, Santa Fe, New Mexico, *Physica B* **136**, 371.

Williams, J. M. (1985c), *in* "Progress in Inorganic Chemistry" (S. J. Lippard, ed.), Vol. 33, Chapter 4, pp. 183–220, Wiley (Interscience), New York.

Wudl, F. (1984), *Acc. Chem. Res.* **17,** 227.

Wudl, F., and Aharon-Shalom, E. (1982), *J. Am. Chem. Soc.* **104**, 1154.

Wudl, F., and Nalewajek, D. (1980), *J. Chem. Soc. Chem. Commun.,* p. 866.

Wudl, F., Smith, G. M., and Hufnagel, E. J. (1970), *J. Chem. Soc. Chem. Commun.,* 1453.

Wudl, F., Aharon-Shalom, E., and Bertz, S. H. (1981), *J. Org. Chem.* **46**, 4612.

Wudl, F., Aharon-Shalom, E., Nalewajek, D. Waszczak, J. V., Walsh, W. M., Rupp,

L. W., Chaikin, P., Lacoe, R., Burns, M., Poehler, T. O., Beno, M. A., and Williams, J. M. (1982), *J. Chem. Phys.* **76**, 5497.

Yagubskii, E. B., Shchegolev, I. F., Pesotskii, S. I., Laukhin, V. N., Kononovich, P. A., Kartsovnik, M. W., and Zvarykina, A. V. (1984a), *JETP Lett. (Engl. Transl.)* **39**, 329.

Yagubskii, E. B., Shchegolev, I. F., Laukhin, V. N., Kononovich, P. A., Karatsovnik, M. W., Zvarykina, A. V., and Buravov, L. I. (1984b), *JETP Lett. (Engl. Transl.)* **39**, 12.

Review Articles

Bechgaard, K., and Jérome, D. (1982), Organic superconductors, *Sci. Am.* **247**, 51–61.

Bechgaard, K., and Jérome, D. (1985), Superconducting organic solids, *Chem. Tech.* (November), 682–685.

Berlinsky, A. J. (1979), One-dimensional metals and charge density wave effects in these materials, *Rep. Prog. Phys.* **42**, 1243–1283.

Bryce, M. R. (1985), Tetrathiafulvalenes (TTF) and their selenium and tellurium analogs (TSF and TTeF): electron donors for organic metals, *Aldrichimica Acta* **18**, 73–77.

Bryce, M. R., and Murphy, L. C. (1984), Organic metals, *Nature* **309**, 119–126.

Epstein, A. J., and Miller, J. S. (1979), Linear-chain conductors, *Sci. Am.* **241**, 52–61.

Greene, R. L., and Street, G. B. (1984), Conducting organic materials, *Science* **226**, 651–656.

Jérome, D., and Schulz, H. J. (1982), Organic conductors and superconductors, *Adv. Phys.* **31**, 299–490 and references therein.

Maugh, T. H. (1983), Number of organic superconductors grows, *Science* **222**, 606.

Metz, W. D. (1975), New materials: a growing list of nonmetallic metals, *Science (Washington, D. C.)* **220**, 450.

Robinson, A. L. (1983), Nanocomputers from organic molecules, *Science* **220**, 940.

Torrance, J. B. (1985), An overview of organic charge-transfer solids: insulators, metals and the neutral-ionic transitions, *Mol. Cryst. Liq. Cryst.* **126**, 55–67.

Williams, J. M., and Carneiro, K. (1985), *Adv. Inorg. Chem. Radiochem.* **29**, 249–296.

Williams, J. M., Beno, M. A., Wang, H. H., Emge, T. J., Copps, P. T., Hall, L. N., Carlson, K. D., and Crabtree, G. W. (1985), Organic superconductors: structure-property relations and materials, *Philos. Trans. R. Soc. London Ser. A* **314**, 83–95.

Williams, J. M. (1985), Organic superconductors, *in* "Progress in Inorganic Chemistry" (S. J. Lippard, ed.), Vol. 33, pp. 183–220, Wiley, New York.

Williams, J. M., Beno, M. A., Wang, H. H., Leung, P. C. W., Emge, T. J., Geiser, U., and Carlson, K. D. (1985), Organic superconductors: structural aspects and design of new materials, *Acc. Chem. Res.* **18**, 261–267.

Williams, J. M., Wang, H. H., Emge, T. J., Geiser, U., Beno, M. A., Leung, P. C. W., Carlson, K. D., Thorn, R. J., and Schultz, A. J. (1987), The rational design of synthetic metal superconductors, *in* "Progress in Inorganic Chemistry" (S. J. Lippard, ed.), Wiley, New York.

Wudl, F. (1984), From organic metals to superconductors: managing conduction electrons in organic solids, *Acc. Chem. Res.* **17**, 227–232.

Books

Alcacer, L., ed. (1980), "The Physics and Chemistry of Low-Dimensional Solids," pp. 1–436, Reidel Publ., Dordrecht.

Devreese, J. T., Evrard, R. P., and Van Doven, V. E. (1977), "Highly Conducting One-Dimensional Solids," pp. 1–422, Plenum, New York.

Hatfield, W. E., ed. (1979), "Molecular Metals," pp. 1–555, Plenum, New York.

Keller, H. J., ed. (1977), "Chemistry and Physics of One-Dimensional Metals," pp. 1–426, Plenum, New York.

Keller, H. J., ed. (1975), "Low Dimensional Cooperative Phenomena," pp. 1–350, Plenum, New York.

Lehn, J. M., and Rees, Ch. W., eds. (1985), "Molecular Semiconductors—Photoelectrical Properties and Solar Cells," pp. 1–288, Springer-Verlag, Berlin.

Miller, J. S., ed. (1981–1983), "Extended Linear Chain Compounds," Vols. 1–3, Plenum, New York.

Miller, J. S., and Epstein, A. J., eds. (1978), "Synthesis and Properties of Low-Dimensional Materials," *Ann. N. Y. Acad. Sci.* **313**, 1–828.

Pope, M., and Swenberg, C. E. (1982), "Electronic Processes in Organic Crystals," pp. 1–799, Clarendon, New York.

Proc. Int. Conf. Synth. Met., (1983), *J. Phys. (Paris) Colloq. C3,* **44**.

Schuster, H. G., ed. (1975), "One Dimensional Conductors," pp. 1–371, Springer-Verlag, Berlin and New York.

3 CONDUCTIVE ORGANIC POLYMERS

I. CONDUCTING ORGANIC POLYMERS

In the mid-1970s considerable attention was directed toward the study of conducting organic polymers following the discovery of Shirakawa *et al.* (1973) and Ito *et al.* (1974), who found that polyacetylene $(CH)_x$ could be prepared as films having a metallic luster. Earlier Hatano *et al.* (1961) had found that polyacetylene conducted electrical current. In 1977 it was discovered that the conductivity of polyacetylene could be increased by 13 orders of magnitude by doping it with various donor or acceptor species to give *p*-type or *n*-type semiconductors and conductors (Shirakawa *et al.*, 1977, 1978; Chiang *et al.*, 1977, 1978a–c; Park *et al.*, 1979; Fincher *et al.*, 1978). Conductivities approaching 10^3 (ohm cm)$^{-1}$ were found in doped $(CH)_x$ (e.g., $[(CH)(AsF_6)_{0.1}]_x$). Although variants of $(CH)_x$ have been since investigated, none have approached the high conductivity of doped $(CH)_x$. Doped polyacetylene remains the most promising conductive organic polymer discovered to date, and is also the most extensively studied and understood of the organic polymers (Chien, 1984). As a consequence, the bulk of the discussion in this chapter is directed toward polyactylene and, in particular, doped polyacetylene.

A. Polyacetylene

1. Synthesis

The synthesis of $(CH)_x$ is beset by numerous side reactions depending on the method of synthesis. This is rather typical of the synthesis of other conductive polymers as well. One major side reaction involves cyclization to benzene. Other reactions lead to branched and cross-linked polymers. The degree of polymerization also varies and several methods of polymerization are possible. Polymerization can be initiated by radiation and by other initiators (e.g., anionic, cationic, and radical).

A catalyst is needed for the polymerization and the preferred catalyst is $Ti(OBu)_4$–$AlEt_3$ since it produces almost linear, high molecular weight polymers in the form of free-standing films of high mechanical strength. If a temperature of ~78°C is employed, the film is almost totally the cis isomer. If the temperature is allowed to reach ~150°C, in decane as solvent, the film that forms is the trans isomer. The cis isomer may be converted to the trans isomer by heating at ~200°C for ~1 h (MacDiarmid and Heeger, 1979/1980). The cis and trans forms of $(CH)_x$ are shown as

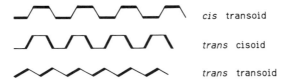

Polyacetylene may be found to reach aggregation numbers of 500 to 220,000. Unfortunately $(CH)_x$ is insoluble in most solvents and conventional molecular weight studies are impossible.

There are four energetically favorable backbone geometries that can be envisioned for $(CH)_x$. Three of these are

                                                             cis transoid

                                                             trans cisoid

                                                             trans transoid

In these structures, the heavy lines represent short bonds, slightly longer than the C=C bond in ethylene; the light lines represent short bonds (somewhat shorter than the C—C bond in ethane); the fourth cis cisoid is thought to be sterically too unfavorable to exist, except in diblock copolymers (Chien, 1984).

2. Doping Polyacetylene

Pure *cis*- and *trans*-polyacetylene have room temperature conductivities of 10^{-10} and 10^{-5} (ohm cm)$^{-1}$, respectively, which can be increased to 10^3 (ohm cm)$^{-1}$ with doping. Conductivity can be achieved by oxidation or by reduction as

$$(CH)_x + \tfrac{3}{2}yI_2 \rightarrow (CH)_x^{y+} + yI_3^- \tag{1}$$

$$(CH)_x + Li \rightarrow (CH)_x^{y-} + yLi^+ \tag{2}$$

and these are basically redox reactions.

The doping process may also be seen to involve a charge-transfer reaction. The two types of doping can be p-type or n-type. In p-type doping, doping is conducted with an electron acceptor (e.g., I_2) whereby a monovalent anion and a stabilized polycarbonium ion are created. The dopant removes the free spin (neutral soliton) for $(CH)_x$ and converts it to a positive soliton [Eq. (1)]. Here conduction involves positive charge carriers. In n-type doping, doping is conducted with a donor specie (e.g., Li). A monovalent cation and a stabilized polycarbanion are formed. In this case an electron is added to the neutral soliton converting it to a negative soliton [Eq. (2)]. Conduction then involves negative charge carriers. For a further discussion on theory related to solitons see Chapter 7.

There are various methods available for the doping of polyacetylene. Unfortunately because $(CH)_x$ is insoluble in most solvents, most methods involve heterogeneous doping and create inhomogeneously doped materials which are somewhat difficult to characterize precisely.

Both *cis*- and *trans*-polyacetylene can be doped. Doping the cis isomer is generally preferred because it is mechanically stronger, flexible, and dopes to give higher conductivities (Chien, 1984).

a. Doping with Gaseous Vapors. Doping of $(CH)_x$ may be achieved by exposure to vapors (e.g., AsF_5, SbF_5, I_2, and Br_2). The usual technique is to simultaneously measure conductivity as the doping process occurs. To diminish the inhomogeneous doping it is generally recommended that the procedure be carried out slowly. Doping with I_2 vapor requires 24 h to reach a maximum conductivity. Doping with Br_2 or Cl_2 can cause complications such as halogenation reactions. Figure 1 shows the electrical conductivity of bromine-doped polyacetylene measured during doping. Figure 2 shows the conductivity of iodine-doped polyacetylene as a function of dopant concentration.

Doping with H_2SO_4 and $HClO_4$ are also possible (Gau et al., 1979), although care must be taken with $HClO_4$. Elsenbaumer and Miller (1985), reported explosions using $HClO_4$ doped polyacetylene and the products are apparently unstable [see also Varyu et al. (1985)]. Österholm and Passiniemi (1986) have found films of polythiophene doped with $HClO_4$ to be stable at room temperature, but are unstable when heated to temperatures above 100°C.

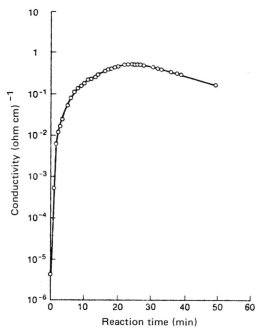

Fig. 1.   Electrical conductivity of *trans*-polyacetylene during bromine doping. [From Chiang *et al.* (1978c).]

b.   Doping by Oxidizing Cations.   Salts that contain NO^+ or NO_2^+ ions can be used to provide *p*-doping of polyacetylene. The salts oxidize the $(CH)_x$ and introduce anions in the lattice, which stabilize the polycarbocation. The oxidizing cation is liberated as the reduced neutral gas in the process according to Gau *et al.*, 1979, and MacDiarmid and Heeger, 1979/1980.

$$(CH)_x + 0.05x NO_2^+ SbF_6^- \rightarrow [CH(SbF_6)_{0.05}]_x + 0.05x NO_2 \qquad (3)$$

c.   Photo-Initiated Doping.   Polyacetylene may also be doped by using triarylsulfonium and diaryliodonium salts which decompose under ultraviolet radiation to produce the corresponding protonic acids (Chien, 1984).

d.   Solution Doping.   This method is used mostly to produce *n*-doped polyacetylene. Sodium naphthalide in THF is added to polyacetylene and the conductivity monitored.

e.   Electrical Doping.   The electrochemical method of doping appears to be the preferred method. In a typical experiment *cis*-polyacetylene film is used as the anode in the electrolysis of aqueous 0.5M KI solution and the conductivity is monitored during the electrolysis. Both *p*- and *n*-type

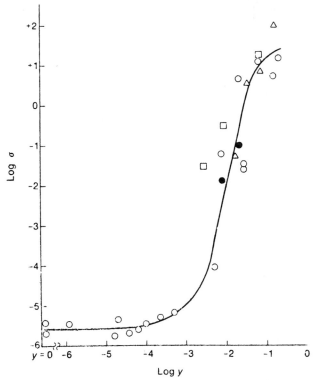

Fig. 2.   Variation of conductivity with y for *trans*-(CHI$_y$)$_x$: △, Chiang *et al.* (1978c); □, Epstein *et al.* (1981b); ○, Warakomski *et al.* (1983); ●, Sichel *et al.* (1982b). [From Chien (1984).]

doping are possible by use of this method. Some polymers (e.g., polypyrrole and polythiophene) may be simultaneously oxidized and polymerized electrochemically. Several advantages of the electrochemical doping process are apparent:

(1)   doping levels can be precisely controlled;
(2)   the approach to equilibrium can be determined by the current level; and
(3)   electrochemical undoping is cleaner, forming no chemical products which require removal, and the sample can be returned to a neutral state.

Table I shows the conductivities for various *p*-type doped polyacetylenes and Table II lists the conductivities for various *n*-type doped polyacetylenes. Table III lists the various dopants that have been used with polyacetylene.

As has been alluded to earlier in this chapter, one of the major problems

TABLE I

p-Type Dopants for Polyacetylene[a,b,c]

Sample	Conductivity [(ohm cm)$^{-1}$ at 298 K]	Sample	Conductivity [(ohm cm)$^{-1}$ at 298 K]
cis-$(CH)_x$	1.7×10^{-9}	cis-$[CH(SbF_5)_{0.06}]$	5×10^1
$trans$-$(CH)_x$	4.4×10^{-5}	cis-$[CH(SbCl_6)_{0.009}]_x$	1×10^{-1}
cis-$[CH(HF)_{0.115}]_x{}^d$	4.9	cis-$[CH(SbCl_8)_{0.0095}]_x$	1×10^1
$trans$-$[CH(HBr)_{0.04}]_x{}^d$	7×10^{-4}	cis-$[CH(SbCl_5)_{0.022}]_x$	2
$trans$-$(CHCl_{0.02})_x{}^d$	1×10^{-4}	cis-$[CH(BF_2)_{0.09}]_x$	1×10^2
$trans$-$(CHBr_{0.23})_x$	4×10^{-1}	cis-$[CH(IF_{5.63})_{0.096}]_x$	1.5×10^2
cis-$[CH(ICl)_{0.14}]_x$	5×10^1	cis-$[CH(SO_3F)_y]_x{}^e$	7×10^2
cis-$(CHI_{0.30})_x$	5.5×10^2	cis-$[CH(ClO_4)_{0.0645}]_x$	9.7×10^2
$trans$-$(CHI_{0.30})_x$	1.6×10^2	cis-$[CH_{1.11}(AsF_5OH)_{0.011}]_x$	$\sim 7 \times 10^2$
cis-$[CH(IBr)_{0.15}]_x$	4.0×10^2	cis-$[CH_{1.058}(PF_5OH)_{0.058}]_x{}^f$	$\sim 3 \times 10^1$
$trans$-$[CH(AsF_5)_{0.15}]_x$	4.0×10^2	cis-$[CH(HSO_4)_{0.12}(CH_3NO_2)_{0.02}]_x$	4.3×10^2
cis-$[CH(AsF_5)_{0.10}]_x{}^d$	1.2×10^3	cis-$[CH(H_2SO_4)_{0.106}(H_2O)_{0.070}]_x$	1.2×10^3
cis-$[CH_{1.1}(AsF_6)_{0.10}]_x$	$\sim 7 \times 10^2$	cis-$[CH(HClO_4)_{0.127}(H_2O)_{0.297}]_x$	1.2×10^3
cis-$[CH(AsF_4)_{0.077}]_x$	2.0×10^2	cis-$[CH(CF_3SO_3H)_y]_x{}^g$	—
cis-$[CH(SbF_6)_{0.05}]_x$	4.0×10^2	cis-$[CH(XeOF_4)_{0.025}]_x$	5×10^1

[a] Taken from MacDiarmid and Heeger (1979/1980) and Chien (1984).

[b] Cis or trans refers to the principal isomeric composition before doping. [After Chiang *et al.* (1978b).]

[c] Composition by elemental analysis unless stated otherwise.

[d] Composition by weight uptake.

[e] Dopant was $(SO_3F)_2$; no composition or analysis was given (Anderson *et al.*, 1978).

[f] By electrochemical doping.

[g] Resistivity as a function of y and temperature given, but not conductivity. [Rolland *et al.* (1980).]

TABLE II

n-Type Dopants for Polacetylene[a,b,c]

Sample	Conductivity [(ohm cm)$^{-1}$ at 298 K]
cis-$[Li_{0.30}(CH)]_x$	2.0×10^2
cis-$[Na_{0.21}(CH)]_x$	2.5×10^1
cis-$[K_{0.16}(CH)]_x$	5.0×10^1
$trans$-$[Na_{0.28}(CH)]_x$	8.0×10^1
cis-$[CH(Bu_4N)_{0.08}]_x{}^d$	—

[a] Cis and trans refers to principal isomeric condition before doping. [After Chiang *et al.* (1978b).]

[b] Composition by weight uptake.

[c] Taken from Chien (1984).

[d] Electrochemical doping.

TABLE III

Various Dopants Used with Polyacetylene

Dopant	References
I_2	Sichel et al. (1982a), Chiang et al. (1977)
Br_2	Kletter et al. (1980), Sichel et al. (1982)
Cl_2	Chien (1984)
ICl	MacDiarmid and Heeger (1979/1980)
IBr	MacDiarmid and Heeger (1979/1980)
$(NO_2)BF_4$	Chien (1984)
$(NO_2)PF_6$, $(NO_2)SbF_6$	Gau et al. (1979)
$HClO_4$	Gau et al. (1979)
H_2SO_4	Gau et al. (1979)
HNO_3	Billaud et al. (1983a)
HSO_4^-	Przluski et al. (1983)
$H(H_2O)_y^+(BF_3OH)^-$	McAndrew and MacDiarmid (1986)
$AgClO_4$	Clarke et al. (1978)
$Fe(ClO_4)_3$	Reynolds et al. (1983)
Rare earth nitrates	Boils et al. (1983)
BF_3	Tanaka et al. (1980)
$FeCl_3$	Sichel et al. (1982a), Prón et al. (1981a,c), Azakura et al. (1985), Jones et al. (1985)
$FeBr_3$	Prón et al. (1981b)
$AlCl_3$	Kúlszewicz et al. (1981)
$InCl_3$	Billaud et al. (1983b)
InI_3	Billaud et al. (1983b)
$ZrCl_4$	Shirakawa and Kobayashi (1983), Billaud et al. (1983b)
$HfCl_4$	Shirakawa and Kobayashi (1983); Billaud et al. (1983b)
$TeCl_4$	Selig et al. (1983)
$TeBr_4$	Shirakawa and Kobayashi (1983)
TeI_4	Shirakawa and Kobayashi (1983)
$SnCl_4$	Shirakawa and Kobayashi (1983), Billaud et al. (1983b)
SnI_4	Shirakawa and Kobayashi (1983)
$SeCl_4$	Shirakawa and Kobayashi (1983)
$TiCl_4$	Shirakawa and Kobayashi (1983)
TiI_4	Billaud et al. (1983b)
$FeCl_4^-$	Przyluski et al. (1983)
$AlCl_4^-$	Przyluski et al. (1983)
AsF_5	Chiang et al. (1977), Shirakawa et al. (1977)
SbF_5	Chiang et al. (1977), Soderholm et al. (1985)
$NbCl_5$	Shirakawa and Kobayashi (1983)
NbF_5	Shirakawa and Kobayashi (1983)
TaF_5	Shirakawa and Kobayashi (1983)
$TaCl_5$	Shirakawa and Kobayashi (1983)
$TaBr_5$	Shirakawa and Kobayashi (1983)
TaI_5	Shirakawa and Kobayashi (1983)
$MoCl_5$	Shirakawa and Kobayashi (1983), Galthier et al. (1983)
ReF_6	Selig et al. (1983)

continued

TABLE III *(continued)*

Dopant	References
$IrCl_6$	Sichel *et al.* (1982a)
InF_6	Selig *et al.* (1983)
UF_6	Selig *et al.* (1983)
OsF_6	Selig *et al.* (1983)
XeF_6	Selig *et al.* (1983)
TeF_6	Selig *et al.* (1983)
SF_6	Selig *et al.* (1983)
SeF_6	Selig *et al.* (1983)
WF_6	Shirakawa and Kobayashi (1983)
WCl_6	Shirakawa and Kobayashi (1983)
ReF_7	Selig *et al.* (1983)
Alkali metal naphthalides	MacDiarmid and Heeger (1979/1980)

in the doping of $(CH)_x$, or for that matter most other polymers, is the inhomogenous doping that results. There are at least three types of inhomogeneity that result in the doping process of polyacetylene, namely,

(1) dopant concentration gradients from the surface to the center of sample (Moses *et al.*, 1982);

(2) radial concentration gradients of dopant from the surface to the center of the $(CH)_x$ fibrils (~200 Å diameter) (Epstein *et al.*, 1981a); and

(3) aggregation of dopant so that it remains undispersed through the sample (Mortensen *et al.*, 1980; Tomkiewicz *et al.*, 1981).

Sichel *et al.* (1982a) have suggested that the dopant molecular size and shape determines how well the dopant is dispersed in the polymer. A comparison of dopant sizes is shown in Fig. 3. In the case of iodine-doped polyacetylene, the I_3^- ion appears to intercalate the polymer causing minimum disruption of the polymer chains (Street and Clarke, 1981). The larger "spherical" dopants may cause greater disturbance to the $(CH)_x$ chains and inhomogeneity may be greater when using such dopants.

3. Nature of Doped Polyacetylene

Only the nature of the iodine specie in iodine-doped polyacetylene is known with certainty. By a combination of resonance Raman and Mössbauer spectroscopy it was found that it exists as I_3^- and I_5^- anions. Raman bands at $105-107$ cm^{-1} were indicative of I_3^- and assigned to the symmetric stretching vibration of the I_3^- ion. Bands at $150-160$ cm^{-1} were indicative of I_5^- (Hsu *et al.*, 1978; Lefrant *et al.*, 1979) and assigned to the I_5^- ion. Figure 4 shows the Raman spectrum of trans-rich polyacetylene (Hsu *et al.*, 1978). The Mössbauer spectra showed that no I^- ions were present.

Dopant	Structure		Dimensions
I I_3^-	Linear chain 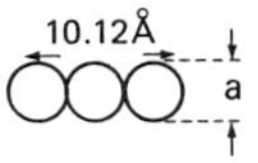	10.12Å ⟷ a	a = 3.37Å
Br Br_3^-	Linear chain 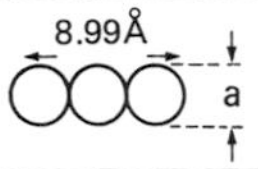	8.99Å ⟷ a	a = 3.00Å
$IrCl_6^{-2}$	Octahedron 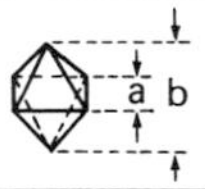	a b	a = 6.08Å b = 8.6Å
AsF_6^{-1}	Octahedron 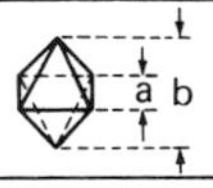	a b	a = 4.41Å b = 6.24Å
$FeCl_4^{-2}$	Tetrahedron 	a	a = 7.05Å

Fig. 3. Dopant molecule sizes and shapes. [From Sichel *et al.* (1982a).]

The I_5^- moiety forms from the reaction $I_2 \cdots I_3^- \rightarrow I_5^-$. Pumping on heavily iodine-doped polyacetylene causes the I_5^- band in the Raman spectrum to disappear (Faulques and Lefrant, 1983). Results with bromine-doped polyacetylene appear to show similar polyhalide moieties (e.g., Br_3^- and Br_5^-) (Faulques and Lefrant, 1983). Further support for these results came from x-ray photoelectron spectroscopy (XPS) (Salaneck *et al.*, 1980) and mass spectroscopy (Allen *et al.*, 1979). Table IV summarizes vibrational data for various materials including iodine-doped $(CH)_x$ containing I_3^- and I_5^- moieties.

For other dopants the nature of the dopant moiety is not entirely clear. For example, in AsF_5-doped materials, several conflicting conclusions have been made because of the problems arising from the multioxidation states of As. AsF_5 is very reactive and in the presence of water forms HF. The first step in AsF_5 doping is thought to follow the equation

$$(CH)_x + yAsF_5 \rightarrow [CH^{y+}(AsF_5^- \cdot)_y]_x \qquad (4)$$

However, EPR studies do not show any paramagnetism, which rules out radical formation. It has been proposed that radicals dimerize to form As_2F_{10} species according to

$$[CH^{y+}(AsF_5^- \cdot)_y]_x \rightarrow [CH^{y+}(As_2F_{10})_{y/2}^{2-}]_x \qquad (5)$$

(MacDiarmid and Heeger, 1979/1980). Another proposal is that of Clarke *et al.* (1979), where an AsF_6^- moiety was postulated according to

$$2(CH)_x + 3xyAsF_5 \rightarrow 2[CH^+(AsF_6)_y^-]_x + xyAsF_5 \qquad (6)$$

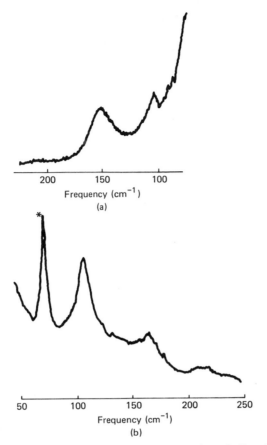

Fig. 4. Raman spectrum of the trans-rich polyacetylene–iodine derivative at room temperature. (a) Laser excitation is 15 mW at 4765 Å; bandpass is 1–2 cm^{-1} at 5100 Å. (b) Laser excitation is 15 mW at 6471 Å; bandpass is 2 cm^{-1} at 6471 Å; asterisk indicates ghost line of the spectrometer. [From Hsu *et al.* (1978).]

Other mechanisms are, of course, possible. Ebert and Selig (1982) have proposed, as in graphite intercalation with AsF$_5$ and SbF$_5$, the following reactions:

$$AsF_6^- + AsF_5 \rightarrow As_2F_{11}^- \tag{7}$$

$$SbF_6^- + SbF_5 \rightarrow Sb_2F_{11}^- \tag{8}$$

The evidence for the dimetallic species was provided by NMR measurements (e.g., ^{121}Sb). With FeCl$_3$ doping, the following reaction has been proposed:

$$(CH)_x + 0.2\ FeCl_3 \rightarrow [CH^{0.2+}(FeCl_4)_{0.1}^-]_x + 0.1\ FeCl_2 \tag{9}$$

[see Prón *et al.* (1981c)].

TABLE IV

Comparison of Raman and FIR for Various Polyiodine Compounds[a,b]

	Raman (cm^{-1})		Far infrared (cm^{-1})			Geometry of polyiodides
	$\nu_1(I_2)$	$\nu_3(I_3^-)$[c]	$\nu_1(I_3^-)$	$\nu_3(I_3^-)$	$\nu_2(I_3^-)$	
Iodine (benzene)[d]	209(S)					
Iodine (Me$_2$SO)[d]	189(S)					
[(ϕCH)$^{0.4-2.0+}$(I$_5^-$)$_{0.4-2.0}$]	166(S)(p)	143(VW)	112(M–W)	143(M)[e]	n.a.	$(I_2)_2 \cdots I-I-I^{(-)}$
2 Perylene·3I$_2$	175(S)	145(W)	115(M)		n.a.	$I_2 \cdots I-I-I^{(-)}$
Ni(OEP)(I$_5$)$_{0.24}$	167(S)		113(sh)		n.a.	
Pd(dpg)$_2$(I$_5$)$_{0.2}$	160(VS)		104(M–W)		n.a.	$I_2 \cdots I-I=I^{(-)}$
Ni(dpg)$_2$(I$_5$)$_{0.2}$	162(VS)		107(W)		n.a.	$I_2 \cdots I-I-I^{(-)}$
(Phenancetin)$_2$H$^+$I$_3^-$I$_2$	187(M)		120(S)		n.a.	$I_2(I_3^-)^-$
Cs$_2$I$_8$	172(S)	150(W)	105(S)		n.a.	$I_2(I_3^-)_2^-$
[Poly(acetylene)$^{0.07+}$(I$_3^-$)$_{0.07}$]			107(S)		n.a.	I_3^-
Ni(Pc)(I$_3$)$_{0.33}$			107(S)		n.a.	I_3^-
I$_3^-$ (Solution)[d]		~149	114	145	52	I_3^-
CsI$_3$		~140	~107, ~100	149[f]	69	I_3^-

Right of $(I_2)_2 \cdots I-I-I^{(-)}$ through $I_2(I_3^-)_2^-$: All contain I$_3^-$ and I$_2$ units and possess Raman spectra of these species

Right of the I_3^- entries: All contain I$_3^-$ units and possess Raman spectra of species

[a] Taken from J. R. Ferraro *et al.* (1984).
[b] Abbreviations: n.a. = not available; S = strong; M = medium; W = weak; sh = shoulder; p = polarized Raman band.
[c] Normally forbidden in Raman; selection rules relax when ion becomes bound to I$_2$.
[d] From K. Nakamoto (1978).
[e] 166 cm^{-1} band also observed, which may be the ν_1 (I$_2$) vibration.
[f] 103 cm^{-1} band also observed and assigned to ν_1 (I$_3^-$) vibration.

4. Crystal Structures of Polyacetylene and Doped Polyacetylene

Many difficulties are involved in attempting to determine the crystal structure of $(CH)_x$. Fewer than 10 reflections are observed when using x-ray diffraction methods and the polymer is randomly oriented causing errors in assigning the chain-axis direction. The crystallites may not provide an atomic arrangement that is most favorable thermodynamically. Three structural studies of *cis*-polyacetylene have been made. The x-ray study by Baughman *et al.* (1978) and two electron diffraction studies by Chien *et al.* (1982) and Lieser *et al.* (1980) are compared in Table V. Figure 5 depicts the molecular configuration of *cis*-polyacetylene in the crystal.

Several crystal studies of *trans*-polyacetylene have also been reported. Table VI shows the derived unit cell parameters and Fig. 6 shows the packing arrangement. Discrepancies are noted in Table VI, which may be due to more than one structural modification of *trans*-$(CH)_x$. For additional discussion on the crystal structures of $(CH)_x$ see Chien (1984).

Iodine-doped polyacetylene has been studied by Baughman *et al.* (1978) and Hsu *et al.* (1978) using x-ray diffraction techniques. A sample with the stoichiometry as $(CHI_{0.16})_x$ had a d spacing of 7.6–7.9 Å. The iodine intercalates between the (100) planes of $(CH)_x$, and an expansion occurs in the interplanar spacing to accommodate the iodine atoms. Figure 7 shows a structural model for the iodine-intercalated polyacetylenes.

TABLE V

Unit Cell Parameters for *cis*-Polyacetylene[a]

	E.D.[b,c] aligned fibrils	X-ray[d] randomly oriented film	E.D.[b,e] polymer suspension
Lattice type	Orthorhombic	Orthorhombic	Orthorhombic
a (Å)	7.68	7.61	7.74
b (Å)	4.46	4.47	4.32
c (Å)	4.38[f]	4.39	4.47
Density (g cm^{-3})	1.15	1.16	1.16
C--C--C	125°	127°	
Setting angle ϕ	32°[g]	59°	

[a] Taken from Chien (1984).
[b] E.D. = electron diffraction.
[c] Chien *et al.* (1982).
[d] Baughman *et al.* (1978) took the b axis to be along the molecular axis.
[e] Lieser *et al.* (1980).
[f] Molecular chain axis and fiber axis.
[g] Value for the cis–transoid structure.

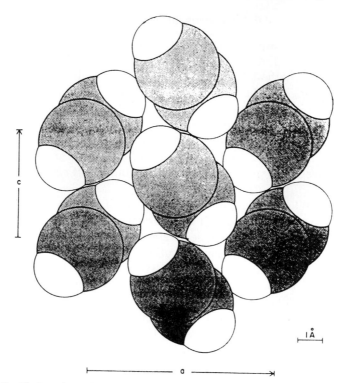

Fig. 5. Chain axis projection of the structure of *cis*-polyacetylene. Hydrogen atoms are white and carbon atoms are grey. The atom radii correspond to the van der Waals radii. [From Baughman *et al.* (1978).]

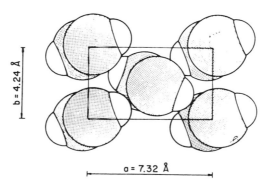

Fig. 6. Molecular conformation of *trans*-polyacetylene in the crystal. [From Chien (1984).]

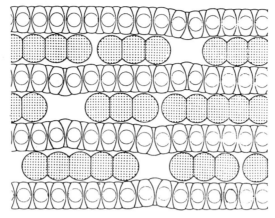

Fig. 7. Structured model for iodine-doped polyacetylene. [From Chien (1984).]

TABLE VI

Unit Cell Parameters for *trans*-Polyacetylene[a]

	E.D.[b,c] aligned fibrils	E.D.[b,d] model I unoriented	E.D.[b,d] model II	X-ray[e]
Lattice type	Orthorhombic[f]	Orthorhombic	Monoclinic $(\gamma = 98°)$	Monoclinic $(\beta = 91–93°)$
a (Å)	7.32	5.62	3.73	4.24
b (Å)	4.24	4.92	3.73	7.32
c (Å)	2.46	2.59	2.44	2.46
Density (g cm^{-3})	1.13	1.20	1.27	
C--C--C	122°	135°	120°	
Setting angle ϕ	24–28°			55°

[a] Taken from Chien (1984).
[b] E.D. = electron diffraction.
[c] Shimamura *et al.* (1981).
[d] Lieser *et al.* (1980).
[e] Fincher *et al.* (1982).
[f] Approximate lattice type.

The crystal structure of Na-, K-, Rb-, and Cs-doped polyacetylene have been determined by crystal packing and x-ray diffraction techniques [see Baughman *et al.* (1983)]. Each metallic complex is tetragonal and the polyacetylene chains form a host lattice in which the alkali-metal ions are present in channels. Lithium forms an amorphous structure, possibly because of its smaller size compared to other alkali metals. Figure 8 shows the coordination of the metal ion in alkali-metal-doped polyacetylene. Table VII lists a comparison of the observed and calculated unit cell

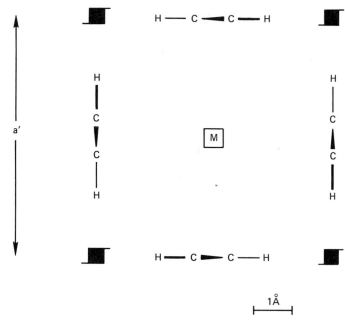

Fig. 8. Coordination of a metal ion (M) in alkali-metal-doped polyacetylene. [From Baughman *et al.* (1983).]

TABLE VII

Comparison of the Observed Unit Cell Parameters (a') for Alkali-Metal-Polyace-tylene Complexes with the Minimum Parameters Calculated From the Require-ment That the Metal Ions Fit in the Channels and the Minimum Parameter Calculated From the Requirements of Nonoverlapping Polyacetylene Chains (6.04 ± 0.36 Å)[a]

Metal	Calculated a'	Observed a'
Li	4.45 Å[b]	Amorphous
Na	5.52[b]	6.0 ± 0.1
K	5.94	5.98 ± 0.05
Rb	6.16	6.19 ± 0.07
Cs	6.43	6.43 ± 0.07

[a] From Baughman *et al.* (1983).

[b] Since these values of a' calculated using the carbon–metal distances are smaller than the 6.04 Å parameter required for the host phase, the latter is predicted for a'.

parameters for alkali-metal $(CH)_x$ polymers. Crystallographic parameter similarities with alkali-metal-doped graphite were observed. The results suggest that hybridization between carbon p_z orbitals and metal s orbitals are involved in the alkali-metal-doped polyacetylene.

5. Stability

If the doped polyacetylene is to become valuable as an electrical conductor it must have reasonable stability to air, moisture, and heat. Doped polyacetylene is relatively stable to auto-oxidation unlike the undoped polymer (Chien, 1984). For example, iodine-doped $(CH)_x$ loses its silvery luster and the relative resistance increases from 1.5 to 4.5 ohm with pumping in a vacuum system for 24 h and some loss in iodine is noted. The iodine-doped polymer is relatively stable to both oxygen and moisture. The stability in oxygen may be due to the lack of sites in heavily doped polymers preventing the reaction

$$(CHIy)_x + O_2 \leftrightharpoons [CHI_{y'} O_{2y''}] \rightleftarrows [CHO_{2y}]_x \tag{10}$$

from occurring, where y' and $y'' < y$, and leading to a nonconductive material.

Prón et al. (1981b) studied the thermal properties of $(CHI_y)_x$ using TGA and DSC. Exotherms were observed at 145 and 267°C and the main products are HI and I_2. The I_5^- Raman band at 162 cm^{-1} decreases in intensity as the doped $(CH)_x$ is heated to 100°C, whereas the I_3^- band intensity at 107 cm^{-1} increases. The initial reaction may be

$$[CH(I_5)_{y'}]_x \xrightarrow{\Delta} I_2 + [CH(I_3)_{y''}]_x \xrightarrow{\Delta} I_2 + (CHI'_{y''})_x \tag{11}$$

with further degradation occurring with additional heat. The bromine- and chlorine-doped $(CH)_x$ behave similarly. The bromine-doped $(CH)_x$ loses principally HBr upon heating and the chlorine-doped $(CH)_x$ loses HCl upon heating (Allen et al., 1979; Chien, 1984).

The AsF_5-doped $(CH)_x$ is less stable than the halogen-doped material, losing 14% of the dopant very rapidly under vacuum at room temperature. It is relatively insensitive to oxygen, like other p-doped polyacetylenes and it is also stable toward moisture.

6. Electrical Properties

The electrical properties of iodine-doped *cis*- and *trans*-polyacetylene have been extensively studied. Figure 9 shows the variation of conductivity with y (dopant concentration) for *trans*-$(CHI_y)_x$ and $[CH(AsF_5)_y]_x$. The σ_{RT} remains constant between $y = 10^{-6}$ and $y = 10^{-4}$, and then increases gradually. From $y = 10^{-3}$ and higher, σ increases rapidly and reaches 10^2

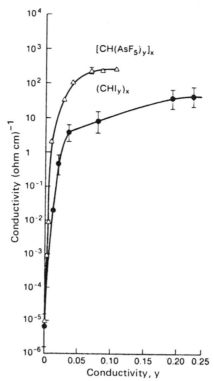

Fig. 9.   Electrical conductivity of halogen-doped polyacetylene as a function of halogen concentration: ($\bullet$) iodine and ($\triangle$) AsF$_5$. [From Chiang *et al.* (1978a).]

(ohm cm)$^{-1}$ at $y = 0.32$. The variation of σ with y for *cis*-(CHI$_y$)$_x$ is illustrated in Fig. 10. The maximum σ is $\sim$300 (ohm cm)$^{-1}$.

Figure 11 shows the linear variation of σ with temperature for (CHI$_y$)$_x$ at various y levels. The general temperature behavior is that a decrease in conductivity occurs with decreasing temperatures.

Only a few studies of the behavior of doped polyacetylene with pressure have been made to date. Measurements on (CHI$_{0.0094}$)$_x$ showed a decrease of resistivity by 22× with pressure. The decrease diminishes as one increases the amount of dopant. For example, for (CHI$_{0.03}$)$_x$, the decrease is 3.5×; for (CHI$_{0.219}$)$_x$ the decrease is 2× (Ferraris *et al.*, 1980). The pressures reached were $\sim$20 kbar. The decrease has been attributed to either reduced interfibril resistance or an increase in density of conducting fibrils per unit volume. At pressures higher than 20 kbar the resistivity of iodine-doped polyacetylene increases and the increase is greater for the more highly doped samples. It was believed that this was due to either iodine being ejected from the polymer or iodination of polyacetylene occurring with pressure. Cross linking and scission with pressure are also possible.

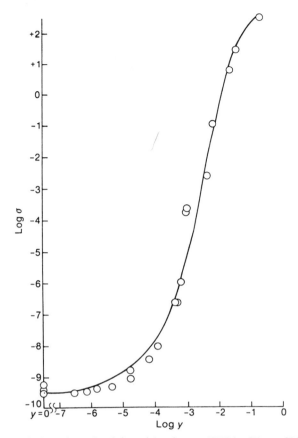

Fig. 10.   Variation of conductivity with y for cis-$(CHI_y)_x$. [From Chien (1984).]

7. Visible, Infrared, and Raman Spectra

The visible spectrum of $trans$-polyacetylene shows a strong absorption with the edge at 1.4 eV and peak maximum at 1.0 eV. Cis-polyacetylene shows a similar band at higher energies. These bands have been attributed to the π–π* transition. With iodine-doping the trans polymer shows a slight decrease in the magnitude of this transition and a new absorption appears at lower frequency. Fincher et $al.$ (1979) showed that the doping of the cis polymer with iodine caused the absorption to change progressively to that of the trans isomer with the absorption intensity, at lower frequency, increasing with doping. Absorption spectra for $trans$-$[CH(AsF_5)_y]_x$ are shown in Fig. 12.

When either cis- or $trans$-polyacetylene is doped, new infrared bands appear. Figure 13 shows results with cis-$[CH(AsF_5)_y]_x$. An intense and broad band at ~900 cm^{-1} (0.11 eV) appears. Similar results are obtained

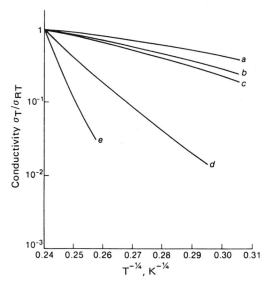

Fig. 11. Ln σ versus $T^{-1/4}$ for $(CHI_y)_x$, for y: (a) 0.19, (b) 3.7×10^{-2}, (c) 2.9×10^{-2}, (d) 1.3×10^{-2}, and (e) 0. [From Chiang *et al.* (1978c).]

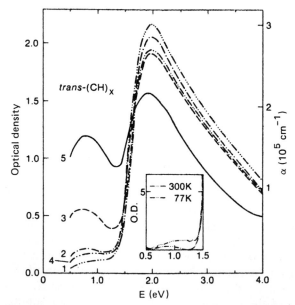

Fig. 12. Absorption spectra of *trans*-$[CH(AsF_5)_y]_x$: (1) undoped, (2) $y = 10^{-4}$, (3) $y = 10^{-3}$, (4) compensated with NH_3, and (5) $y = 5 \times 10^{-3}$. The inset shows the temperature dependence of undoped *trans*-polyacetylene. [From Suzuki *et al.* (1980).]

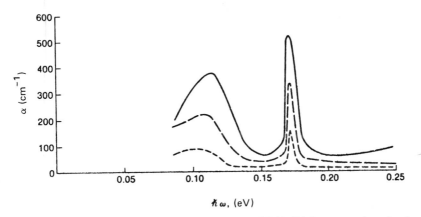

Fig. 13. Additional infrared absorption of *cis*-[CH(AsF₅)ᵧ]ₓ over undoped polymer: (——), $y = 3 \times 10^{-4}$; (---), $y = 6 \times 10^{-4}$; (——), $y = 10^{-3}$. [From Fincher *et al.* (1979).]

for the iodine-doped sample with the addition of a new absorption at ~1280 cm⁻¹ (0.16 eV).

Raman spectra of iodine-doped $(CH)_x$ are diagnostic for the presence of I_3^- and I_5^- species (see Fig. 4). Figure 14 shows Raman results for AsF₅ doped and undoped polyacetylene. The resonance enhanced C=C stretching vibration at ~1500 cm⁻¹ in *cis*-$(CH)_x$ appears to broaden upon doping and shifts toward higher frequency (~1630 cm⁻¹).

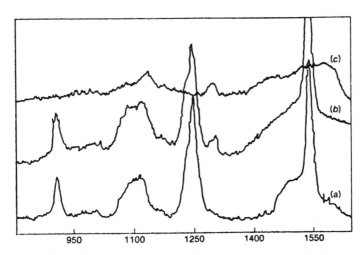

Fig. 14. Raman spectra of (a) *cis*-polyacetylene, (b) lightly doped with AsF₅, and (c) heavily doped with AsF₅, $y = 0.12$. [From Kuzmany (1980).]

8. Derivative Polymers of Poly(acetylene)

Several derivatives of polyacetylene have been prepared. These are poly(phenylacetylene) (I), poly(propiolonitrile) (II), poly(3-chloro-1-propyne) (III), and poly(3,3,3-trifluoro-1-propyne) (IV).

$$+CH{=}C+_x \qquad +CH{=}C+_x \qquad +CH{=}C+_x \qquad +CH{=}C+_x$$

(benzene ring)	C ≡≡≡ N	CH$_2$ Cl	CF$_3$

(I)   (II)   (III)   (IV)

These polymers retain the conjugated backbone of polyacetylene but the hydrogen atoms are replaced by different pendant groups which tends to increase solubility in nonpolar solvents. When doped with iodine vapor, the conductivities rose to values ranging from 6×10^{-6} to 2×10^{-3} (ohm cm)$^{-1}$ (Deits *et al.*, 1981). Table VIII lists the properties of these materials and compares them to (CH)$_x$ and poly(phenylacetylene). A more extensive discussion of poly(phenylacetylene) follows in Section B.2.

9. Block Copolymers—Soluble Polyacetylene

A soluble form of polyacetylene has been synthesized by attaching a solubilizing group to (CH)$_x$. This attachment occurs on the (CH)$_x$ chain and can be varied to provide the desired polymer–solvent interactions (Baker and Bates, 1983). It is preferable to attach the second polymer at the terminus of (CH)$_x$ if possible. This tends to preserve the symmetry and desirable properties of the (CH)$_x$ segment. Some of the grafted copolymers used are polystyrene and polyisoprene. The presence of (CH)$_x$ in the block copolymers has been verified by infrared and NMR spectroscopy. The average molecular weight in solution is estimated to be 3×10^4 and 6×10^4 g/mole. Tubino has measured the vibrational spectra of soluble polyacetylene (Tubino, 1985). Polybutadiene* has also been used to solubilize (CH)$_x$ and molecular weights of 210,000 in solution have been achieved (Bolognesi *et al.*, 1985). Higher molecular weights of ∼900,000 were reached in a gel state using 97.7% 1,4-*cis*-polyisoprene with 2.3% 1,3-cis form (Bolognesi *et al,*. 1985). It is of interest to determine to what extent these block copolymers can be doped and what their conductivities

---

*95% polybutadiene 1,4-cis, 4% 1,2-trans, and 1% 1,4-trans side groups.

TABLE VIII

Properties of Analogues of Polyacetylene (Iodine Doped)[a]

	Polyacetylene	Poly(phenyl-acetylene)	Poly(propio-lonitrile)	Poly(3-chloro-1-propyne)	Poly(3,3,3-tri-fluoro-1-propyne)
Maximum level of conductivity (ohm^{-1} cm^{-1})	~500	~10^{-5}	~10^{-7}	~10^{-3}	~10^{-4} [b]
Semiconductor to insulator transition	Present	Absent	Absent	f	Absent
Solvent extraction with MeOH	Iodine retained	Iodine retained	Iodine removed	f	Iodine removed
Pumping in dynamic vacuum	Iodine retained	Iodine removed	Iodine removed	f	Iodine removed
Passing current through sample	Steady current	Declining current	Declining current	f	e
Onset of significant weight loss (°C)	400[d]	130[e]	175	f	200–240
Solubility	Insoluble	Soluble	Insoluble	Insoluble	High MW insoluble

[a] From Deits et al. (1981).
[b] Probably due to adsorbed I$_2$.
[c] Onset of chain degradation without accompanying weight loss.
[d] Ito et al. (1975).
[e] Insufficient mechanical integrity to press pellet.
[f] Limited quantity of material available.

are. It would appear that homogeneous doping would be possible if soluble $(CH)_x$ is doped in a solvent. Preliminary studies with bromine showed that Br_2 adds quantitatively to the double bonds of $(CH)_x$ and fails to dope (Baker and Bates, 1983). For a review on soluble polymers see Frommer (1986).

10. Applications

A number of potential applications for the doped polyacetylene polymer have been suggested. Some of these are used as antistatic coatings, fuel cell catalysts, solar electrical cells, photoelectrodes in a photogalvanic cell, protective coatings on electrodes in photoelectrochemical cells, and as lightweight, inexpensive batteries. For reviews on potential uses of these materials see Yoshimura (1979) and Engler et al. (1979).

The most promising application has been the potential use as a recharg-able battery. Most current batteries are heavy in weight and have limita-tions in their charging and discharging characteristics. MacDiarmid and co-workers have demonstrated and developed $(CH)_x$ based, lightweight rechargeable batteries (Nigrey et al., 1979, 1981a,b; MacInnes et al., 1981; Kaneto et al., 1982; MacDiarmid et al., 1983).

Several different types of polyacetylene batteries have been made and they are of three types:

(1) Polyacetylene acts as the anode. In this battery, polyacetylene film is immersed in a tetrahydrofuran (THF) solution of lithium naphthalide; n-doping occurs according to

$$(CH)_x + xy\text{Li}(\text{Naph})^- \rightarrow [(CH)^{y-}\text{Li}^+y]_x + \text{Naph} \tag{12}$$

The battery consists of a strip of $(CH)_x$ and lithium metal which are placed in the THF solution and connected through an ammeter and an electrical current is produced.

(2) The use of $(CH)_x$ as the cathode is possible whereby it becomes p-doped in the process. This type battery consists of a $(CH)_x$ cathode with a lithium anode immersed in 1.0 M LiClO_4 in THF. A spontaneous elec-trochemical reaction occurs when $(CH)_x$ and Li electrodes are connected by a wire external to the cell. The discharge reactions are

$$\text{Anode:} \quad xy\text{Li} + xy\text{Li}^+ + xy\text{e}^- \tag{13}$$

$$\text{Cathode:} \quad (CH)_x + xy\text{e}^- \rightarrow (CH^{-y})_x \tag{14}$$

$$\text{Overall:} \quad xy\text{Li} + (CH)_x + \text{Li}^+(CH^{-y})_x \tag{15}$$

[see MacDiarmid et al. (1983)].

(3) The use of $(CH)_x$ as both anode and cathode is the third type of battery that has been fabricated. In one such battery the p-doping of $(CH)_x$ is done at various levels. The more highly doped material serves as the cathode and the lightly doped material as the anode.

Solid state batteries are also possible using $(CH)_x$. Several of these types of batteries have been fabricated (Béniére et al., 1983). One such battery incorporates

$$Ag\,|\,RbAg_4I_5\,|\,(CHI_y)_x$$

The design of this battery is shown in Fig. 15. The advantages of this type of battery are that they

(1) are waterproof,
(2) are operational at room temperature,
(3) are shock-resistant,
(4) have extended life, probably >10 years in the discharge state,
(5) may be recharged (possibly many times),
(6) produce a current density of up to 50 mA cm^{-2},
(7) develop a voltage of 0.65 V,
(8) have an internal resistance of 10 ohm, and
(9) yield an energy density of 10 watt hour/kg.

Another solid state battery incorporates all polymeric ingredients (see Chiang, 1981). A schematic of the all-polymer solid state battery is seen in Fig. 16. For more extensive discussion of $(CH)_x$ batteries see Chien, (1984).

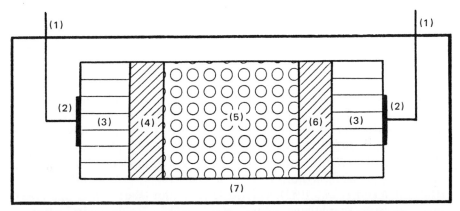

Fig. 15. Solid state battery schematic for $Ag\,|\,I_2\,|$ polyacetylene system: (1) copper, (2) silver lacquer, (3) graphite, (4) silver, (5) Rb Ag_4I_5, (6) $(CH)_x$, (7) Resin. [From Béniere et al. (1983).]

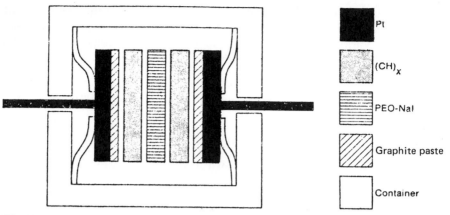

Fig. 16.   Schematic structure for an all-polymer solid state battery. [From Chiang (1981).]

If $(CH)_x$ is doped to semiconducting levels it has application as a semiconductor comparable to many inorganic semiconductors.

The potential use of $(CH)_x$ in solar cells is promising although some technical problems must be solved. It holds great promise because $(CH)_x$ is inexpensive and easy to make, and estimates indicate that it could possibly provide electrical energy at a cost well below the present cost goals set by the U.S. government. $(CH)_x$ is also attractive because of its interesting properties. Its optical band gap is 1.5 eV and has an absorption coefficient of 10^5 cm^{-1} over the photon energy range of 1.5–3.0 eV, matching the solar spectrum very well (see Chien, 1984).

B.   Other Conducting Organic Polymers

A number of derivatives of $(CH)_x$, other organic polymers, and their doped derivatives have been synthesized and evaluated as conductors. Some of these include, poly(diacetylenes) (Greene *et al.*, 1979; Bloor *et al.*, 1979a), bis(aromatic sulfonate) diacetylenes (Bloor *et al.*, 1979b), poly(phenylacetylene) (Diaconu *et al.*, 1979), poly(1,6-heptadiyne) (Pochan *et al.*, 1980), polypyrrole (Diaz *et al.*, 1981; Kanazawa *et al.*, 1979, 1980), poly(*p*-phenylene)vinylene (Wnek *et al.*, 1979), poly(*p*-phenylene) (Ivory *et al.*, 1979; Shacklette *et al.*, 1979), poly(*p*-phenylene sulfide) (Rabolt *et al.*, 1980; Chance *et al.*, 1980), polythiophene (Yamamoto *et al.*, 1980), poly(ethynylferrocene) (Pittman, 1977), poly(propiolic anhydride) (Gibson *et al.*, 1983), poly(azophenylene) (Barbarin *et al.*, 1983), and polyisothianaphthene (Wudl *et al.*, 1985; Brédas, 1985; Brédas and Street, 1985). Many of these polymers are based on aromatic ring systems, such as polypyrrole, polythiophene, poly(*p*-phenylene), and poly(phenylacetylene) (Brédas *et al.,* 1985). Unless doped, all of these conjugated polymers remain insulators, having a band gap >1.5 eV; $(SN)_x$ being an exception. Polyisothianapthene being an exception demonstrating a band gap of ~ 1 eV (Brédas and Street, 1985; Wudl *et al.,* 1985). However, none of these polymers, when doped, approach the conductivities of doped $(CH)_x$ (e.g., $[(CH)_x(AsF_6)_{0.1}]_x$, which exhibits a conductivity of 10^3 (ohm cm)$^{-1}$ (Druy *et al.,* 1980).

Structural formulas of the various conducting polymers are illustrated in Fig. 17.

1. Polypyrrole

Polypyrrole is a promising synthetic metallic polymer when doped with *p*-type dopants. It shows promise as an organic electrode (Diaz and Kanazawa, 1979; Diaz *et al.,* 1981; Skotheim *et al.,* 1983; Devreux *et al.,*

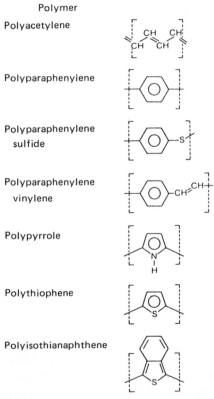

Fig. 17. Structures of various conducting polymers.

1983; Kanazawa *et al.*, 1979), for information storage (Meyer *et al.*, 1985), as well as a conducting polymer (Kanazawa *et al.*, 1979, 1980).

The polymerization of pyrrole takes place electrochemically on platinum, carbon, or silicon surfaces (Diaz *et al.*, 1981; Skotheim *et al.*, 1983). If the polymerization is accomplished in acetonitrile solutions containing $(ET_4N)BF_4$, the oxidized form of polypyrrole is produced according to the reaction

$$\tag{16}$$

The polymer is believed to form by linking the pyrrole units via the α-carbon atoms, with the pyrrole rings remaining intact. Evidence for this comes from reflection infrared and transmission infrared data (Street *et al.*, 1985), Raman data (Street *et al.*, 1981; Kreutzberger and Kalten, 1961), and from NMR experiments (Street *et al.*, 1981).

The shiny, blue–black films produced are flexible and highly stable in air and water, and may be heated to 250°C with little effect on the conductivity properties. The pyrrole units carry the positive charge when doped with $(Et_4N)BF_4$ and the charge is balanced by the BF_4^- ions. The similarity to *p*-type doped $(CH)_x$ is obvious (e.g., doping with $AgClO_4$ or $AgBF_4$). Conductivity values measured by a four-point probe method (Kanazawa *et al.*, 1979; Street *et al.*, 1981), show values of $\sim 100 \,(ohm \, cm)^{-1}$ at room temperature and decrease to $\sim 30 \,(ohm \, cm)^{-1}$ at $-193°C$. Figure 18 shows a comparison of conductivity versus temperature for polypyrrole and other polymeric conductors (Kanazawa *et al.*, 1979). If the electropolymerization is carried out in a solvent containing $LiClO_4$, $[pyr^{y+}(ClO_4)_y^-]_x$ is formed ($y = 0.25$–0.4), which shows reduced conductivities in the ~ 1–$20 \,(ohm \, cm)^{-1}$ range (Devreau *et al.*, 1983). The corresponding poly-*n*-methylpyrrole-BF_4 polymer shows much lower room temperature conductivities of $\sim 1 \times 10^{-3} \,(ohm \, cm)^{-1}$ (Genies and Syed, 1984; Kanazawa *et al.*, 1980). Copolymers of doped pyrrole and *n*-methylpyrrole show electrical conductivities between that of pyrrole and *n*-methylpyrrole. Grafting other polymers to polypyrrole, or electrochemical deposition of polypyrrole in the matrix of polymers that can swell such

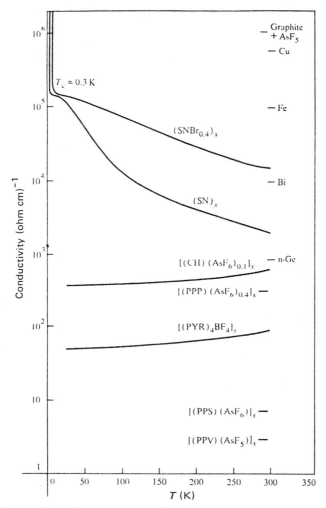

Fig. 18. Conductivity versus temperature from various conducting polymers and classical conductors: PPP = poly(p-phenylene), PYR = pyrrole, PPS = poly(p-phenylene sulfide), and PPV = poly(p-phenylene vinylene). [Modified from Street and Clarke (1981).]

as styrene (Street *et al.,* 1985), can modify the mechanical properties significantly. Polyvinyl alcohol or polyvinyl chloride have served to transform polypyrrole into a tough flexible plastic (Street, unpublished data).

Some effort has been devoted to improving the mechanical properties and stability of polypyrrole. For example, this has been accomplished when toluene sulfonate ion has been incorporated as the anion (Diaz and Hall, 1983).

The mechanism of electrical conduction in these systems has been studied by ESR methods, and is thought to proceed via bipolarons as the charge carriers (Scott *et al.*, 1984, 1985; Kaufman *et al.*, 1985).

The [(pyr)(BF$_4$)]$_x$ films show various changes upon storage. These changes are due to the removal of BF$_4^-$ anions and a subsequent decrease in conductivity. A strong synergistic effect was noted with oxygen and water on the degradation kinetics (Erlandsson *et al.*, 1985).

2. Poly(phenylacetylene)

Poly(phenylacetylene) is a derivative of poly(acetylene) and the replacement of a hydrogen atom by a phenyl group increases the solubility of this polymer. Whereas poly(acetylene) is insoluble unless a solubilizing group is attached, poly(phenylacetylene) is soluble in most nonpolar solvents. This offers several advantages insofar as availability of methods of study, which may require solutions of the polymer. It is apparent that one can easily dope the polymer in a solvent. Additionally, the polymer becomes more processable because of its solubility. The polymer is stable in air and in water.

Poly(phenylacetylene) polymerization has been achieved by the use of Zeigler-type catalysts. Table IX lists the various catalysts that have been used. The average molecular weight obtained falls in the range of ~300–15,000 (Masuda *et al.*, 1974). Polymerization solvents and the nature of the catalysts formed have a large influence on the size of the polymer formed (Masuda *et al.*, 1982). Table X shows the states of average molecular weight as a function of the catalyst used in the polymerization in 1,4-dioxane solvent.

Poly(phenylacetylene) has been prepared in amorphous (cis and trans) and crystalline (cis) forms. Using MoCl$_5$ catalyst at temperatures below 40°C, produces the cis isomer almost exclusively (Masuda *et al.*, 1979). A trans-rich poly(phenylacetylene) results from a WCl$_6$ catalyst solution. The amorphous cis material was separated into high and low molecular weight portions by fractional precipitation, using benzene as a solvent and methanol as a nonsolvent. The cis material is an orange powder, while the trans material has a reddish-brown color. Both forms are readily soluble in nonpolar solvents. Kern (1969) polymerized poly(phenylacetylene) in benzene using Fe(acac)$_3$·Al(*t*-Bu)$_3$ catalyst, and obtained an orange solid, which was insoluble in common organic solvents, but which was converted to the soluble amorphous trans isomer by heating to 120–130°C. Kern (1969) has also reported on the various possible isomers in poly(phenylacetylene). For the trans isomer it is possible to have two

TABLE IX

Some Catalysts Used for the Polymerization of Poly(phenylacetylene)

Catalyst	Reference
WCl_6	Percec (1983), Masuda et al. (1975, 1979)
$MoCl_5$	Percec (1983), Masuda et al. (1975, 1979)
$\Phi M(CO)_3$ (M = Cr, Mo, W)	Woon and Farona (1974)
$Co(NO_3)_2$–$NaBH_4$	Byrd (1969)
$(iBu)_2AlH$	Kern (1969)
$RhCl(\Phi_3P)_3$	Kern (1969)
$WCl_6 \cdot \Phi_4Sn$	Masuda et al. (1982)
$TiCl_4$	Simionescu and Percec (1980)
$Re(CO)_5X$ (X = Cl, Br)	Tsonis and Farona (1979)
$Fe(acac)_3$–$Al(Et)_3$	Nguyen et al. (1978), Simionescu et al. (1977)
$Fe(acac)_3$–$Al(i\text{-}Bu)_2H$	Kern (1969)
$(Et)_3Al$–$Ti(OEt)_4$	Chiang et al. (1982)
$AlEt_3$–$TiCl_4$	Simionescu et al. (1977)
Φ_3P–Pt compounds	Cordischi et al. (1971)
$[Rh(bipy)cyclooctadiene]PF_6$	Mestroni et al. (1982)
$[Pt(C\equiv C\text{-}Ph)_2(PPh_3)_2]$	Furlani et al. (1969)
$(P\Phi_3)_2PtCl_2$	Furlani et al. (1969)
$[Ni(NCS)(C\equiv CR)L_2]$ where $\quad L = Bu_3P, \Phi_3P$ and R = Ph $\quad$ or $C(CH_3)_2OH$	Bicev et al. (1980)
L_2NiXY where $L = P\Phi_3$, $\quad PBu_3\Phi_2PC_2P\Phi_2/2$, and X = $\quad Y = Cl, Br, I, NO_3, SCN$	Furlani et al. (1977)

TABLE X

Polymerization of Phenylacetylene in 1,4-Dioxane with Different Catalysts—
Degree of Polymerization[a]

Catalyst	$\overline{M}_n$ [b]
WCl_6	86,000
$WCl_6 \cdot \frac{1}{2}CH_3OH$	91,000
$WCl_6 \cdot \Phi_4Sn$	92,000
$MoCl_5$	5,300
$MoCl_5 \cdot \frac{1}{2}CH_3OH$	5,000
$MoCl_5 \cdot \Phi_4Sn$	3,900

[a] Taken from Masuda et al. (1982).
[b] $\overline{M}$ is average molecular weight; highest number.

forms. A 180° rotation about the C—C single bonds transforms the trans transoid form to trans cisoid.

trans transoid   *trans cisoid*

For the cis polymers a sequence of three double bonds in the cis cisoid structure forms a six-membered ring. In a chain of alternating cis double bonds this limits rotation about C—C bonds. The formation of a helix becomes possible and from the cis transoid form this is illustrated as

cis transoid

A number of poly(phenylacetylene) polymers can form depending on the catalyst. Table XI tabulates the properties of some of these polymers (Kern, 1969).

The conductivity of undoped poly(phenylacetylene) is $\sim 10^{-12}$ (ohm cm)$^{-1}$. Doping with iodine increases the conductivity to 10^{-5}–10^{-4} (ohm cm)$^{-1}$. Heavily iodine doped poly(phenylacetylene) has been prepared, which demonstrates conductivity values of 10^{-3} (ohm cm)$^{-1}$ (Furlani and Russo, unpublished data). At best, doped poly(phenylacetylene) is a semiconductor (Deits *et al.,* 1981; Ferraro *et al.,* 1984). The nature of the incorporated ion was shown to be I_5^-. Ferraro *et al.* (1984), using Raman, polarized Raman, and infrared spectroscopy, found polarized maxima occurring at 166 and 112 cm^{-1} in the Raman spectrum (see Fig. 19). The far infrared data showed maxima at 143 and 112 cm^{-1} and the vibrational spectroscopy was consistent with the I_5^- moiety (I_2–I_3^-) even down to low concentrations of dopant (see Fig. 20). Table IV summarizes these results and compares them with vibrational data for other compounds containing I_3^- and I_5^- moieties.

TABLE XI

Properties of Phenylacetylene Polymers[a]

Species	Initiator	Color	Softening temperature (°C)[b]	Ultraviolet fluorescence	$\overline{M}_n$[c]	Characteristics[d]
I-A	Transition metal-reducing agent	Bright yellow	Transforms	No	4900	Strong 13.5 μM; broad 11.0–11.5 μM cluster
I-B	Transition metal-reducing agent	Orange	203–208	No	4200(Fe)[e] 5400(Rh)[f]	11.25–11.36 μM doublet
I-C	Transition metal-reducing agent	Dark red	198–203	No	—	XRD periodicity
I-D	Transition metal-reducing agent	Yellow	216–219	Yellow	4900	Obtained by pyridine treatment
II	RhCl[(C₆H₅)₃P]₃	Tan	152–157	Yellow-orange	1100	Absence of 11.0–11.5 μM; 6.77–6.89 μM doublet
II-Py	RhCl[(C₆H₅)₃P]₃		213–217	Yellow-orange	1100	Pyridine treatment of 11
III	Thermal	Yellow	158–163	Yellow	700–900	Thermal initiation
III-Py	Thermal	Yellow	213–217	Yellow	700–900	Pyridine treatment of III

[a] Taken from Kern (1969).
[b] Softening temperature determined by using Fisher–Johns apparatus.
[c] Vapor-phase osmometry in benzene.
[d] Numbers refer to infrared bands; XRD = x-ray diffraction.
[e] Using iron catalyst in polymerization.
[f] Using rhodium catalyst in polymerization.

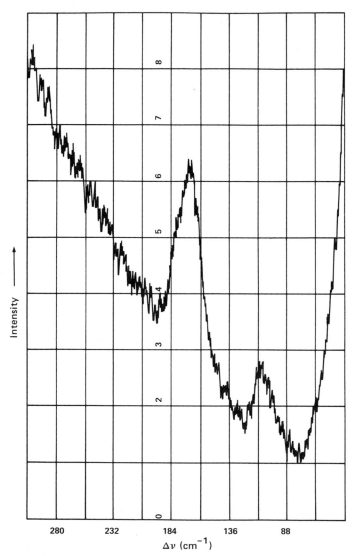

Fig. 19. Low frequency spectrum of doped poly(phenylacetylene) (30% iodine). [From Ferraro *et al.* (1984).]

Some controversy exists as to the mechanism of conductivity in poly(phenylacetylene). Diets *et al.* (1981) and Cukor *et al.* (1981) suggest that poly(phenylacetylene) is an ionic conductor. Baughman *et al.* (1982) have proposed that the conduction is based on hole carriers and the low conductivity is due to low mobility of hole carriers.

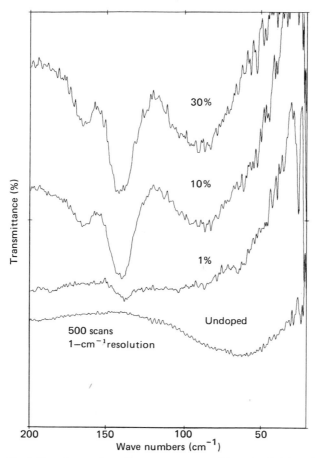

Fig. 20. Far infrared specular reflectance spectra of doped poly(phenyl)acetylene on Al foil for 30% I_2, 10% I_2, 1% I_2, and undoped polymer from 200–20 cm^{-1}. Vertical scale in arbitrary transmittance units. Undoped polymer spectrum was a Nujol mull spectrum. [From Ferraro *et al.* (1984).]

3. Poly(*p*-phenylene sulfide)

Poly(*p*-phenylene) (PPS) sulfide has garnered interest because it is the only melt and solution-processible polymer that can be doped to produce a conducting polymer. The polymer is stable in a dry oxygen atmosphere, but deteriorates in water vapor. It can be annealed to its melting point of 285°C with no change. Another advantage of this polymer is that it is commercially available. It is also of interest from a theoretical viewpoint, since it is the first polymeric system without a continuous system of overlapping C-π orbitals.

Undoped poly(p-phenylene sulfide) has a conductivity of 10^{-6} (ohm cm)$^{-1}$ (Chance *et al.*, 1980). When doped with AsF$_5$, conductivities increase by 6 orders of magnitude to ~1–3 (ohm cm)$^{-1}$ (Chance *et al.*, 1980; Rabolt *et al.*, 1980; Shacklette *et al.*, 1981). The conductivity can be controlled by varying the dopant concentration or by adding a compensating agent such as dimethylamine. Figure 21 shows the conductivity of AsF$_5$ doped PPS versus T. Better fits are found if σ is plotted versus $T^{-1/2}$.

Related polymers have also been synthesized and exposed to AsF$_5$. Poly(m-phenylene sulphide) has a conductivity of 10^{-1} (ohm cm)$^{-1}$ (Chance *et al.*, 1980; Rabolt *et al.*, 1980). Poly(m-phenylene oxide) shows a conductivity of 10^{-2} (ohm cm)$^{-1}$ when doped with AsF$_5$ (Chance *et al.*, 1980).

Shacklette *et al.* (1981) have provided evidence that the doping process with AsF$_5$ or SbF$_5$ promotes the formation of C–C links bridging the sulfur linkages to form thiophene rings leading to a system of overlapping orbitals. The polymer also swells and loses both crystallinity and chain orientation (Murthy *et al.*, 1984). These structural changes may account for the insensitivity of the conductivity of the doped polymer to the degree of crystallinity and orientation of the polymer.

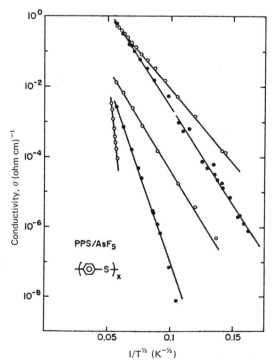

Fig. 21. Conductivity versus $T^{-1/2}$ for AsF$_5$-doped levels (in PPS). Doping levels increase from left to right on figure. [From Shacklette *et al.* (1981).]

In these systems, the sulfur atoms contribute to a delocalized electronic structure due to significant overlap of sulfur p orbitals and aromatic π orbitals on the neighboring rings even where the configurations are non-planar. This illustrates the fact that the highly conducting polymers are not restricted to conjugated polymers with extended electronic structure if the doping process alters the polymeric structure.

4. Poly(p-phenylene)

This polymer is the first example of a nonacetylenic, hydrocarbon polymer that can be doped, either with electron donors or acceptors to give polymers with metallic properties (Shacklette et al., 1979; Ivory et al., 1979).

The polymer can be synthesized by the oxidative cationic polymerization of benzene using $AlCl_3$–$CuCl_2$ catalyst at 35°C. When doped with AsF_5 the material becomes green–black in appearance (Kovacic and Oziomek, 1964; Kovacic and Kyriakis, 1963). Doping with potassium naphthalide in THF provides a reddish-gold solid (Shacklette et al., 1979). Doping with iodine or bromine does not provide highly conductive materials (Chiang et al., 1978b).

Poly(p-phenylene) and cis-polyacetylene have the same crystal packing arrangement with the same projection symmetry (Pgg). The space group for poly(p-phenylene) is monoclinic ($P2_1/a$) while that for cis-polyacetylene is orthorhombic ($Pmma$). Figure 22 shows the chain-axis projections for

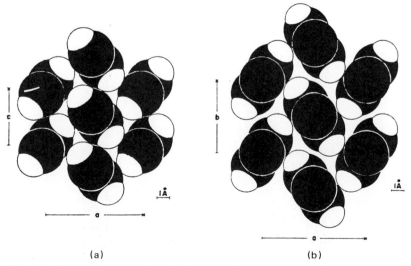

(a)   (b)

Fig. 22.   Chain-axis projections for the crystal structures of (a) cis-polyacetylene and (b) poly(p-phenylene). Hydrogen atoms are white and carbon atoms are black. [From Shacklette et al. (1979).]

the two polymers (Baughman *et al.*, 1978, 1979; Baughman and Hsu, 1979). See also Pradere *et al.* (1985) and Stamm *et al.* (1985) for the crystalline structures of the doped-poly(*p*-phenylene) polymer.

The undoped-poly(*p*-phenylene) polymer shows conductivities of 10^{-9}–10^{-10} (ohm cm)$^{-1}$ at room temperature. After doping with AsF$_5$, the conductivity reaches 145 (ohm cm)$^{-1}$. Iodine-doped poly(*p*-phenylene) shows conductivities of 4×10^{-3} (ohm cm)$^{-1}$ while bromine-doped samples have conductivities of 10^{-4} (ohm cm)$^{-1}$ (Chiang *et al.* 1978b). With *n*-doping K naphthalide gives solids which show conductivities of ~7 (ohm cm)$^{-1}$. Figure 23 shows the conductivity behavior versus *T* of the AsF$_5$-doped poly(*p*-phenylene). The behavior is similar to that observed for doped (CH)$_x$.

The AsF$_5$-doped poly(*p*-phenylene) polymer is more stable under atmospheric conditions. Only a slow decrease in conductivity is noted with time (Ivory *et al.*, 1979). The doped polymer can also be dipped in water without changing the electrical conductivity. The potential advantage of this polymer is in the high stability (>450°C in air and 550°C in an inert atmosphere).

A new poly(*p*-phenylene) polymer has been synthesized by electropolymerization (Froyer *et al.*, 1985). When doped with AsF$_5$, the conductivity was found to range from 2–15 (ohm cm)$^{-1}$.

The EPR technique has been used to characterize the electronic states of doped poly(*p*-phenylene) polymer. Evidence is provided that most of the charge in the polymer is due to bipolarons (Kispert *et al.*, 1985).

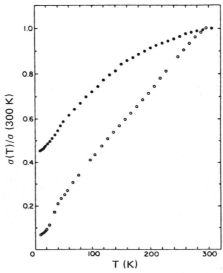

Fig. 23.   The temperature dependence of σ of AsF$_5$ doped poly(*p*-phenylene). Upper curve is for a freshly doped sample and lower curve was for a sample after aging in poor vacuum.

5. Poly(thiophene)

Poly(thiophene) is another polymer of the type poly(heterocycles) and is thus related to poly(pyrrole). It may be synthesized both by chemical (Street *et al.*, 1982; Kobayashi *et al.*, 1984) and electrochemical methods (Tourillon and Garnier, 1982; Hotta *et al.*, 1983; Kaneto *et al.*, 1983). The chemical method involves the polymerization of 2,5- diiodothiophene in a solution of ether/anisole using nickel catalysis of Grignard reagents (Kobayashi *et al.*, 1984). Polymers of ~4000 molecular weight consisting of 46–47 thiophene rings (~184–188 carbon atoms along the backbone) have been prepared. The fact that the pyrrole rings remain intact in the polymerization is confirmed by infrared spectroscopy (Kobayashi *et al.*, 1984).

Partial oxidation with AsF_5 causes an enhancement of the conductivity of 10 orders of magnitude to occur. Conductivity values of >10 (ohm cm)$^{-1}$ are achieved. The conductivity decreases with a decrease in temperature and the data correspond to a $T^{-1/4}$ dependency.

The doped poly(thiophene) polymer shows good stability toward O_2 and moisture like poly(pyrrole), and in this respect is superior to $(CH)_x$ (Tourillon and Garnier, 1983, 1984; Waltman *et al.*, 1983). It also is heat stable to ~250°C.

Mermilliod-Thevenin and Bidan (1985) have synthesized related thiophene polymers such as poly(2,2'-bithiophene) and poly(3-methylthiophene). These materials were doped with $Fe(ClO_4)_3 \cdot 9H_2O$. The poly(2,2'-bithiophene) showed a conductivity of 17 (ohm cm)$^{-1}$ and poly(3-methylthiophene) had a conductivity of 7×10^{-3} (ohm cm)$^{-1}$. Another variant of thiophene is polyisothianaphthene (PITM) (Wudl *et al.*, 1985). PITN has its thiophene ring fused to a benzene ring along the β—β' thiophene bond (see Fig. 17). The benzene ring stabilizes the quinoid structure in the ground-state geometry of PITN (Brédas, 1985; Brédas and Street, 1985) and its bandgap is found to be ~1 eV lower than thiophene (~2 eV) and other conductive polymers. PITN can be produced electrochemically as a film and it can be reversibly oxidized. In its oxidized form it becomes a transparent, yellow–green color and has a σ of 1–50 (ohm cm)$^{-1}$ (Brédas and Street, 1985). It is transparent in that when highly doped the absorption appears in the infrared region.

C. Discussion

None of the polymers synthesized and doped since the discovery of doped polyacetylene have reached the high conductivities of doped $(CH)_x$. Table XII intercompares the conductivity values, stabilities in air, water, and heat for various conducting polymers. For a more extensive discussion on conducting polymers see recent symposium reports found in Molecular

TABLE XII

Summary of Conductivities and Stabilities of Various Conducting Organic Polymers

Polymer	Dopant	σ (ohm cm)$^{-1}$	Stability in air (O$_2$) and H$_2$O	Thermal stability
Poly(acetylene)cis	AsF$_5$	1.2×10^3	Less than I$_2$-doped polymers	Sl. decrease in σ with T to 100°C. AsF$_5$-doped conductors better heat resistant than the I$_2$-doped polymers
Poly(acetylene)trans	AsF$_5$	4×10^2		
trans-(CHIy)$_x$, y = 3.2	I$_2$	10^2	Good	
cis-(CHIy)$_x$, y = 0.2	I$_2$	300	Good	Stable to 250°C
Poly(pyrrole)	BF$_4^-$	100	Highly stable, better than (CH)$_x$ polymers	
	ClO$_4^-$	1–20	Highly stable, better than (CH)$_x$ polymers	
Poly(thiophene)	I$_2$	3×10^{-4}	Highly stable, better than (CH)$_x$ polymers	
	AsF$_5$	$2 \times 10^{-2}, \sim 10^a$	Highly stable, better than (CH)$_x$ polymers	
Poly(2,2'-bithiophene)	Fe(ClO$_4$)$_3 \cdot$ 9H$_2$O	17		
Poly(3-methylthiophene)	Fe(ClO$_4$)$_3 \cdot$ 9H$_2$O	7×10^{-3}		
Poly(phenylacetylene)	I$_2$	10^{-4}–10^{-5}	Stable	
Poly(p-phenylene)	AsF$_5$	145	Stable	High stability to 550°C in inert atms. and 450°C in air.
Poly(m-phenylene)	K	7		
	AsF$_5$	10^{-3}		
Poly(phenylene sulfide)	AsF$_5$	1–3	Stable in dry O$_2$, deteriorates in H$_2$O	Melt processible, can be annealed to M.P. at 285°C.
Poly(phenylene oxide)	AsF$_5$	10^{-3}		
Poly(thienylene)	AsF$_5$	~ 10		
Poly(phenylene vinylene)	AsF$_5$	1.2×10^{-1}		

[a] Depends on degree of doping. High value corresponds to 24 mol. % of AsF$_5$.

Crystals and Liquid Crystals, **118**, Parts A and B (1985), and the Journal de Physique, Colloque C3 (1983) (see reference and book sections at end of chapter).

Baughman *et al.* (1982) have considered the various factors relative to structure which lead to enhanced electrical conductivities in conducting polymers. Polymers which have the most homogeneous chain structures appear to be the best conductors. Table XIII shows that regular copolymers which are doped lead to lower conductivities than the homopolymers containing one of the constituent chain elements. There are exceptions in the cases where doping causes chemical transformations or where the copolymerization causes the removal of steric hindrances to planarity which are present in the homopolymers.

Heterogeniety can yield carrier localization on the chain unit which provides the lowest potential for holes (or electrons in the case of donor doping). Brédas *et al.* (1981, 1982) have shown that a link exists between homogeneous character of the polymer backbone and the width of the highest occupied π band. The width of the π bands can be correlated to the degree of delocalization of the π system along the polymer backbone and also to the mobility of the carriers in these bands. Table XIV shows calculated band widths, ionization potential, and band gaps for various polymers. As one notes from Table XIV as the ionization potential decreases and the width of the highest occupied π band increases, the band gap decreases, thus creating a better conductor (Baughman *et al.*, 1982). These results show why polymer research is primarily centered on π-bonded unsaturated systems since these materials have low ionization potentials and/or large electron affinities. These π electrons can be easily removed or added to these systems and little effect is noted on the σ bonds which are necessary to hold the basic backbone of the polymer intact (Brédas and Street, 1985).

TABLE XIII

Observed Conductivities (ohm cm)$^{-1}$ for Unoriented AsF$_5$-doped Polymers and Copolymers[a]

A	B	$(A)_x$	$(B)_x$	$(AB)_x$	$(AB_2)_x$
—HC=CH—	—C$_6$H$_4$—	1200	500	3	
—C$_6$H$_4$—	—C$_6$H$_4$S—	500	1	0.3	0.02
—C$_6$H$_4$S—	—C$_6$H$_4$O—	1	10^{-3}	10^{-4}	5×10^{-6}

[a] Taken from Baughman *et al.* (1982).

TABLE XIV

Calculations Using the Ab Initio Quality Valence Effective Hamiltonian Method[a]

	Ionization potential (eV)[b]	Width of highest occupied π band (eV)	Band gap (eV)
Polyacetylene all-trans	4.7	6.5	1.4
Cis, transoid	4.8	6.4	1.5
Trans, cisoid	4.7	6.5	1.3
Polydiacetylene acetylenic	5.1	3.9	2.1
Butatriene	4.3	4.5	
Poly(p-phenylene)			
coplanar phenyls	5.5	3.9	3.2
twisted (22° between phenyls)[c]	5.6	3.5	
perpendicular phenyls	6.9	0.2	
Poly(m-phenylene)			
coplanar	6.1	0.7	4.5
twisted (28° between phenyls)[c]	6.2	0.2	
Poly(p-phenylenevinylene)	5.1	2.8	2.5
Poly(p-phenylenexylylidene)	5.6	2.5	3.4
Polybenzyl	6.5	0.6	

[a] Taken from Baughman et al. (1982).

[b] The ionization potential has been corrected for lattice polarization energy by subtracting 1.9 eV from the calculated single chain value. The correction was chosen to provide good agreement between experimental and theoretical ionization potential.

[c] Rotation angles of 22 and 28° between phenyls in poly(p-phenylene) and poly(m-phenylene), respectively, are suggested by the geometry of model compounds.

II. SUPERCONDUCTIVE INORGANIC POLYMER (SN)$_x$

Polythiazyl was first synthesized in 1910 (Burt, 1910). Walatka *et al.* (1973) discovered that the material was electrically conductive [$\sigma = 2 \times 10^3$ (ohm cm)$^{-1}$ at room temperature]. This caused increased interest in other polymeric inorganic materials as well as organic electrical conductors. In 1975, Greene *et al.* showed that (SN)$_x$ was superconductive at 0.3 K. Thus far, (SN)$_x$ is the only superconductive inorganic polymer discovered. Replacement of sulfur with selenium has been unsuccessful (Wolmershäuser *et al.*, 1978), as have all efforts to prepare other analogs.

Greene and Street (1977) studied the effects of various donors and acceptors on the properties of (SN)$_x$. Bromine was found to increase the electrical conductivity, which was first noted by Yoffe (1976) to be 2–4 $\times$ 10^4 (ohm cm)$^{-1}$ at 300 K (SNBr$_{0.4}$)$_x$. The intercalation of both bromine or iodine into the (SN)$_x$ lattice causes an increase in electrical conductivity of an order of magnitude, without destroying the superconducting transition. The (SN)$_x$ polymer with the highest conductivity was (SNBr$_{0.4}$)$_x$, which is one order of magnitude less than copper metal at low temperature. The relationship between conductivity and temperature for (SN)$_x$ and bromine-doped (SN)$_x$ and comparison with various polymeric organic conductors is illustrated in Fig. 18. An increase in conductivity with decreasing temperature is observed. Both (SN)$_x$ and bromine-doped (SN)$_x$ show a linear increase of conductivity with a lowering of temperature. The T_c for (SNBr$_{0.4}$)$_x$ is ~0.36° K.

A large number of studies have led to the conclusion that (SN)$_x$ is an anisotropic three-dimensional semimetal and exhibits bulk type II superconductivity explained by the BCS mechanism (Greene and Street, 1977). The material demonstrates a partial Meissner effect indicative that it is an intrinsic superconductor (Dee *et al.*, 1977). The intrinsic nature arises from the presence of one unpaired electron associated with each S–N unit. Under the effect of an applied field unpaired electrons move and carry the charge giving rise to the observed electrical conductivity.

The structure of (SN)$_x$ has been determined by x-ray (Cohen *et al.*, 1976) and electron diffraction techniques (Heeger *et al.*, 1978; Mikulski *et al.*, 1975). The chain structure in (SN)$_x$ is shown in Fig. 24. The polymer chains are parallel with good overlap between π orbitals creating a pathway for conduction through the length of the chain. The conductivity ratio, $\sigma_\parallel/\sigma_\perp$ at 298 K is ~15, indicative of significant orbital overlap between chains. The structure of the brominated (SN)$_x$ is also illustrated in Fig. 25 with the Br$_3^-$ ions located between the (SN)$_x$ chains. The bromination reaction involves a charge-transfer process, delocalizing the polymeric

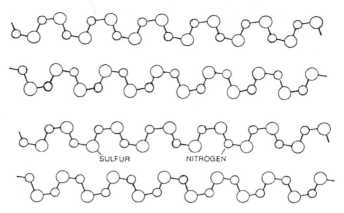

Fig. 24.   Parallel chains found in $(SN)_x$.

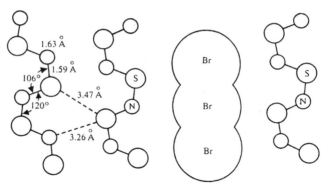

Fig. 25.   Structure of brominated $(SN)_x$ showing the orientation of Br_3^- ions relative to the $(SN)_x$ chains. [From Street and Gill (1979).]

cation and the counter anion of the dopant. The reaction can be depicted as

$$(SN)_x + 1.5Br_2 \rightarrow (SN)^{y+} + 3Br_y^- \tag{17}$$

It has been established by Raman spectroscopy (Temkin *et al.*, 1978; Iqbal *et al.*, 1978; Temkin and Street, 1978) that the bromine moiety is present as Br_3^- although Br_5^- $(Br_2-Br_3)^-$ is not excluded. Strong Raman bands are found at ~ 230 and 150 cm^{-1}. Doping with iodine is also possible and the iodine moiety has been interpreted as I_5^- $(I_2-I_3)^-$ (I_3^- ions linked to a distorted I_2 unit). Doping of $(SN)_x$ with ICl vapor was not successful, as no polyhalide species were observed (Iqbal *et al.*, 1978; Wolmerhäuser and Street, 1978).

Both $(SN)_x$ and $(SNBr_{0.4})_x$ show an increase in conductivity with pressure, with the latter compound showing a smaller effect. T_c increases with pressure (Gill *et al.*, 1975; Müller et al., 1978), while T_c for brominated $(SN)_x$ decreases monotonically to 150 mK under 10 kbar pressure indicative of a three-dimensional $s-p$ superconductor (Dee *et al.*, 1977).

The applications of $(SN)_x$ and doped $(SN)_x$ have been limited because of the difficulties encountered in the synthesis of the material. Finally, it should be pointed out that $(SN)_x$ is explosive and is synthesized from explosive intermediates (Greene and Street, 1984).

LIST OF SYMBOLS AND ABBREVIATIONS

σ	Conductivity is $(ohm\ cm)^{-1}$ units
$\sigma_\parallel$	Conductivity parallel to polymer chain
$\sigma_\perp$	Conductivity perpendicular to polymer chain
p-type	Semiconductor where doping is conducted with an electron acceptor
n-type	Semiconductor where doping is conducted with a donor specie
XPS	X-ray photoelectron spectroscopy
XRD	X-ray diffraction
TGA	Thermal gravimetric analysis
DSC	Differential scanning calorimetry
EPR	Electron paramagnetic resonance

REFERENCES

Allen, W. N., Decorpo, J. J., Saalfeld, F. E., Wyatt, J. R., and Weber, D. C. (1979), *Synth. Met.* **1**, 370.

Anderson, L. R., Pez, G. P., and Hsu, S. L. (1978), *J. Chem. Soc. Chem. Commun.*, p. 1066.

Azakura, K., Ikimoto, I., Kuroda, H., Kobayashi, T., and Shirakawa, H. (1985), *Bull. Soc. Jpn.* **58**, 2113.

Baker, G. L., and Bates, F. S. (1983), *J. Phys. Colloq.* **44**, C3-11.

Barberin, E., Berthet, G., Blanc, J. P., Dugay, M., Fabre, C., Germain, J. P., and Robert, H. (1983), *J. Phys. Colloq.* **44**, C3-749.

Baughman, R. H., and Hsu, S. L. (1979), *J. Polym. Sci. Polym. Lett. Ed.* **17**, 185.

Baughman, R. H., Hsu, S. L., Pez, G. P., and Signorelli, A. J. (1978), *J. Chem. Phys.* **68**, 5405.

Baughman, R. H., Hsu, S. L., Anderson, L. R. Pez, G. P., and Signorelli, A. J. (1979), *in* "Molecular Metals" (W. E. Hatfield, ed.), pp. 187–202, Plenum, New York.

Baughman, R. H., Brédas, J. L., Chance, R. R., Elsenbaumer, R. L., and Shacklette, L. W. (1982), *Chem. Rev.* **82**, 209.

Baughman, R. H., Murthy, N. S., and Miller, G. G. (1983), *J. Chem. Phys.* **79**, 515.

Béniére, F., Boils, D., Cánepa, H., Franco, J., Le Corre, A., and Louboutin, J. P. (1983), *J. Phys. Colloq.* **44**, C3-567.

Bicev, P., Furlani, A., and Russo, M. V. (1980), *Gazz. Chim. Ital.* **110**, 25.

Billaud, D., Haleem, M. A., Lefrant, S., and Krichine, S. (1983a), *J. Phys. Colloq.* **44**, C3-175.

Billaud, D., Kulszewicz, I., Prón, A., Bernier, P., and Lefrant, S. (1983b), *J. Phys. Colloq.* **44**, C3-33.

Bloor, D., Hubble, C. L., and Ando, D. J. (1979a), *in* "Molecular Metals" (W. C. Hatfield, ed.), p. 243, Plenum, New York.

Bloor, D., Ando, D. J., Fisher, D. A., and Hubble, C. L. (1979b) *in* "Molecular Metals" (W. C. Hatfield, ed.), p. 255, Plenum, New York.

Boils, D., Schué, F., Sledz, J., and Giral, L. (1983), *J. Phys. Colloq.* **44**, C3-189.

Bolognesi, A., Catellini, M., and Desdri, S. (1985), *Mol. Cryst. Liq. Cryst.* **117**, 29.

Brédas, J. L. (1985), *J. Chem. Phys.* **82**, 3808.

Brédas, J. L., and Street, G. B. (1985), *Acc. Chem. Res.* **18**, 308.

Brédas, J. L., Chance, R. R., Baughman, R. H., and Silbey, R. (1981), *Int. J. Quantum Chem. Symp.* **15**, 30; *J. Chem. Phys.* **76**, 3673 (1982).

Brédas, J. L., Street, G. B., Thémans, B., and Andre, J. M. (1985), *J. Chem. Phys.* **83**, 1323.

Burt, F. P. (1910), *J. Chem. Soc.* 1171.

Byrd, N. R. (1969), *J. Polym. Sci. Part A-1* **7**, 3419.

Chance, R. R., Shacklette, L. W., Miller, G. G., Ivory, D. M., Sowa, J. M., Elsenbaumer, R. L., and Baughman, R. H. (1980), *J. Chem. Soc. Chem. Commun.*, p. 349.

Chiang, C. K. (1981), *Polymer* **22**, 1454.

Chiang, C. K., Fincher, C. R., Park, Y. W., Heeger, A. J., Shirakawa, H., Louis, E. J., Gau, S. C., and MacDiarmid, A. G. (1977), *Phys. Rev. Lett.* **39**, 1098.

Chiang, C. K., Gau, S. C., Fincher, C. R., Park, Y. W., MacDiarmid, A. G., and Heeger, A. J. (1978a), *Appl. Phys. Lett.* **33**, 18.

Chiang, C. K., Druy, M. A., Gau, S. C., Heeger, A. J., Louis, E. J., MacDiarmid, A. G., Park, Y. W., and Shirakawa, H. (1978b), *J. Am. Chem. Soc.* **100**, 1013.

Chiang, C. K., Park, Y. W., Heeger, A. J., Shirakawa, H., Louis, E. J., and MacDiarmid, A. G. (1978c), *J. Chem. Phys.* **69**, 5098.

Chiang, A. C., Waters, P. F., and Aldridge, M. H. (1982), *J. Polym, Sci. Polym. Chem. Ed.* **20**, 1807.

Chien, J. C. W. (1984), "Polyacetylene—Chemistry and Physics and Material Science," pp. 1–634, Academic Press, New York.

Chien, J. C. W., Karasz, F. E., and Shimamura, K. (1982), *Macromolecules* **15**, 1012; *J. Polym. Sci. Polym. Lett. Ed.* **20**, 97 (1982).

Clarke, T. C., Geiss, R. H., Kwak, J. F., and Street, G. B. (1978), *J. Chem. Soc. Chem. Commun.*, p. 489.

Clarke, T. C., Geiss, R. H., Gill, W. D., Grant, P. M., Morawitz, H., and Street, G. E. (1979), *Synth. Met.* **1**, 21.

Cohen, M. J., Garito, S. F., Heeger, A. J., MacDiarmid, A. G., Mikulski, C. M., Saran, M. S., and Kleppinger, J. (1976), *J. Am. Chem. Soc.* **98**, 3844.

Cordischi, D., Furlani, A., Bicev, P., Russo, M. V., and Carusi, P. (1971), *Gazz. Chim. Ital.* **101**, 526.

Cukor, P., Krugler, J. I., and Rubner, M. F. (1981), *Makromol. Chem.* **182**, 165.

Dee, R. H., Dollard, D. H., Turrell, B. G., and Carolan, J. F. (1977), *Solid State Commun.* **24**, 469.

Deits, W., Cukor, P., Rubner, M., and Jopson, H. (1981), *I&EC Prod. Res. Dev.* **20**, 696.

Devreux, F., Genoud, F., Nechtschein, M., Travers, J. P., and Bidan, G. (1983), *J. Phys. Colloq.* **44**, C3-621.

Diaconu, I., Dumitreschu, S., and Simionescu, C. (1979), *Eur. Polym. J.* **15**, 1155.

Diaz, A. F., and Hall, B. (1983), *IBM J. Res. Dev.* **27**, 342.

Diaz, A. F., and Kanazawa, K. K. (1979), *J. Chem. Soc. Chem. Commun.*, p. 635.

Diaz, A. F., Vasquez Vallejo, J. M., and Martinez Duran, A. (1981), *IBM J. Res. Dev.* **25**, 42.

Druy, M. A., Tsang, C. H., Brown, N., Heeger, A. J., and MacDiarmid, A. G. (1980), *J. Polym. Sci. Phys. Ed.* **18**, 429.

Ebert, E. B., and Selig, H. (1982), *Chem. Eng. News,* October 4, p. 27.

Elsenbaumer, R. L., and Miller, G. G. (1985), *Chem. Eng. News* **63** (25), 4.

Engler, E. M., Fox, W. B., Interrante, L. V., Miller, J. S., Wudl, F., Yoshimura, S., Heeger, A. J., and Baughman, R. H. (1979), "Molecular Metals" (W. E. Hatfield, ed.), pp. 541–543, Plenum, New York.

Epstein, A. J., Rommelmann, H., Druy, M. A., Heeger, A. J., and MacDiarmid, A. G. (1981a), *Solid State Commun.* **38**, 683.

Epstein, A. J., Gibson, H. W., Chaikin, P. M., Clark, W. G., and Grüner, G. (1981b), *Chem. Sci.* **17**, 135.

Erlandsson, R., Inganäs, O., Lundström, I., and Salaneck, W. R. (1985), *Synth. Met.* **10**, 303.

Faulques, E., and Lefrant, S. (1983), *J. Phys. Colloq. (Orsay, Fr.) Suppl.* **44**(6), C3-337.

Ferraris, J. P., Webb, A. W., Weber, D. C., Fox, W. B., Carpenter, E. R., and Brant, P. (1980), *Solid State Commun.* **35**, 15.

Ferraro, J. R., Martin, K., Furlani, A., and Russo, M. V. (1984), *Appl. Spectrosc.* **38**, 267.

Fincher, C. R., Peebles, D. L., Heeger, A. J., Druy, M. A., Matsumura, Y., MacDiarmid, A. J., Shirakawa, H., and Ikeda, S. (1978), *Solid State Commun.* **27**, 489.

Fincher, C. R., Ozaki, M., Heeger, A. J., and MacDiarmid, A. G. (1979), *Phys. Rev. B* **19**, 4140.

Fincher, C. R., Chen, C. E., Heeger, A. J., MacDiarmid, A. G., and Hastings, J. B. (1982), *Phys. Rev. Lett.* **48**, 100.

Frommer, J. E. (1986), *Accts. Chem. Res.* **19**, 20.

Froyer, G., Maurice, F., Goblot, J. Y., Fauvarque, J. F., Petit, M. A., and Diqua, A. (1985), *Mol. Cryst. Liq. Cryst.* **118**, 267.

Furlani, A., and Russo, M. V. (1985), unpublished data.

Furlani, A., Collamati, I., and Sartori, G. (1969), *J. Organomet. Chem.* **17**, 463.

Furlani, A., Bicev, P., Russo, M. V., and Fiorentino, M. (1977), *Gazz. Chim. Ital.* **107**, 373.

Galthier, M., Gay, J. M., Montaner, A., and Ribet, J. L. (1983), *J. Phys. Colloq.* **44**, C3-107.

Gau, S. C., Milliken, J., Prón, A., MacDiarmid, A. G., and Heeger, A. J. (1979), *J. Chem. Soc. Chem. Commun.,* p. 662.

Genies, E. M., and Syed, A. A. (1984), *Synth. Met.* **10**, 21.

Gibson, H. W., Epstein, A. J., Rommelmann, H., Tanner, D. B., Yang, X.-Q., and Pochan, J. M. (1983), *J. Phys. Colloq.* **44**, C3-651.

Gill, W. D., Greene, R. L., Street, G. B., and Little, W. A. (1975), *Phys. Rev. Lett.* **35**, 1732.

Greene, R. L., and Street, G. B. (1977), "Chemistry and Physics of One-Dimensional Metals" (H. J. Keller, ed.), p. 167, Plenum, New York.

Greene, R. L., and Street, G. B. (1984), *Science* **226**, 651.

Greene, R. L., Street, G. B., and Suter, L. J. (1975), *Phys. Rev. Lett.* **34**, 577.

Greene, R. L., Clarke, T. C., Gill, W. D., Grant, P. M., Kwak, J. F., and Street, G. B. (1979), *in* "Molecular Metals" (W. E. Hatfield, ed.), p. 203, Plenum, New York.

Hatano, M., Kambara, S., and Okamoto, S. (1961), *J. Polym. Sci.* **51**, 526.

Heeger, G. Kleu, S., Pintschovious, L., and Kahlert, H. (1978), *Solid State Chem.* **23**, 341.

Hotta, S., Hosaka, T., and Shimotsuma, W. (1983), *Synth. Met.* **6**, 381.

Hsu, S. L., Signorelli, A. J., Pez, G. P., and Baughman, R. H. (1978), *J. Chem. Phys.* **69**, 106.

Iqbal, Z., Baughman, R. H., Kleppinger, J., and MacDiarmid, A. G. (1978), *Solid State Commun.* **25**, 409; "Synthesis and Properties of Low-Dimensional Materials" (J. S. Miller and A. J. Epstein, eds.), Vol. 313, p. 775, Anal. N. Y. Acad. Sci.

Ito, T., Shirakawa, H., and Ikeda, S. (1974), *J. Polym. Sci. Polym. Chem. Ed.* **12**, 11.

Ito, T., Shirakawa, H., and Ikeda, S. (1975), *J. Polym. Sci. Polym. Chem. Ed.* **13**, 1943.

Ivory, D. M., Miller, G. G., Sowa, J. M., Shacklette, L. W., Chance, R. R., and Baughman, R. H. (1979), *J. Chem. Phys.* **71**, 1506.

Jones, T. E., Ogden, T. R., McGinnis, W. C., Butler, W. F., and Gottfredson, D. M. (1985), *J. Chem. Phys.* **83**, 2532.

Kanazawa, K. K., Diaz, A. F., Geiss, R. H., Gill, W. D., Kwak, J. F., Logan, J. A., Rabolt, J. F., and Street, G. B. (1979), *J. Chem. Soc. Chem. Commun.*, p. 854.

Kanazawa, K. K., Diaz, A. F., Gill, W. D., Grant, P. M., Kwak, J. F., Street, G. B., Gardini, G. P., and Kwak, J. F. (1980), *Synth. Met.* **1**, 329.

Kaneto, K., Maxfield, M., Nairns, D. P., MacDiarmid, A. G., and Heeger, A. J. (1982), *J. Chem. Soc. Faraday Trans. 1* **78**, 3417.

Kaneto, K., Kohno, Y., Yoshimo, K., and Inushi, Y. (1983), *J. Chem. Soc. Chem. Commun.*, p. 382.

Kaufman, J. H., Colameri, N., Scott, J. C., Kanazawa, K. K., and Street, G. B. (1985), *Mol. Cryst. Liq. Cryst.* **118**, 171.

Kern, R. J. (1969), *J. Polym. Sci.* **7**, 621.

Kispert, L. D., Joseph, J., Miller, G. G., and Baughman, R. H. (1985), *Mol. Cryst. Liq. Cryst.* **118**, 313.

Kletter, M. J., Worner, T., Prón, A., MacDiarmid, A. G., Heeger, A. J., and Park, Y. W. (1980), *J. Chem. Soc. Chem. Commun.*, p. 426.

Kobayashi, M., Chen, T., Chung, T-C., Moraes, F., Heeger, A. J., and Wudl, F. (1984), *Synth. Met.* **9**, 77.

Kovacic, P., and Kyriakis, A. (1963), *J. Am. Chem. Soc.* **85**, 454.

Kovacic, P., and Oziomek, J. (1964), *J. Org. Chem.* **29**, 100.

Kreutzberger, A., and Kalten, P. A. (1961), *J. Phys. Chem.* **65**, 624.

Kúlszewicz, I., Billaud, D., Prón, A., Bernier, P., and Przyluski (1981), Proc. Int. Conf. Low-Dimensional Conductors, Part E, Boulder, Colorado, p. 1191.

Kuzmany, H. (1980), *Phys. Status Solidi* **97**, 521.

Lefrant, S., Lichtmann, L. S., Temkin, H., Fitchen, D. B., Miller, D. C., Whitwell, G. E., and Burlitch, J. M. (1979), *Solid State Commun.* **29**, 191.

Lieser, G., Wegner, G., Müller, W., and Enkelmann, V. (1980), *Makrmol. Chem. Rapid Commun.* **1**, 621.

MacDiarmid, A. G., and Heeger, A. J. (1979/1980), *Synth. Met.* **1**, 101.

MacDiarmid, A. G., Kaner, R. B., Mammone, R. J., and Heeger, A. J. (1983), *J. Phys. Colloq.* **44**, C3-543.

MacInnes, D., Druy, M. A., Nigrey, P. J., Nairns, D. P., MacDiarmid, A. G., and Heeger, A. J. (1981), *J. Chem. Soc. Chem. Commun.*, p. 317.

Masuda, T., Ohtori, T., and Higashimura, T. (1974), *Makromolecules* **7**, 728.

Masuda, T., Sasaki, N., and Higashimura, T. (1975), *Makromolecules* **8**, 717.

Masuda, T., Ohtori, T., and Higashimura, T. (1979), *Polym. J.* **11**, 849; *Polym. Chem.*, p. 781 (1979).

Masuda, T., Takahashi, T. Yamamoto, K., and Higashimura, T. (1982), *J. Polym. Sci.* **20**, 2603.

McAndrew, T. P., and MacDiarmid, A. G. (1986), *ACS Nat. Meet. Div. Inorg. Chem., 191st,* New York, 1986.

Mermilliod-Thevenin, N., and Bidan, G. (1985), *Mol. Cryst. Liq. Cryst.* **118**, 227.

Mestroni, G., Camus, A., Furlani, A., and Russo, M. V. (1982), *Gazz. Chim. Ital.* **112**, 1435.

Meyer, W. H., Kiess, H., Binggeli, B., Meier, E., and Harbeke, G. (1985), *Synth. Met.* **10**, 255.

Mikulski, C. M., Russo, P. J., Saran, M. S., MacDiarmid, A. G., Garito, A. F., and Heeger, A. J. (1975), *J. Am. Chem. Soc.* **97**, 6358.

Mortensen, K., Thewalt, M. L. W., and Tomkiewicz, Y. (1980), *Phys. Rev. Lett.* **45**, 490.

Moses, D., Denerstein, A., Chen, J., Heeger, A. J., McAndrew, P., Woerner, T., MacDiarmid, A. G., and Park, Y. W. (1982), *Phys. Rev. B* **25**, 7652.

Müller, W. H. G., Baumann, F., Dammer, G., and Pintschovius, P. L. (1978), *Solid State Commun.* **25**, 119.

Murthy, N. S., Elsenbaumer, R. L., Frommer, J. E., and Baughman, R. H. (1984), *Synth. Met.* **9**, 91.

Nakamoto, K. (1978), *in* "Infrared and Raman Spectra of Inorganic and Coordination Compounds," 3rd ed., pp. 116–117, Wiley, New York.

Nguyen, N. Y., Amdur, S., Ehrlich, P., and Allendoerfer, R. D. (1978), *J. Polym, Sci. Polym. Syn.* **65**, 63.

Nigrey, P. J., MacDiarmid, A. G., and Heeger, A. J. (1979), *J. Chem. Soc. Chem. Commun.,* p. 594.

Nigrey, P. J., MacInnes, D., Nairns, D. P., MacDiarmid, A. G., and Heeger, A. J. (1981a), *J. Electrochem. Soc.* **128**, 1651.

Nigrey, P. J., MacInnes, D., Nairns, D. P., MacDiarmid, A. G., and Heeger, A. J. (1981b), *in* "Conductive Polymers" (R. D. Seymour, ed.), p. 227, Plenum, New York.

Österholm, J. E., and Passiniemi, P. (1986), *Chem. Eng. News* **64** (10), 2.

Park, Y. W., Denenstein, A., Chiang, C. K., Heeger, A. J., and MacDiarmid, A. G. (1979), *Solid State Commun.* **29**, 747.

Percec, V. (1983), *Polym. Bull.* **10**, 1.

Pittman, C. U. (1977), *in* "Organometallic Reactions" (E. Becker and M. Tsutoni, eds.), Vol. 6, Plenum, New York.

Pochan, J. M., Gibson, H. W., and Bailey, F. C. (1980), *J. Polym. Sci. Polym. Lett.* **18**, 447.

Pradere, P., Boudet, A., Goblot, J. V., Froyer, G., and Maurice, F. (1985), *Mol. Cryst. Liq. Cryst.* **118**, 277.

Prón, A., Billaud, D., Kulszewicz, I., Budrowski, C., Przyluski, J., and Suwalaki, J. (1981a), *Mat. Res. Bull.* **16**, 1229.

Prón, A. Bernier, P., Rolland, M., Lefrant, S., Aldissi, M., Pachdi, F., and MacDiarmid, A. G. (1981b), *J. Mater. Sci.* **7**, 305.

Prón, A., Kulszewicz, I., Billaud, D., and Przyluski, J. (1981c), *J. Chem. Soc. Chem. Commun.,* p. 784.

Przyluski, J., Billaud, D., Zagorska, M., Budrowski, C., and Prón, A. (1983), *J. Phys. Colloq.* **44**, C3-29.

Rabolt, J. F., Clarke, T. C., Kanazawa, K. K., Reynolds, J. R., and Street, G. B. (1980), *J. Chem. Soc. Chem. Commun.,* p. 347.

Reynolds, J. R., Chien, J. C. W., Karasz, F. E., Lillya, C. P., and Curran, D. J. (1983), *J. Phys. Colloq.* **44**, C3-171.

Rolland, M., Bernier, R., and Aldissi, M. (1980), *Phys. Status Solidi* **62**, K5.

Salaneck, W. R., Thomas, H. R., Bigelow, R. W., Duke, C. B., Plummer, E. W., Heeger, A. J., and MacDiarmid, A. G. (1980), *J. Chem. Phys.* **72**, 3674.

Scott, J. C., Brédas, J. L., Yakushi, K., Pfluger, P., and Street, G. B. (1984), *Synth. Met.* **9**, 165.

Scott, J. C., Brédas, J. L., Kaufman, J. H., Pfluger, P., Street, G. B., and Yakushi, K. (1985), *Mol. Cryst. Liq. Cryst.* **118**, 163.

Selig, H., Holloway, J. H., Prón, A., and Billaud, D. (1983), *J. Phys. Colloq.* **44**, C3-179.

Shacklette, L. W., Chance, R. R., Ivory, D. M., Miller, G. G., and Baughman, R. H. (1979), *Synth. Met.* **1**, 307.

Shacklette, L. W., Elsenbaumer, R. L., Chance, R. R., Eckhardt, H., Frommer, J. E., and Baughman, R. H. (1981), *J. Chem. Phys.* **75**, 1919.

Shimamura, K., Karasz, F. E., Hirsch, J. A., and Chien, J. C. W. (1981), *Makromol. Chem. Rapid Commun.* **2**, 1473.

Shirakawa, H., and Kobayashi, T. (1983), *J. Phys. Colloq.* **44**, C3-3.

Shirakawa, H., Ito, T., and Ikeda, S. (1973), *Polym. J. (Tokyo)* **4**, 460.

Shirakawa, H., Louis, E. J., MacDiarmid, A. G., Chiang, C. K., and Heeger, A. J. (1977), *J. Chem. Soc. Chem. Commun.*, p. 578.

Shirakawa, H., Ito, T., and Ikeda, S. (1978), *Makromol. Chem.* **179**, 1565.

Sichel, E. K., Rubner, M. F., and Tripathy, S. K. (1982a), *Phys. Rev. B* **26**, 6719.

Sichel, E. K., Knowles, M., Rubner, M., and Georges, J. (1982b), *Phys. Rev. B* **25**, 5574.

Simionscu, C. I., and Percec, V. (1980), *J. Polym. Sci. Chem. Ed.* **18**, 147.

Simionscu, C. I., Percec, V., and Dumitrescu, S. (1977), *J. Polym. Sci. Polym. Chem. Ed.* **15**, 2497.

Skotheim, T. A., Feldberg, S. W., and Armand, M. B. (1983), *J. Phys. Colloq. (Orsay, Fr.)* **44**(3), C3-615.

Soderholm, L., Mathis, C., and Francois, B. (1985), *Synth. Met.* **10**, 261.

Stamm, M., Fink, J., and Tieke, B. (1985), *Mol. Cryst. Liq. Cryst.* **118**, 281.

Street, G. B., unpublished data.

Street, G. B., and Clarke, T. C. (1981), *IBM J. Res. Dev.* **25**, 51.

Street, G. B., and Gill, W. D. (1979), *in* "Molecular Metals" (W. E. Hatfield, ed.), pp. 301–326, Plenum, New York.

Street, G. B., Clarke, T. C., Krounbi, M., Kanazawa, K., Lee, V., Pfluger, P., Scott, J. C., and Weiser, G. (1981), *IBM Res. Div. Rep.* RJ3267 (39723), 1–12.

Street, G. B., Clarke, T. C., Krounbi, M., Pfluger, P., Rabolt, R. J., and Geiss, R. H. (1982), *Polym. Prepr. (Am. Chem. Soc. Div. Polym. Chem.)* **23**, 117.

Street, G. B., Lindsey, S. E., Nazzal, A. I., and Wynne, K. J. (1985), *Mol. Cryst. Liq. Cryst.* **118**, 137.

Suzuki, N., Ozaki, M., Etemad, S., Heeger, A. J., and MacDiarmid, A. G. (1980), *Phys. Rev. Lett.* **45**, 1209.

Tanaka, M., Watanabe, A., and Tanaka, J. (1980), *Bull. Chem. Soc. Jpn.* **53**, 645.

Temkin, H., and Street, G. B. (1978), *Solid State Commun.* **25**, 455.

Temkin, H., Fitchen, D. B., Street, G. B., and Gill, W. D. (1978), *Anal. N. Y. Acad. Sci.* **313**, 771.

Tomkiewicz, Y., Schultz, T. D., Broom, H. B., Taranko, A. R., Clarke, T. C., and Street, G. B. (1981), *Phys. Rev. B* **24**, 4348.

Tourillon, G., and Garnier, F. (1982), *J. Electroanal. Chem.* **135**, 173.

Tourillon, G., and Garnier, F. (1983), *J. Electroanal. Chem.* **130**, 2042.

Tourillon, G., and Garnier, F. (1984), *J. Electroanal. Chem.* **161**, 51.

Tsonis, C. P., and Farona, M. F. (1979), *J. Polym. Sci. Polym. Chem. Ed.* **17**, 1779.

Tubino, R. (1985), *Mol. Cryst. Liq. Cryst.* **117**, 319.

Varyu, M. E., Schlenoff, J. B., and Dabkowski, G. M. (1985), *Chem. Eng. News* **63**, 4.

Walatka, V. V., Labes, M. M., and Perlstein, J. H. (1973), *Phys. Rev. Lett.* **31**, 1139.

Waltman, R. J., Bargon, J., and Diaz, A. F. (1983), *J. Phys. Chem.* **87**, 1459.

Warakomski, J. M. (1983), *in* "Polyacetylene Chemistry, Physics, and Material Science," J. C. W. Chien (1984), p. 504, Academic Press, New York.

Wnek, D. E., Chien, J. C. W., Karasz, F. E., and Lillya, C. P. (1979), *Polymer.* **20**, 1441.
Wolmershäuser, A., and Street, G. B. (1978), *Inorg. Chem.* **17**, 2685.
Wolmershäuser, A., Brulet, C. R., and Street, G. B. (1978), *Inorg. Chem.* **17**, 3586.
Woon, P. S., and Farona, M. F. (1974), *J. Polym. Sci. Polym. Chem. Ed.* **12**, 1749.
Wudl, F., Kobayashi, M., Colaneri, N., Boysel, M., and Heeger, A. J. (1985), *Mol. Cryst. Liq. Cryst.* **118**, 199.
Yamamoto, T., Sanechika, K., and Yamamoto, A. (1980), *J. Polym. Sci. Polym. Lett. Ed.* **18**, 9.
Yoffe, A. D. (1976), *Chem. Soc. Rev.* **5**, 51.
Yoshimura, S. (1979), *in* "Molecular Metals" (W. E. Hatfield, ed.), pp. 471–489, Plenum, New York.

Review Articles

Baughman, R. H., Murthy, N. S., and Miller, G. G. (1985), The structure of metallic complexes of polyacetylene with alkali metals, *J. Chem. Phys.* **79**, 515.
Baughman, R. H., Brédas, J. L., Chance, R. R., Elsenbaumer, R. L., and Shacklette, L. W. (1982), Structural basis for semiconducting and metallic polymer/dopant systems, *Chem. Rev.* **82**, 209.
Chidsey, C. E. D., and Murray, R. W. (1986), Electroactive polymers and macromolecular electronics, *Science* **231**, 25.
Conwell, E. M. (1985), The differences between one-dimensional and three-dimensional semiconductors, *Phys. Today* **38**, 46.
Deits, W., Cukor, P., Rubner, M., and Jopson, H. (1981), Analogues of polyacetylene, preparation and properties, *I&EC Prod. Res. Dev.* **20**, 696.
Frommer, J. (1986), Conducting polymer solutions, *Acc. Chem. Res.* **19**, 2–9.
Greene, R. L., and Street, G. B. (1984), Conducting organic materials, *Science* **226**, 651.
Hayes, W. (1985), Conducting polymers, *Contemp. Phys.* **26**, 421.
MacDiarmid, A. G., and Heeger, A. J. (1980), Organic metals and semiconductors: the chemistry of polyacetylene, $(CH)_x$ and its derivatives, *Synth. Met.* **1**, 101.
Perlstein, J. H. (1977), Organic metals—the intermolecular migration of aromaticity, *Angew. Chem. Int. Ed. Engl.* **16**, 519.
Street, G. B., and Clarke, T. C. (1981), Conducting polymers: a review of recent work, *IBM J. Res. Dev.* **25**, 51.
Street, G. B., and Gill, W. D. (1979), The chemistry and physics of polythiazyl $(SN)_x$, and the polythiazyl halides, *in* "Molecular Metals" (W. E. Hatfield, ed.), pp. 301–326, Plenum, New York.

Books

Bernier, P., Andre, J. J., and Comes, R., eds. (1985). "Proceedings of the International Conference on the Physics and Chemistry of Polymeric Conductors," *J. Phys. Colloq.* **44**, C3-1–C3-598.
Chien, J. C. W. (1984), "Polyacetylene—Chemistry, Physics and Material Science," pp. 1–634, Academic Press, New York.
Davydov, A. S. (1985), "Solitons in Molecular Systems," pp. 1–319, D. Reidel Publ., Dordrect, Holland.
Engler, E. M., and Tanaka, J., eds. (1985), "The 1984 International Chemical Congress of Pacific Basin Societies," Honolulu, Hawaii, December 16–21, 1984, *Mol. Cryst. Liq. Cryst.* **126**, 1–128.
Hatfield, W. E., ed. (1979), "Molecular Metals," pp. 1–555, Plenum, New York.
Lehn, J. M., and Rees, Ch. W. (1985), "Molecular Semiconductors," pp. 150–200, Springer-Verlag, Berlin and New York.

Miller, J. S., and Epstein, A. J., eds. (1978), "Synthesis and Properties of Low-Dimensional Materials," pp. 1–828, N. Y. Acad. Sci., New York.

Pecile, C., Zerbi, G., Bozio, R., and Girlando, A., eds. (1985), "Proceedings of the International Conference on the Physics and Chemistry of Low-Dimensional Synthetic Metals," Abano, Terme, Italy, June 17–22, 1984, Part A, *Mol. Cryst. Liq. Cryst.* **117**, 1–486.

Pecile, C., Zerbi, G., Bozio, R., and Girlando, A., eds. (1985), "Proceedings of the International Conference on the Physics and Chemistry of Low-Dimensional Synthetic Metals," Abano, Terme, Italy, June 17–22, 1984, Part B, *Mol. Cryst. Liq. Cryst.* **118**, 1–450.

Pope, M., and Swenberg, C. E. (1982), "Electronic Processes in Organic Crystals," pp. 1–799, Clarendon Press, New York.

Seanor, D. (1982), "Electrical Properties of Polymers," pp. 1–360, Academic Press, New York.

Skotheim, T. A., ed. (1986), "Handbook of Conducting Polymers," pp. 1–1464, Marcel Dekker, New York.

Van der Berg, E. J., ed. (1984), "Contemporary Topics in Polymer Science," pp. 281–319, Plenum Press, New York.

4 KROGMANN SALTS—PARTIALLY OXIDIZED PLATINUM-CHAIN METALS AND RELATED MATERIALS

I. INTRODUCTION

The quest for materials with metallic properties has provided another new area of research centered on inorganic "linear chain" or "pseudo-one-dimensional" (1D) compounds. One-dimensional metal-chain compounds may be synthesized from inorganic materials either as stacks of individual molecules or as covalently bonded polymers. It is in the former class of complexes that major interest has developed.

The most extensive research has centered on one-dimensional inorganic systems that are based on tetracyanoplatinate ions. As early as 1842 Knop prepared a complex of platinum and cyanide anions. The complex had a gold–bronze color, but was not too well understood. Further work by Knop and Schnedermann (1845) and Wilm (1888) was undertaken, but it was not until 1912 that Levy (1912) found that the complex consisted of a "mixed" valence state of platinum. Originally, these materials were of interest for their flourescent properties. Krogmann and Hausen (1968) reinitiated a study of the tetracyanoplatinates and deduced that the stoichiometry of the potassium salt was $K_2[Pt(CN)_4]X_{0.3}(H_2O)_n$, where X could be Cl or Br, with platinum exhibiting the nonintegral oxidation state of $\sim 2.3^+$. Therefore, these materials were found to be partially oxidized tetracyanoplatinate (POTCP) salts and *not* consisting of a mixture of valence states of platinum. These materials exhibit metallic properties at room temperature. They were capable of stacking of the planar portions of the molecule, which resulted in linear metal–metal atom chains with short separations of less than the van der Waal radius sum of the metal atom. Thus, strong metal–metal interactions led to band formation. The subsequent research in this area, in an attempt to create new materials, resulted in the finding of 10–12 new and unusual POTCP compounds.

Since these POTCP salts were originally synthesized by mixing divalent and tetravalent tetracyanoplatinates, it appears appropriate to discuss the properties of both these starting materials.

II. DIVALENT TETRACYANOPLATINATES

Since their discovery over 150 years ago, the tetracyanoplatinate (TCP) salts of simple monovalent and divalent cations have been interesting materials because of their broad range of colors and their highly anisotropic optical properties [see Gmelin (1939–1940)]. Tables I and II tabulate some well characterized TCP compounds, which are obtainable as single crystals which greatly facilitates their study.

A. Structural Studies

An x-ray study of $Mg[Pt(CN)_4] \cdot 7H_2O$ was carried out by Bozorth and Pauling (1932), and a columnar stacked-chain structure for the $[Pt(CN)_4]^{2-}$ ion was proposed with d_{Pt-Pt} of 3.23 Å. This distance may be compared with that in Pt metal (2.78 Å) at room temperature. Many related TCP salts have similarly been examined structurally and Table I lists several of these. In most cases the d_{Pt-Pt} has been determined from the unit cell dimension in the platinum atom-chain direction (usually the needle axis of a crystal) assuming that the platinum atom chains are linear and that all the intrachain Pt–Pt separations are identical. These assumptions are reasonable, although detailed x-ray and neutron diffraction studies have revealed that some nonlinear platinum atom chains are present with unequal Pt–Pt separations in some TCP salts. Table II lists the structural details of compounds on which full x-ray or neutron crystal structure measurements have been performed.

The data in Table I demonstrates that the d_{Pt-Pt} ranges from a minimum of about 3.09 Å to over 3.6 Å. The Pt–Pt separation appears to be influenced by cation size and charge and the degree of hydration. For example, $Sr[Pt(CN)_4] \cdot 3H_2O$ has the shortest d_{Pt-Pt} (3.09 Å), whereas the pentahydrate has one of the longest d_{Pt-Pt} separations. The stacking of the $[Pt(CN)_4]^{2-}$ groups occurs in the divalent TCP and although the intrachain Pt–Pt distances can be varied by changing the cation and the degree of hydration, the distances are not sufficiently short to give rise to metallic properties and these molecules do not exhibit high electrical conductivities.

TABLE I

Crystal and Spectral Properties of Divalent (Pt^{2+}) Tetracyanoplatinates[a]

Compound	Color	Crystal system	d_{Pt-Pt} (Å)	Reflection band $E_\parallel c$ (cm^{-1})	Emission band[b] $E_\parallel c$ (cm^{-1})
$Sr[Pt(CN)_4]\cdot 3H_2O$	Violet	Monoclinic	3.09		
$Mg[Pt(CN)_4]\cdot 7H_2O$	Red	Tetragonal	3.155	18,020	17,600
$Ba[Pt(CN)_4]\cdot 2H_2O$	Dark red	Orthorhombic	3.16		
$Er[Pt(CN)_4]_3\cdot 21H_2O$			3.17		
$Li_2[Pt(CN)_4]\cdot xH_2O$			3.18	20,400	
$Dy_2[Pt(CN)_4]_3\cdot 21H_2O$		Orthorhombic	3.18		
$Tb_2[Pt(CN)_4]_3\cdot 21H_2O$			3.18		17,800
$Y_2[Pt(CN)_4]_3\cdot 21H_2O$		Orthorhombic	3.18	19,420	17,800
$KLi[Pt(CN)_4]\cdot 2H_2O$			3.20		
$K_2Sr[Pt(CN)_4]\cdot 2H_2O$	Violet–red	Monoclinic	3.21	20,830	
$KNa[Pt(CN)_4]\cdot 3H_2O$		Monoclinic	3.26(2)		
$(NH_4)_2[Pt(CN)_4]\cdot 2H_2O$			3.26	21,280	
$Ba[Pt(CN)_4]\cdot 4H_2O$	Yellow–green	Monoclinic	3.321(3)	21,930	21,000
$K_2Sr[Pt(CN)_4]\cdot 6H_2O$	Yellow–green	Monoclinic	3.33		
$Sm_2[Pt(CN)_4]_3\cdot 18H_2O$			3.35[b]		21,650
$Mg[Pt(CN)_4]\cdot 4.5H_2O$	Yellow	Triclinic	3.36		
$Eu_2[Pt(CN)_4]\cdot 18H_2O$			3.37[b]		22,000
$Ca[Pt(CN)_4]\cdot 5H_2O$	Yellow	Orthorhombic	3.38	22,900	22,250
$Rb_2[Pt(CN)_4]\cdot 4H_2O$	Green	Monoclinic	3.421(2)	24,390	
$K_2[Pt(CN)_4]\cdot 3H_2O$		Orthorhombic	3.478(1)	24,690	
$Cs_2[Pt(CN)_4]\cdot H_2O$		Hexagonal	3.545(1)		
$Sr[Pt(CN)_4]\cdot 5H_2O$	Colorless	Monoclinic	3.60	27,170	
$Na_2[Pt(CN)_4]\cdot 3H_2O$	Colorless	Triclinic	3.71[c]		26,750

[a] Taken from Williams *et al.* (1982).
[b] Deduced from position of emission band.
[c] Four independent spacings exist in this salt (3.65 Å, 3.69 Å, 3.75 Å, and 3.75 Å).

Detailed structural studies as tabulated in Table II have shown that several of these compounds have linear or approximately linear chains of equally spaced platinum atoms. A typical example of this type of compound is $Ba[Pt(CN)_4]\cdot 4H_2O$ (Maffly *et al.,* 1977). The structure (see Fig. 1) of this compound has been shown by neutron diffraction at room temperature to consist of columnar stacks of $[Pt(CN)_4]^{2-}$ ions with a d_{Pt-Pt} of 3.321 Å (Williams *et al.,* 1982). The $[Pt(CN)_4]^{2-}$ ions were found to be tilted by 3° with respect to the c axis of the crystal. Adjacent $[Pt(CN)_4]^{2-}$ groups are in an almost totally staggered configuration with a torsion angle of ~45°. The platinum chains are bound together by Ba^{2+} and water molecule hydrogen-bond interactions. The Ba^{2+} ions

TABLE II

Summary of Structural Data for Divalent Tetracyanoplatinates[a]

Compound	Neutron (N) or x-ray (X)	Space group	Intrachain Pt–Pt separation (Å)	Torsion angle (degrees)	Comments
KNa[Pt(CN)$_4$]·3H$_2$O	N	$Cc(C_s^4)$	3.263(2)	34.5(8) 34.5(8) 34.5(16) 34.4(16)	The Pt atom chain is nearly linear (angles 176.4°); K$^+$ and Na$^+$ have 6-fold coordination.
Ba[Pt(CN)$_4$]·4H$_2$O	N	$C2/c(C_{2h}^6)$	3.321(3)	46.27(7) 42.44(4) 48.85(7)	The Pt atom chain is linear; Ba^{2+} surrounded by 6 H$_2$O and 4 CN$^-$.
Cs$_2$[Pt(CN)$_4$]·4H$_2$O	N X	$P6_1$ or $P6_5$ $(C_6^2$ or $C_6^3)$	3.545(1) 3.543(1)	32.79(31)	Unusual helical chain of Pt atoms and Pt–Pt angle is 156.01°; 7- and 8-fold coordination around Cs$^+$.
Na$_2$[Pt(CN)$_4$]·3H$_2$O	N	$P\bar{1}(C_i^1)$	3.651(4) 3.691(4) 3.745(4) 3.754(2)	Eclipsed average angle 0°	Four independent intrachain Pt–Pt separations. The Pt atom chain is bent with Pt–Pt angles of 166.04(8) and 162.48(8)°; 6-fold coordination about Na$^+$.
Rb$_2$[Pt(CN)$_4$]·1.5H$_2$O	N	$C2/c(C_{2h}^6)$	3.421(2)	Varies from 29.5 to 32.2	Pt atom chain bent and Pt–Pt–Pt angle is 170.97(4)°.
K$_2$[Pt(CN)$_4$]·3H$_2$O	N	$Pbcn$	3.478(1)	15.7(1) 16.7(1)	Linear Pt atom chain; 7-fold coordination about K$^+$.

[a] Taken from Williams *et al.* (1982).

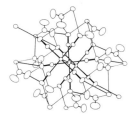 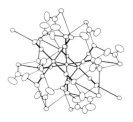

Fig. 1. A stereoview of the Pt–Pt chain (c axis) and the water molecule interactions between the cyanide groups and the Ba ions in $Ba_2[Pt(CN)_4] \cdot 4H_2O$. Ellipsoids are shown at 50% probability. [From Williams *et al.* (1982).]

are coordinated by six water molecules and four CN^- ions. On the other hand, the $Rb_2[Pt(CN)_4] \cdot 1.5H_2O$ salt exemplifies a compound having a nonlinear platinum atom chain (Koch *et al.*, 1977a). The Pt–Pt–Pt angle is 170.97(4)°, and the platinum atoms are displaced from linearity by 0.270(1) Å. For $Cs_2[Pt(CN)_4] \cdot H_2O$ an unusual helical chain of platinum atoms has been observed by both x-ray and neutron-diffraction techniques (see Fig. 2) (Johnson *et al.*, 1977; Otto *et al.*, 1977). The helix has a repeat unit of 19.34 Å and the Pt atom distance from the origin (at 000) is 1.47 Å. The Pt–Pt–Pt angle is 156.01°. Both seven- and eight-fold coordination exist around the Cs^+ ions.

The torsion angle between $[Pt(CN)_4]^{2-}$ ions in the Pt^{2+} materials has been found to vary from ~45°, corresponding to a staggered configuration, to a totally eclipsed configuration of 0°. Compounds with very short d_{Pt-Pt} have a torsion angle of 45°, which minimizes the CN^- group repulsive interactions. Compounds with long d_{Pt-Pt} have the eclipsed configuration due to steric effects. Compounds with an intermediate d_{Pt-Pt} possess torsion angles between 0 and 45° and appear to show no correlation between d_{Pt-Pt} and the torsion angle.

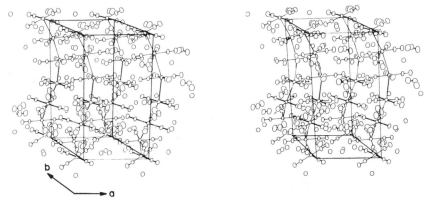

Fig. 2. A stereoview of the unit cell of $Cs_2[Pt(CN)_4] \cdot 4H_2O$ showing the "helical" Pt atom chain. [From Williams *et al.* (1982).]

B. Optical Studies

1. $[Pt(CN)_4]^{2-}$ Ions in Solution

The isolated $[Pt(CN)_4]^{2-}$ ion demonstrates multiple absorption spectra in dilute aqueous and acetonitrile solutions (Moreau-Colin, 1972; Piepho et al., 1969; Mason and Gray, 1968; Isci and Mason, 1975; Marsh and Miller, 1976). In general, in dilute aqueous solutions, five complex band systems occur at about 35,800 cm^{-1} ($\epsilon = 1500\, M^{-1}\, cm^{-1}$), 39,000 (10,000), 41,000 (2000), 46,000 (20,000), and 51,000 (10,000). The bands have been assigned to metal–ligand charge transfer bands. The lowest lying empty orbital will be the a_{2u} (π^*) orbital which is chiefly localized on the N atoms of the CN groups (Interrante and Messmer, 1974). However, controversy exists over the assignments for several reasons:

(a)　The ordering of the d orbitals in this ion is not known with any degree of certainty. Mason and Gray (1968) and Interrante and Messmer (1974), based on X_α scattered wave calculations, have proposed the order sequence

$$b_{2g}(d_{xy}) > e_g(d_{xz}, d_{yz}) > a_{1g}(d_{z^2}) \tag{1}$$

However, Piepho et al. (1969) and Isci and Mason (1975) (based on MCD studies) suggest the order

$$a_{1g}(d_{z^2}) > e_g(d_{xz}, d_{yz}) > b_{2g}(d_{xy}) \tag{2}$$

(b)　The one electron spin–orbit coupling constant for Pt is 4060 cm^{-1} so that spin forbidden bands will be expected to have quite large intensities (McClure, 1959).

(c)　The inclusion of spin–orbit coupling will increase greatly the number of energy levels and consequently the number of transitions possible for the $[Pt(CN)_4]^{2-}$ ion. The number of observed bands is generally less than the number of theoretically possible transitions. Deconvolution of the spectra suggest seven transitions below 50,000 cm^{-1} (Marsh and Miller, 1976). However, general agreement exists in that the low energy transition at ~35,800 cm^{-1} is due to the a_{2u} (π^*) ligand orbital, and the small extinction coefficient is indicative of a single–triplet transition. There is agreement that the band at 39,000 cm^{-1} arises from the $e_g \rightarrow a_{2u}$ transition giving rise to E_u^1 and A_{2u}^1 states (Moreau-Colin, 1972; Piepho et al., 1969; Mason and Gray, 1968; Isci and Mason, 1975).

C. Solid State Optical Properties

The very intense absorbance of the TCP salts make absorption measurements very difficult to obtain. However, reflectivity studies by Moncuit and Poulet (1962) have allowed a more precise determination of the position of the absorption bands in this class of compounds. Moreau-Colin (1972) has reviewed these results, as well as results of fluorescent spectra of TCP salts (Moreau-Colin, 1965). Yersin and Gliemann (1978) have reported on extensive studies of the polarized emission spectra of TCP complexes (See Table I). The TCP compounds show two differently polarized transitions which vary with d_{Pt-Pt}. As d_{Pt-Pt} decreases the transitions shift to lower energy similar to observations cited for the absorption spectra. The experimental results follow an R^{-n} power low (with $n = 3.0 \pm 0.3$), where $R = d_{Pt-Pt}$ (Yersin and Gliemann, 1975a).

The reflectivity and emission peak energies with E parallel to c ($E \parallel c$) both undergo the same shift to lower energies with decreasing R, showing they are connected with the same transitions (see Fig. 3). Extrapolation of the data to $d_{Pt-Pt}^{-3} = 0$ yields a peak energy of 45,000 and 44,000 cm^{-1}, respectively, suggesting that this low energy band is connected with an intense band system observed at ~46,000 cm^{-1} in the free ions. Since it has a high oscillator strength and is polarized parallel to the platinum atom chain direction it has been suggested that it arises from a transition between the one-electron hybrid molecular state $(5d_{z^2} 6_s) \rightarrow (6p_z, CN\pi^*)$ (Yersin et al., 1977; Yersin and Gliemann, 1975b).

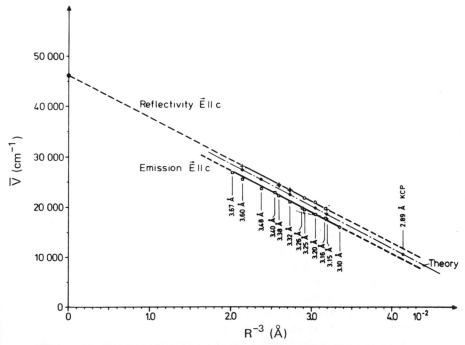

Fig. 3.   Emission and reflectivity peak energies ($E \parallel c$) for various (Pt^{2+}) tetracyano-platinates versus R^{-3}; $R = Pt-Pt$ distance (300 K). The theoretical points are the band gap energies. [From Williams et al. (1982).]

The oscillator strength of the transitions with E perpendicular to c ($E \perp c$) are orders of magnitude smaller than those with $E \parallel c$ (Yersin and Gliemann, 1978). Additionally, the emission lifetime for $E \parallel c$ is at least two orders of magnitude larger than the lifetime of the $E \perp c$ component at room temperature (Yersin and Gliemann, 1978), suggestive of a spin forbidden transition and connected with an excited triplet state.

The variation of the emission spectra with temperature has been studied for $KLi[Pt(CN)_4] \cdot 2H_2O$ and $Mg[Pt(CN)_4] \cdot 7H_2O)$ (Holzapfel et al., 1978; Yersin and Gliemann, 1975b). The red shift observed on lowering the temperature can be attributed mainly to a large thermal contraction in d_{Pt-Pt} (Yersin and Gliemann, 1978). The relative intensities of the polarized emission transitions vary strongly with temperature and this has been explained in part in terms of a Boltzmann distribution between the two emitting states.

Recently the polarized emission spectra of rare earth TCP compounds have also been studied (Yersin 1976, 1978; Yersin et al., 1979). It has been observed that some of these rare earth ions (e.g., Sm^{3+} and Eu^{3+}) produce pronounced changes in the column emission properties compared with those previously observed for the TCP compounds containing alkali and alkaline earth metal ions (Yersin, 1978). For example, in $Sm_2[Pt(CN)_4]_3 \cdot 18H_2O$, the intensity ratio $I(E \perp c)/I(E \parallel c)$ is 10^2 smaller compared with $Ba[Pt(CN)_4] \cdot 4H_2O$ and the $E \perp c$ lifetime is drastically reduced. In addition, emission from the Sm^{3+} ion is also observed which results from a radiationless transfer process from the TCP (donor) stack to the rare earth (acceptor) ion. Energy transfer can only be detected if the energy of the donor state is just above that of the acceptor state. Thus in $Sm_2[Pt(CN)_4]_3 \cdot 18H_2O$ the lowest excited state of Sm^{3+} has a lower energy than the lowest column excited state. Energy transfer takes place and line emission from the $^4G_{5/2}$ excited state to the ^6H_J ground state is observed. Similar behavior is observed for $Eu_2[Pt(CN)_4]_3 \cdot 18H_2O$ at 80 K. However, $Tb_2[Pt(CN)_4]_3 \cdot 21H_2O$ crystallizes with a shorter d_{Pt-Pt} and this reduces the transition energies of the stacks below that of the lowest excited Tb^{3+} ion state. The emission spectrum of this rare earth compound is similar to that of the alkaline earth TCP compounds and no emission spectrum from the Tb^{3+} ion is observed. The energy transfer mechanism between the TCP stack and the rare earth ion has been discussed (Yersin, 1978).

Energy bands within a columnar stack of TCP units have been calculated based on the relatively strong overlap of the Pt $6p_z$, $CN\pi^*$, and Pt $5d_{z^2}$, $6s$) molecular orbitals in the chain direction (Yersin et al., 1977). The variation of these bands with decreasing d_{Pt-Pt} was calculated and it was found that as d_{Pt-Pt} decreased the band splitting increased and the energy gap between the top of the valence (Pt d_{z^2}, $6s$) band and the bottom of the conduction (Pt $6p_z$, $CN\pi^*$) band decreased. The calculated energy gaps agreed well with the d_{Pt-Pt} dependence obtained from optical measurements. However, Day (1975) has suggested that the lowest excited states in these crystals should be described as neutral Frenkel excitons propagating along the stacks and formed from simple molecular transitions coupled by the intermolecular interaction potential. Since the molecular transition dipole vectors within each stack are parallel to one another, the resulting crystal transition would be expected to suffer a Davydov shift to lower energy. This also leads to the $(d_{Pt-Pt})^{-3}$ relationship observed experimentally (Day, 1975). A study of the time-resolved emission spectra of $Ba[Pt(CN)_4] \cdot 4H_2O$ revealed the presence of several localized and delocalized excited states with different behaviors. At high excitation intensities, indications of cooperative exiton effects were observed (Gerhardt *et al.*, 1979).

D. Electrical Conduction Studies

The dc electrical conductivities parallel to the platinum atom chain of several divalent TCP salts have been reported over the temperature range of 100–270 K (O'Neil *et al.*, 1979). The $\sigma_\parallel$ at room temperature was found to be strongly dependent on the cation of the salt and $\sigma_\parallel$ varied with $1/d_{Pt-Pt}$. For $Li_2[Pt(CN)_4] \cdot 3H_2O$ and $Ba[Pt(CN)_4] \cdot 4H_2O$ the anisotropy $\sigma_\parallel/\sigma_\perp$ was $\sim 10^2$. It was suggested that the electrical conductivity properties arise extrinsically due to presence of Pt^{4+} impurities which act as electron acceptors promoted from the top of the $(5d_{z^2}, 6s)$ valence band. The increase of $\sigma_\parallel$ with decreasing d_{Pt-Pt} is principally due to an increasing mobility of carriers (O'Neil *et al.*, 1979). It should be noted that divalent TCP salts are, at best, semiconductors. For example, $K_2[Pt(CN)_4] \cdot 3H_2O$ ($d_{Pt-Pt} = 3.50$) has a conductivity of 5×10^{-1} (ohm cm)$^{-1}$ (Minot and Perlstein, 1971), whereas $Li_2[Pt(CN)_4] \cdot xH_2O$ with $d_{Pt-Pt} = 3.18$ has a conductivity of 3×10^{-1} (ohm cm)$^{-1}$ (Underhill, 1974).

E. Pressure Studies

Few pressure studies have been performed in the case of the divalent TCP salts. Several studies demonstrate the strong dependence of the position of the emission peaks of d_{Pt-Pt} based on the effects of pressure (Yersin, 1978; Stock and Yersin, 1976; Yersin *et al.*, 1979; Stock and Yersin, 1978). Studies on $Na_2[Pt(CN)_4] \cdot 3H_2O$, $Ca[Pt(CN)_4] \cdot 5H_2O$, and $Mg[Pt(CN)_4] \cdot 7H_2O$ demonstrated that it is possible to tune $E \parallel c$ continuously from 27,000 cm^{-1} to 13,000 cm^{-1} by using a suitable combination of TCP salt and pressure (Yersin, 1978; Yersin *et al.*, 1979).

Hara *et al.* (1975) measured the pressure sensitivity of platinum salts such as $Sr[Pt(CN)_4] \cdot 2H_2O$, $Mg[Pt(CN)_4] \cdot 7H_2O$, $Ba[Pt(CN)_4] \cdot 4H_2O$, and $Pt(NH_3)_4 \cdot Pt(CN)_4$. No resistance minimum was observed, but electrical conductivity increased with pressure by a factor of 10^4. Results were interpreted as due to a considerable contribution of the π^* orbitals of the CN^- ligand to the conduction process.

III. TETRAVALENT TETRACYANOPLATINATES

Some of the most common of the tetravalent platinum complexes are those containing halogen ligands coordinated trans to each other, as is the case in the $M_2[Pt(CN)_4X_2]$ salts where M = alkali metal and X = Cl, Br, or I. These octahedrally coordinated platinum salts are prepared by adding an excess of the oxidant in the presence of the Pt^{2+} salt (Blomstrand, 1971). Presumably M is any alkali metal cation, but the full range of constituents representing M is not known. The chemistry of these species can actually be quite complex, e.g., when $K_2[Pt(CN)_4] \cdot 3H_2O$ is treated with an oxidant other than a halogen it does not give definite compounds of Pt^{4+} (Wilm, 1888).

For the sake of brevity we will only consider tetracyanodihalo salts of Pt^{4+} since they are intrinsically relevant to the preparation of the important one-dimensional POTCP salts. Because the $[Pt(CN)_4X_2]^{2-}$ moiety is octahedrally coordinated, with the halogen atoms trans to each other, these groups do not stack in such a manner as to produce direct M–M d_{z^2} overlap. However, some "mixed valence" (Pt^{2+}, Pt^{4+}) salts do stack to form M–X–M bridges, and some electrical conduction can occur through the metal–halogen–metal bridge (Miller and Epstein, 1976).

A. Structural Studies

Four characterized salts of Pt^{4+}, which serve as excellent examples for pointing out an unusual finding related to POTCP salts, are $Na_2[Pt(CN)_4Br_2] \cdot 2H_2O$ (Maffly *et al.*, 1977), $Rb_2[Pt(CN)_4Br_2]$ (Needham *et al.*, 1977), $Ba[Pt(CN)_4Br_2] \cdot 4H_2O$, and $K_2[Pt(CN)_4Br_2]$ (Washecheck and Williams, unpublished work). The unusual finding is that, to the best of our knowledge, if the Pt^{4+} salt is hydrated, then the partially oxidized salt of the corresponding $M^{1+,2+}$ cation cannot be prepared. As an illustration we point out that whereas numerous POTCP salts of K^+ and Rb^+ have been prepared, no salts of Na^+, Ba^{2+}, or any multivalent cation have been reported. In this regard it is interesting to note (Koch *et al.*, 1977b) that from the limited amount of structural data available, if no halide ions are included in the inner coordination sphere of the cation in the hydrated Pt^{4+} salts (e.g., Ba^{2+} and Na^{+}), then no POTCP derivatives have been reported. However, in the anhydrous K^{+} and Rb^{+} derivatives there are always halide ions in the inner coordination sphere of the Pt^{4+} ion and a POTCP complex of the cation can be prepared. The specific relevance of these findings is not known and we remind the reader that the results are derived from solid state crystallographic studies whereas the preparation of POTCP salts is most frequently carried out in solution.

In concluding this section it should be pointed out that a wide variety of $[Pt(CN)_4XY]^{-n}$ anionic salts can be prepared as illustrated by the few given in Table III (Osso and Rund, 1978). However, whether or not POTCP salts can be derived from these complexes must await further investigation.

B.   Optical Studies in Solution

Whereas Pt^{2+} salts, and those of $[Pt(CN)_4]^{2-}$ in particular, have been studied in great detail, this is not the case for the $[Pt(CN)_4X_2]^{2-}$, X = halide, anion salts. The reason for this lies primarily in the fact that while the $[Pt(CN)_4]^{2-}$ specie can stack to form platinum atom chains, with resulting metallic properties, the $[Pt(CN)_4X_2]^{2-}$ anion cannot stack because the X atoms do not allow the required d_{z^2} orbital overlap. For this reason the optical and electrical properties of the $[Pt(CN)_4X_2]^{2-}$ anion have not been investigated in any great detail.

However, the solution chemistry of cyanide and halide complexes of

TABLE III

Tetravalent Tetracyanoplatinates Containing Different X and Y Groups
$[(n\text{-}C_4H_9)_4N]_2[Pt(CN)_4XY]^a$

X	Y
I	I
I	Cl
I	Br
Br	Br
C_6H_5S	Cl
Br	Cl
Cl	Cl
I	CN
SCN	SCN
NO_2	NO_2

[a] Osso and Rund (1978).

Pt^{2+} and Pt^{4+} has been investigated in some detail and brief mention of the associated chemistry is relevant here because many of the platinum chain conductors can be prepared by mixing solutions containing the two species. It is well known that Pt^{2+} catalyzes Pt^{4+} substitution reactions (Mason, 1972). Chain formation appears to result from a reaction between a square–planar Pt^{2+} donor and a solvated Pt^{4+} acceptor. If the axially bound ligands of the octahedrally coordinated Pt^{4+} complex are replaced in solution by solvent molecules, and if the solvent molecule itself can be displaced by reaction with the donating Pt^{2+} species, then chain formation occurs in a reaction which may be illustrated as follows (Anderjan *et al.*, 1974; Saillant and Jaklevic, 1974):

$$[Cl \cdot Pt^{IV}(CN)_4 \cdot H_2O]^- + [H_2O \cdot Pt^{II}(CN)_4 \cdot H_2O]^{2-} \rightleftarrows$$

$$[H_2O \cdot Pt^{II}(CN)_4 \cdot Pt^{IV}(CN)_4 \cdot H_2O]^{2-} + H_2O + Cl^-$$

$$[H_2O \cdot Pt^{II}(CN)_4 \cdot Pt^{IV}(CN)_4 \cdot H_2O]^{2-} + [H_2O \cdot Pt(CN)_4 \cdot H_2O]^{2-} \rightleftarrows$$

$$[H_2O \cdot Pt^{II}(CN)_4 \cdot Pt^{IV}(CN)_4 \cdot Pt^{II}(CN)_4 \cdot H_2O]^{4-} + 2H_2O$$

As has been pointed out previously, in the solid state, POTCP salts are not generally of the "mixed-valence" type containing discrete Pt^{2+} and Pt^{4+} ions but instead contain $Pt^{2.1-2.4}$ ions which are most probably all in the same valence state. In a preliminary report (Schindler *et al.* (1979), it has been observed that in spectroscopic and photochemical studies of K^+ and Ba^{2+} tetracyanoplatinate salts in solution there was definite evidence for formation of $[Pt(CN)_4]_n^{-2n}$ oligomers which exhibit linear chain pseudo one dimensionality. Previously, such platinum atom chain formation had been thought to exist in the solid state only. Due to the absence of Pt–Pt chain formation in the Pt^{4+} salts one would expect them to be electrical insulators, which they are, and lacking in unusual conduction properties. Therefore, no discussion of their electrical conduction properties will be presented.

In the next section we turn to the partially oxidized tetracyanoplatinate, POTCP, salts which are of immense interest due to their unusual properties, not the least of which is the phenomenon of one-dimensional electrical conductivity.

IV.   PARTIALLY OXIDIZED TETRACYANOPLATINATE METALS

The most avidly studied conducting transition metal complexes have been the partially oxidized tetracyanoplatinate (POTCP) metals. These are appropriately classified as organomctallic compounds due to the pres-

ence of metal–carbon σ bonds. These materials exhibit anisotropic physical properties, with the most important being the one-dimensional metallic
conductivity along the metal-chain axis. In general, the most promising
have been the materials which involve square–planar coordination complexes of platinum, involving columnar stacks of infinite platinum atom
chains, with overlapping $5d_{z^2}$ platinum orbitals (see Fig. 4).

Several aspects of one dimensionality and the highly conducting systems should be considered before discussing the POTCP metal complexes.

(1) *Mixed valency.* The electrical properties of a solid are determined
to a large extent by its electronic energy levels, their occupation, and the
gaps between them. According to band theory when N atoms are brought
together so also are N molecular orbitals covering a range of orbital
energies of finite width. At indefinitely large values of N, these energies
can be considered to form a continuous band. Metallic conduction results
from the presence of partially filled bands in which electrons can be readily

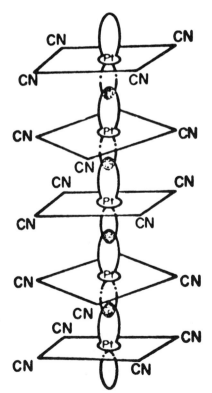

Fig. 4.   A chain of square planar $[Pt(CN)_4]^{x-}$ groups showing the overlapping Pt d_{z^2}
orbitals.

excited to unfilled orbitals close in energy ($E \ll k$T) to the highest filled orbital, the result being electron mobility and electrical conduction. The presence of nonintegral oxidation states and hence, the creation of partially filled bands, thus becomes an important factor in the creation of highly conducting systems. The theory behind charge transport has been treated in detail elsewhere (Andre *et al.*, 1976; Soos and Klein, 1975; Torrance, 1977). Briefly, in systems with integral oxidation states charge is mobilized through the creation of states in which positive and negative ions are separated along the chain, while partially oxidized systems already have holes which have been formed as a result of fractional oxidation. In order for the former to occur, a large activation energy is required. In the latter, however, charge transport can readily occur by movement of the holes between degenerate configurations.

(2) *Steric aspects.* For strong intermolecular interactions to occur, close molecular approach is a necessity. One way to facilitate this is accomplished by square–planar coordination about the metal ion, and by using planar ligand systems. (In the transition metal area efforts to achieve one-dimensional chains by using binuclear complexes M_2L_6 or M_2L_4 have not been as successful). Short metal–metal distances that are only 0.1 to 0.3 Å longer than those in the parent metal have been obtained in the partially oxidized square–planar tetracyanoplatinates by premeditated design (Williams and Schultz, 1978). Even for organic molecular metals, a criterion for creating conductive molecular crystals is (frequently) that the parent compound be a planar complex of metal and ligand (Phillips and Hoffman, 1977; Phillips *et al.*, 1980). The logic behind employing planar systems is evident, i.e., bulky ligands or molecules may not approach each other due to steric repulsions as well as large size. On the other hand some complexes which show relatively large distortions from planarity have been found to exhibit unexpectedly large conductivities. The partially oxidized octamethyltetrabenzporphinatonickel(II), Ni(OMTBP)I, is a case in point (Phillips *et al.*, 1980). In fact, the nickel and palladium *bis*(diphenylgloximates), despite the bulky phenyl substituents, display higher conductivities in their partially oxidized form than the corresponding more planar *bis*(benzoquinonedioximates). These compounds are discussed in greater detail in Chapter 5.

Given that square–planar geometry is a desirable characteristic in one-dimensional transition metal complexes, an investigation into the propensity of various metals for forming this geometry is appropriate. Krogmann has pointed out that metals having accessible d^8 electron configurations have a greater tendency to form square–planar coordinated complexes, which stack to form columnar structures (Krogmann, 1969). Table IV lists metals with accessible d^8 electron configurations. Presently

TABLE IV

Metals with Accessible d^8 Electron Configurations

Configuration	Metal (oxidation state)			
$3d^8$	Fe(0)	Co($+1$)	Ni($+2$)	Cu($+3$)
$4d^8$	Ru(0)	Rh($+1$)	Pd($+2$)	Ag($+3$)
$5d^8$	Os(0)	Ir($+1$)	Pt($+2$)	Au($+3$)

only platinum, iridium, and rhodium (in decreasing order of occurrence) have been observed to form one-dimensional metal chain structures with extremely short metal–metal separations. The compounds formed from zero-valent states of iron, ruthenium, and osmium favor trigonal–bipyramidal geometries or clusters which tends to prevent chain formation and hence extended metal–metal interactions.

(3) *Electronic considerations.* These are also of major importance as can be seen by examining the capacity of the d^8 metals for forming highly conducting one-dimensional systems. If, as is often the case in transition metal complexes, the intermolecular interaction involves overlapping d orbitals, orbital spatial extension is a major consideration. Due to $5d > 4d > 3d$ orbital extension, the most highly conducting complexes would be expected for third row transition metals. Thus both Pt^{2+} and Ir^+ form complexes which exhibit metallic conduction, but Au^{3+} does not, due to a contracted $5d$ orbital as a result of higher effective charge. Accordingly, Ni^{2+} and Pd^{2+} would be expected to form complexes exhibiting lower conductivities. This is observed in one-dimensional systems where d orbital interaction is the major contributing mechanism of conduction. For organic moieties where p_z orbital interactions become important a similar reasoning would suggest that the heavier group 4 elements such as silicon would be a better choice in synthesizing organic one-dimensional compounds, all other aspects being equal. The widely differing chemistry of silicon and other group 4 elements, compared to carbon, makes it difficult to prove this assertion, however.

(4) *Ligand system.* The small, nonbulky ligands which promote a planar configuration and have the correct steric requirements to allow close stacking are limited and mainly involve CN^-, CO, and oxalate $(C_2O_4)^{2-}$. NH_3 offers steric problems, S^{2-} appears to be too large, and F^- does not combine with transition metals in their low oxidation states. The complexes formed with the expanded ligand systems, while sometimes highly conducting, do not exhibit the close intermolecular spacings readily accessible with the less bulky systems. Even in the more congested, extended ligand systems, as long as steric repulsions are minimized, a

close intermolecular approach should be allowed by close packing of the complexes. In general, however, while intermolecular M–M distances in the partially oxidized tetracyanoplatinates are on the order of 2.9 Å or even less, the analogous phthalocyanine systems exhibit spacings of 3.2 Å or longer. The mode of conduction, whether via primarily d character orbitals, as in the former, versus ligand-centered conduction in the latter, may also influence the packing of the molecules. Evidently d orbital interactions necessitate closer approach, at least on the basis of the experimental evidence cited in the subsequent sections.

A. The Synthesis of POTCP Metals

The partial oxidation of a $Pt^{2.0+}$ salt to form a nonintegral oxidation state (NIOS) and a Pt^{2+x} complex with a partially filled band, may be achieved by (i) mixing solutions of the appropriate $Pt^{2.0+}$ and $Pt^{4.0+}$ salt (Williams and Schultz, 1979a) or (ii) by wet chemical oxidation of the Pt^{2+} salt using H_2O_2 or (iii) by electrolysis using a dc voltage source and a potential of $\sim 0.75-1.5$ V (Terry, 1928; Williams, 1979, 1980; Miller, 1979). The present method of preference is the electrolytic procedure, because visible crystal growth on the platinum electrodes often occurs in *seconds* or *minutes*. The growth rate appears to be *exponential* as a function of applied voltage and the highest quality crystals are obtained by using low voltage (slow) growth conditions.

Much longer time periods are required in using the wet chemical methods described in (i), which generally involve slow evaporation of aqueous solutions containing the POTCP salt. However, a much more important finding is that while *all* NIOS–POTCP salts prepared to date can be prepared electrolytically, this does *not* appear to be the case when using the wet chemical procedure. The dramatic difference between the two methods is illustrated in Table V, where it is evident that *different* $(F-H-F)^-$ containing salts may be prepared using method (i) or (ii), even when the initial solution concentrations are *identical*.

The POTCP salts formed using these methods all have unusual metallic lusters ranging from gold to bronze to copper. Although a metallic luster does not always signify metallic behavior, in this case the resulting complexes are characterized by anisotropic electrical, magnetic, and optical properties. In particular, electrical conductivity along the metal chain and insulating behavior in directions orthogonal to the chain are observed. In general it appears that as the degree of partial oxidation (DPO) of the platinum atom *increases* the d_{Pt-Pt} decreases and the metallic luster progressively changes in color from copper to bronze to gold. A gold luster

TABLE V

Synthesis of POTCP Complexes Containing $(FHF)^-$ or F^- Anions

Compound oxidation	Concentration (M)			Oxidation
	[MTCP]	[MF]	[HF]	
$K_2[Pt(CN)_4](FHF)_{0.30} \cdot 3H_2O$	0.4	2.9	9.6	1.5 V or H_2O_2
$Rb_2[Pt(CN)_4](FHF)_{0.40}$	0.2	1.9	8.7	1.5 V
$Rb_2[Pt(CN)_4](FHF)_{0.26} \cdot 1.7H_2O$	0.2	1.9	8.7	H_2O_2
$Cs_2[Pt(CN)_4](FHF)_{0.39}$	0.2	2.6	8.7	1.5 V or H_2O_2
$Cs_2[Pt(CN)_4](FHF)_{0.23}$	0.056	1.0	0.076	H_2O_2
$[C(NH_2)_3]_2[Pt(CN)_4](FHF)_{0.26} \cdot xH_2O$	0.8	0	28.9	1.5 V
$Cs_2[Pt(CN)_4]F_{0.19}$	0.3	1.6	$—^a$	1.5 V

^{a} In this preparation, the pH was maintained at 9.0 ± 0.1 during electrolysis by the addition of CsOH.

is found in the two salts which have the highest known DPO's of ~ 0.40, namely, $Rb_2[Pt(CN)_4](FHF)_{0.40}$, $d_{Pt-Pt} = 2.798(1)$ Å (Schultz *et al.*, 1977b) and $Cs_2[Pt(CN)_4](FHF)_{0.39}$, $d_{Pt-Pt} = 2.833(1)$ Å (Schultz *et al.*, 1978a).

B. The Molecular Structure and Conduction Behavior of POTCP Metals

As pointed out earlier, the crystal structures of the Pt^{2+} starting materials are often similar to the POTCP salts. This is not totally unexpected because the geometries of both starting materials and POTCP complexes are guided by the columnar stacking of the square–planar TCP moieties. The TCP stacking is in turn influenced by the allowed Coulombic and hydrogen bonding interactions which can exist between M^+, X^- (halide ion), CN^-, and H_2O. Having formed a platinum atom chain in a POTCP material, then the main variables governing the chain geometry are the platinum atom intrachain spacings (d_{Pt-Pt}), which are governed largely by the degree of Pt–Pt overlap and the DPO, and the torsion angles between adjacent $[Pt(CN)]_4^{-1.7-}$ groups. Somewhat later we will show that for POTCP salts there is a very good correlation between d_{Pt-Pt} and the DPO.

Diffraction studies indicate that for the limited number of POTCP compounds studied to date two types exist, namely, the "cation deficient" (CD) salts such as $M_{1.75}[Pt(CN)_4] \cdot xH_2O$, $M = Li^+$, K^+, Rb^+, or Cs^+, and the (predominant) "anion deficient" (AD) complexes such as the prototype $K_2[Pt(CN)_4]Br_{0.3} \cdot 3H_2O$, "Krogmann's salt" (Krogmann, 1969) or "KCP." Only one of the cation deficient salts has been well characterized, $K_{1.75}[Pt(CN)_4] \cdot 1.5H_2O$, (Keefer *et al.*, 1976), while numerous anion deficient materials have been studied in some detail, such as KCP, which has been studied (Williams and Schultz, 1979a; Miller and Eptstein, 1976) theoretically, spectroscopically (IR, ESR, and NMR), and by optical reflectivity. All POTCP complexes which have been structurally characterized to date are summarized in Table VI.

C. Anion Deficient POTCP Salts

The anion deficient (AD) salts comprise two main types that differ in that the type P (primitive) are hydrated and the type I (body centered) are usually anhydrous. The two types differ considerably in their electrical conduction behavior at low temperature. The main generalizations regarding the AD salts are as follows:

(1)   Type P hydrated salts form primitive tetragonal lattices (space group $P4mm$) with the M^+ cations in one half of the unit cell while the other half contains H_2O molecules. More importantly, the cations are located between the $Pt(CN)_4^-$ groups. Therefore, different platinum atom chains are parallel and about 10 Å apart, and are bound via a complex network involving, in part, weak hydrogen bonding interactions which are easily broken making crystal decomposition a constant problem. In general, the parallel platinum atom chains in type P salts are ~1 Å further apart than in type I salts. Compounds in this class are KCP(Cl), KCP(Br), RbCP(Cl), ACP(Cl), where $A = NH_4^+$, and KCP(FHF). Implied in the chemical formulas for these salts is $X^+ \cong 0.3$, and the anion sites in the crystal are only partially occupied (~0.6) in an apparently random fashion.

To illustrate the structure-conductivity properties of a type P salt we have chosed $Rb_2[Pt(CN)_4]Cl_{0.3} \cdot H_2O$, studied by neutron diffraction at 298 K (Williams et al., 1978) and 110 K (Brown and Williams, 1979) with the surprising result that the platinum chain is dimerized and the dimerization decreases with temperature (see Fig. 5). This is somewhat contrary to expectation because from variable temperature conductivity studies (Underhill et al., 1976) it was known that the electrical conductivity along the platinum chain also decreases with decreasing temperature with the conductivity trend $(NH_4)CP(Cl) < RbCP(Cl) < KCP(Cl)$ (see Fig. 6). Indeed, KCP also has decreasing conductivity with decreasing temperature but the chain is not dimerized. One would expect that with a decrease in chain dimerization there would be a concomitant decrease in electron localization along the chain; hence higher electrical conductivity would be expected.

The answer to this puzzling question does not lie in the degree of chain dimerization but rather in terms of differing temperatures at which some

TABLE VI

Crystal Structure and Conductivities for Several Krogmann-Type Conductors[a]

Conductor	Space group[b]	d_{Pt-Pt} (Å) at 298 K	Conductivity[c] (ohm cm)$^{-1}$	Color
Pt metal		2.775	9.4×10^4	Metallic
$K_2[Pt(CN)_4]Br_{0.30} \cdot 3H_2O$	$P4mm$	2.89	4–1050	Bronze
$K_2[Pt(CN)_4]Cl_{0.30} \cdot 3H_2O$	$P4mm$	2.87	~200	Bronze
$K_2[Pt(CN)_4]Br_{0.15}Cl_{0.15} \cdot 3H_2O$	$P4mm$			
$Rb_2Pt(CN)_4]Cl_{0.30} \cdot 3H_2O$	$P4mm$	2.877, 2.924	10	Bronze
$Cs_2[Pt(CN)_4]Cl_{0.30}$	$I4/mcm$	2.859	~200	Bronze
$(NH_4)_2(H_3O)_{0.17}[Pt(CN)_4]Cl_{0.42} \cdot 2.83H_2O$	$P4mm$	2.910, 2.930	0.4	Bronze
$Cs_2[Pt(CN)_4](N_3)_{0.25} \cdot 0.5H_2O$	$P4b2$	2.877		Reddish copper
$Rb_3(H_3O)_{0.x}[Pt(CN)_4](O_3SO \cdot H \cdot OSO_3)_{0.49} \cdot (1-x)H_2O$	$P\bar{1}$	2.826		Copper
$K_{1.75}[Pt(CN)_4] \cdot 1.5H_2O$	$P\bar{1}$	2.965, 2.961	115–125	Bronze
$Rb_{1.75}[Pt(CN)_4] \cdot 1.5H_2O$	d	2.94	1	Bronze
$Cs_{1.75}[Pt(CN)_4] \cdot 1.5H_2O$	d	2.88	~25	Bronze
$K_2[Pt(CN)_4](FHF)_{0.30} \cdot 3H_2O$	$P4mm$	2.918, 2.928		Reddish bronze
$Rb_2[Pt(CN)_4](FHF)_{0.40}$	$I4/mcm$	2.798	2300	Gold
$Rb_2Pt(CN)_4](FHF)_{0.26} \cdot 1.7H_2O$	$C2/c$	2.89		Greenish bronze
$Cs_2[Pt(CN)_4](FHF)_{0.39}$	$I4/mcm$	2.833	1600	Reddish gold
$Cs_2[Pt(CN)_4](FHF)_{0.23}$	$I4/mcm$	2.872	250–350	Reddish bronze
$Cs_2[Pt(CN)_4]F_{0.19}$	$Immm$	2.886		Reddish gold
$[C(NH_2)_3]_2Pt(CN)_4](FHF)_{0.26} \cdot xH_2O$		2.90		Bronze
$[C(NH_2)_3]_3[Pt(CN)_4]Br_{0.25} \cdot H_2O$	$I4cm$	2.908		Bronze

[a] Data taken in part from Williams and Schultz (1979a).

[b] For the space group $P4mm$ the Pt–Pt intrachain distances are not required to be equal, but often appear to be so. When they have been determined to be different both distances are tabulated.

[c] Results are for a range of literature values for room temperature and by the four point probe dc conductivity method.

[d] The crystal class is monoclinic but the space group is unknown. The lattice constants for the Cs salt are $a = 18.35$, $b = 5.760$, $c = 19.91$ Å, and $\beta = 109.03°$; for the Rb salt the lattice constants are $a = 10.56$, $b = 33.2$, $c = 11.74$ Å, and $\beta = 114.23°$.

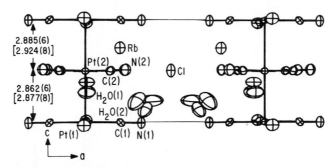

Fig. 5. Drawing of the unit cell (50% probability ellipsoids) of $Rb_2[Pt(CN)_4]Cl_{0.30} \cdot 3H_2O$ showing the linear Pt atom chain, which contains unequal Pt–Pt separations, and the asymmetric location of Rb^+ ions and the H_2O molecules. Distances in angstroms are shown both at 298 K (square brackets) and at 110 K. Note that the Rb ion is in between the planes of the $Pt(CN)_4$ groups. [From Williams *et al.* (1982).]

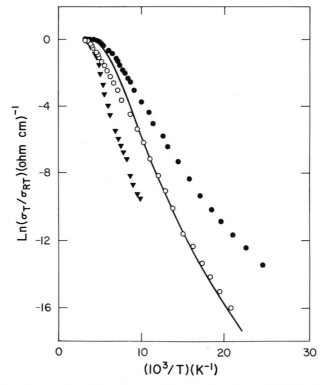

Fig. 6. Temperature dependence of the in-chain dc conductivity for the four isostructural type *P* compounds (space group *P4mm*): full line, KCP(Br); open circles, RbCP(Cl); closed circles, KCP(Cl); and triangles, ACP(Cl). [From Williams *et al.* (1982).]

interchain coupling of one-dimensional distortion (for Peierls distortions see Chapter 7) occurs causing the chains to undergo a three-dimensional ordering transition at T_{3D} (Underhill *et al.*, 1979–1980). For any one-dimensional conductor there is a mean field temperature T_p below which critical fluctuations occur with increasing importance as the temperature is decreased. This behavior in turn creates one-dimensional correlated domains over extensive portions of the platinum chains. In addition, inter-chain Coulombic interactions also increase with decreasing temperature until they are strong enough to cause three-dimensional ordering of the one-dimension distorted chains (at T_{3D}). The interchain coupling parameter η may be described in terms of static Coulombic coupling in the lattice or by electrons hopping from chain to chain depending on the chemical nature of the crystal lattice, etc. In practice T_{3D} is obtained by an analysis of the temperature dependence of the electrical conductivity, i.e., the first deriv-ative of ln $\sigma_\parallel$ versus $1/T$ will exhibit a maximum at T_{3D} (Carneiro, 1979). Since the ratio T_{3D}/T_p is a measure of the interchain coupling η, it depends considerably on the extent of hydrogen bonding between platinum chains.

For KCP(Br), which is the one POTCP salt which has been character-ized in the greatest detail, only partial three-dimensional ordering occurs at even the lowest temperatures and it is thought that this is due to the presence of disordered Br^- anions in the lattice (Carneiro *et al.*, 1979a; Lynn *et al.*, 1975). The observation of a giant Kohn anomaly (Renker and Comés, 1975) and inelastic neutron scattering studies (Lynn *et al.*, 1975) have been proposed as evidence for a Peierls transition in KCP(Br). Since many POTCP salts are isostructural with KCP(Br), it should be pointed out that from the neutron scattering experiments a great deal has been learned regarding the behavior of the platinum atom chains as a function of temperature. At room temperature these studies indicate dynamic sinu-soidal displacements of the platinum atoms along the chain, with no co-herence between chains, and an incommensurate repeat period of 6.67 c' (c' = Pt–Pt intrachain distance), which is exactly that expected for a Peierls distortion. Neutron scattering studies (Lynn *et al.*, 1975) demon-strated that at 77 K and below, a three-dimensional ordering process occurs, the distortion involving the entire $Pt(CN)_4$ group becomes static, and the TCP groups "ride" an electronic charge density wave (CDW) arising from periodic fluctuations in electronic charge due to the Pt d_{z^2} electrons. However, the CDW displacement of the platinum atoms is only 0.025 Å from their perfectly equal spacing derived by classical structure analysis (Williams *et al.*, 1979a). Therefore, the average crystallographic structure of KCP(Br) does not vary with temperature but is "modulated" by a CDW to which the $Pt(CN)_4$ groups respond. In fact, all POTCP salts exhibit "modulated" structures which we will discuss shortly (Williams and Schultz, 1979b). Finally, it should be pointed out than η depends on the extent of hydrogen bonding in AD salts, e.g., T_{3D} can be raised from ~100 to ~195 K if K^+ is replaced by NH_4^+ (Carneiro et al., 1979a) (see Table VII).

Although KCP(Br) never appears to undergo chain dimerization, it is obvious from Table VI that some type P salts do exhibit this behavior even though they have $d_{Pt–Pt}$ which are almost identical to KCP(Br). After numerous diffraction studies of KCP type complexes, in which cations of differing radius (r^+) have been substituted for K^+, the results may be summarized as follows: (i) for a given Pt–Pt spacing, as r^+ increases the likelihood of dimerization increases and (ii) for a given r^+, as the Pt–Pt spacing decreases the likelihood that dimerization will occur decreases. As has been shown previously (Brown and Williams, 1979), for a given r^+, the Pt–Pt spacing tends to depend on the degree of partial oxidation (DPO) of the metal. Therefore, it may be that the degree of platinum chain dimerization is also dependent on the DPO.

(2)   Type I salts are usually anhydrous with body-centered tetragonal lattices (space group $I4/mcm$). Their most important single structural feature is the occupation of the same molecular plane by the M^+ cations and the square–planar $Pt(CN)_4$ groups which maximizes M^+–$^-N{\equiv}C$ interactions. Compounds which fall into this class are $CsCP(Cl)_{0.3}$ (the first anhydrous POTCP halide salt), $RbCP(FHF)_{0.40}$, $CsCP(FHF)_{0.39}$, $CsCP(FHF)_{0.23}$, and $RbCP(N_3)_{0.25}$. As an illustration of this structure type we have chosen $Cs_2[Pt(CN)_4]Cl_{0.3}$ (see Fig. 7) (Brown and Williams, 1979) and $Rb_2[Pt(CN)_4](FHF)_{0.4}$ (see Fig. 8) (Schultz et al., 1977b).

TABLE VII

Three-Dimensional Ordering Temperatures (T_{3D}) and Interchain Coupling Constants (η) in POTCP Compounds[a]

Compound	T_{3D} (K)	η
$(NH_4)(H_3O)_{0.17}[Pt(CN)_4]Cl_{0.42} \cdot 2.83H_2O$	195	0.037
$Rb_2[Pt(CN)_4]Cl_{0.3} \cdot 3H_2O$	110	0.022
$K_2[Pt(CN)_4]Br_{0.3} \cdot 3H_2O$	100	0.021
$K_2[Pt(CN)_4]Cl_{0.3} \cdot 3H_2O$	95	0.019
$Cs_2[Pt(CN)_4]Cl_{0.3}$	90	0.017
$Cs_2[Pt(CN)_4](FHF)_{0.39}$	80	0.017
$Rb_2[Pt(CN)_4](FHF)_{0.40}$	80	0.02
$K_{1.75}[Pt(CN)_4] \cdot 1.5H_2O$	50	0.012
$Rb_{1.75}[Pt(CN)_4] \cdot 1.5H_2O$	50	0.012
$[C(NH_2)_3]_2[Pt(CN)_4]Br_{0.25} \cdot H_2O$	100	0.018

[a] Taken chiefly from Underhill et al. (1979–1980).

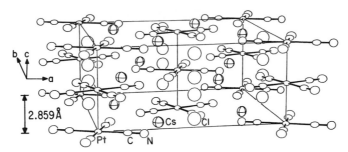

Fig. 7. View of the unit cell of the anhydrous type I salt, $Cs_2[Pt(CN)_4]Cl_{0.3}$, CsCP(Cl). Note that the Cs ion is in the plane of the $Pt(CN)_4$ groups. [From Williams *et al.* (1982).]

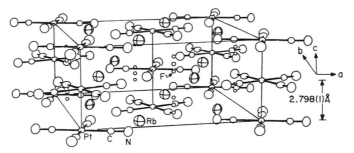

Fig. 8. View of the unit cell of anhydrous $Rb_2[Pt(CN)_4](FHF)_{0.40}$, RbCP $(FHF)_{0.40}$. The Pt–Pt spacing is the shortest reported for any POTCP complex. [From Williams *et al.* (1982).]

Variable temperature electrical conduction studies (Wood *et al.*, 1979a) of $Cs_2[Pt(CN)_4]Cl_{0.3}$ and isostructural type I complexes (Wood *et al.*, 1979b), $Cs[Pt(CN)_4](FHF)_{0.39}$ and $Rb_2[Pt(CN)_4](FHF)_{0.40}$, have been reported and the results are compared with those for KCP in Fig. 9. From Fig. 9, it is obvious that CsCP(Cl) exhibits a conduction behavior intermediate between that of KCP(Br) and RbCP(FHF)$_{0.40}$. For KCP(Br), metallic conduction occurs at room temperature but becomes the semiconducting type (thermally activated conduction) at low temperature (Zeller and Beck, 1974). Previous data indicate that the room temperature value of $\sigma_{\parallel}$ is not sensitive to water content of the cation present, but is dominated by the intrachain Pt–Pt separation. Considerable differences in conductivity do arise, however, at low temperature. From Fig. 9 it can be seen that CsCP(Cl) displays almost temperature-independent conductivity near room temperature, but it does not show a region of negative temperature dependence as does KCP(Br), and that it exhibits conduction behavior intermediate between that of KCP(Br) and RbCP(FHF). In attempting to explain these differences one can compare the interchain

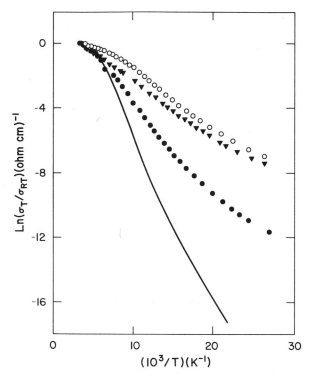

Fig. 9. Temperature dependence of the in-chain dc conductivity for the three iso-structural compounds (space group *I4mcm*): open circles, RbCP(FHF)$_{0.4}$; triangles, CsCP(FHF)$_{0.4}$; and closed circles, CsCP(Cl)$_{0.3}$. The solid curve for KCP(Br) is included for comparison. [From Williams *et al.* (1982).]

Pt–Pt separations of 9.32 Å in CsCP(Cl), 9.23 Å in CsCP(FHF)$_{0.39}$, and 8.97 Å in RbCP(FHF)$_{0.4}$. However, there is little difference in these values and the most obvious answer to the question of differing conductivities would seem to be the difference in anion in these compounds. Thus it is possible that the larger and more easily polarized chloride ion provides a more effective interchain Coulombic coupling mechanism than the bifluoride ion (Wood *et al.*, 1979a). Thus, from Fig. 9, it is clear that the decline in conductivity from the value at room temperature to the onset of semiconducting behavior is considerably less for the (FHF)$^-$ salts than for KCP(Br). An estimation of the activation energies in the semiconductor region provides a value of ~0.02 eV for the bifluoride salts with DPO = 0.40 and ~0.07 eV for KCP(Br) where DPO = 0.30.

Since T_{3D} is a measure of Coulombic interchain coupling, and is partly dependent on hydrogen bonding between the Pt(CN)$_4$ groups, it is not surprising that T_{3D} is lowest in the anhydrous type *I* salts as given in Table VII (Underhill *et al.*, 1979–1980).

D. Cation Deficient POTCP Salts

To date only one cation deficient salt, $K_{1.75}[Pt(CN)_4] \cdot 1.5H_2O$, K(def)TCP, has been fully characterized structurally using diffraction methods (see Fig. 10) (Keefer *et al.*, 1976; Williams *et al.*, 1976). A noteworthy finding is that K(def)TCP possesses the longest intrachain Pt–Pt separations observed in any POTCP salt, namely, 2.961(1) and 2.965(1) Å, including the only bent platinum chain (see Fig. 10). It appears that $d_{Pt-Pt} < 3.0$ Å marks the line of demarcation for POTCP salt formation. As in KCP(X) type complexes the d_{Pt-Pt} are essentially equal in K(def)TCP although not required to be so crystallographically. Even though K(def)TCP has the longest d_{Pt-Pt} reported, it does not have the lowest DPO (see Table VI). X-ray diffuse scattering (XDS) studies (Schultz *et al.*, 1977a) of K(def)TCP establish that it possesses a superlattice which is commensurate with the *c* axis of the crystal with a repeat distance of eight times the average Pt–Pt distance. Since a completely filled band would contain a contribution of two electrons from each platinum atom, the calculated Fermi wave vector is $\mathbf{k}_F = (1.75/2)\pi/c' = 0.875c'$ where c' is the average d_{Pt-Pt}. The wave vector derived from the diffuse scattering

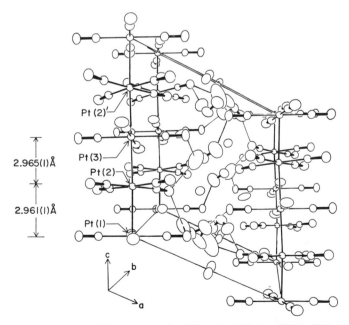

Fig. 10. View of the triclinic unit cell of $K_{1.75}[Pt(CN)_4] \cdot 1.5H_2O$, K(def)TCP. The zigzag Pt-chain structure is shown [Pt(1)-Pt(2)-Pt(3) = 173.25°]. The Pt(2) atom is displaced 0.170 Å perpendicular to the *c* axis. Hydrogen bonding interactions are indicated by faint lines. [From Williams *et al.* (1982).]

superlattice is $\mathbf{k} = 2\pi/8.1c'$, or $\mathbf{k} = 2\pi(1 - 1/8.1)/c'$, which is in excellent agreement with $2\mathbf{k}_F$ predicted from the chemical stoichiometry. Thus for K(def)TCP the DPO = 0.25 derived by chemical analysis was confirmed in the XDS study.

Optical reflectance studies have established that the M(def)TCP salts (M = K$^+$, Rb$^+$, and Cs$^+$), are all one-dimensional metals (Musselman and Williams, 1977), and neutron scattering studies (Carneiro *et al.*, 1976) show a presence of a Kohn anomaly from which Carneiro *et al.* concluded that a half-filled conduction band is present. The well defined Kohn anomaly had little temperature dependence between 80 and 300 K which is similar to the behavior observed in KCP halide salts (Renker *et al.*, 1975).

We have discussed in some detail the many physical measurements reported for K(def)TCP because it is the only CD complex which has been well characterized and because it has a commensurate $(8c')$ superlattice as compared to KCP which is incommensurate $(6.67c')$. The nature of the superlattice may account for the conductivity trend K(def)TCP $\ll$ KCP. From two independent electrical conductivity studies (Epstein and Miller, 1979; Carneiro *et al.*, 1979b) nearly identical experimental data for K(def)TCP were derived ($\sigma_{\parallel} \simeq 100$ ohm^{-1} cm^{-1}, $T = 298$ K); however, the interpretations agree in one aspect only, namely, the electrical conductivity behavior of K(def)TCP is markedly different from that of KCP type P conductors. From their interpretation of the data, Epstein and Miller (1979) derived the zero temperature gap $\Delta(0) = 106$ meV and $T_{3D} = 308$ K while Carneiro *et al.*, (1979b) derived $\Delta(0) = 55$ meV and $T_{3D} = 50$ K. It seems worthwhile to dwell for a moment on this discrepancy between interpretations since it demonstrates how difficult it has been to resolve problems of this kind at a stage where neither crystal growth, physical experiments, nor theory were fully developed. The problem in K(def)TCP was further accentuated because not only does a Peierls transition occur in this compound, but in addition a second one, namely, a non-Peierls distortion occurs, giving rise to two sets of (Δ, T_{3D}) parameters as derived from electrical measurements.

Independent Raman scattering investigations (Steigmeier *et al.*, 1979) showed that $\Delta = 55$ meV and there appeared to be no gap at 106 meV. Furthermore, from the temperature dependence of the oscillator strength of the amplitude mode (vide infra) it could be concluded that $T_{3D} < 100$ K. Consequently, the lower transition with $\Delta = 55$ meV and $T_{3D} = 50$ K is associated with the Peierls instability which leads to an electron–phonon coupling constant of $\lambda = 0.29$. Below we shall show that all parameters of the Peierls instability correlate with the DPO and therefore we require an explanation of why ACP(Cl), i.e., $(NH_4)_4CP(Cl)$, with an identical DPO = 0.25 as in K(def)TCP, has such a different $\lambda = 0.38$. A plausible reason for this difference is that because of the low symmetry (triclinic) of K(def)TCP, the Peierls distortion has both a longitudinal and a transverse part, and this in turn decreases λ. A simple way of accounting for this is to extend the formula for

$$\lambda^{-1} = \frac{(\omega^L_{2k_F})}{g^2 N(\epsilon_F)} \qquad \text{to} \qquad \lambda^{-1} = \frac{\sqrt{[\omega^L_{2k_F}]^2 + [\omega^T_{2k_F}]^2}}{g^2 N(\epsilon_F)}, \qquad (3)$$

where L and T refer to longitudinal and transverse respectively, so that the ratio $\lambda_{ACP}/\lambda_{K(def)TCP} = \sqrt{[\omega^L]^2 + [\omega^T]^2}/\omega^L = 1.12$ using values for the phonon frequencies taken from neutron scattering experiments. However, the experimental ratio derived from the respective Δ values is 1.31 and gives only partial substance to this explanation; hence K(def)TCP does not correlate with the series of seven compounds analyzed by Underhill *et al.* (1979–1980) but it appears to fall into a small class of crystals which possess a particular hydrogen bonding network.

The high temperature of $T_{3D} = 308$ K, which was found by Epstein and Miller (1979) to show a mean field critical behavior, appears to fall in the class of "non-Peierls" (NP) transitions often found in the bisoxalatoplatinates. Here the NP transition is always associated with the occurrence of superlattice spots with an in-chain wavelength component of q^{NP} which is commensurate with $2k_F$ when the NP transition is observed in the conductivity. Indeed, x-ray investigations (Braude *et al.*, unpublished results) showed such spots to occur with intensity $(I_{NP})^2$ below T_{3D} with $q^{NP} = 2k_F$, i.e., similar to γ-$K_{1.81}[Pt(C_2O_4)_2] \cdot 2H_2O$. Regretably, a new batch of K(def)TCP crystals prepared in order to study this phenomena further showed no NP distortion, thereby resembling the situation which occurs in the bisoxalatoplatinates, where preparations under slightly differing conditions may give different types of crystals with different physical properties. However, it is reassuring that a mechanism similar to ligand back bonding in the bisoxalatoplatinates is possibly present in K(def)TCP in that hydrogen bonding between water molecules and cyanide ligand nitrogen atoms occur along chains as observed in the structural study.

A detailed analysis of K(def)TCP is shown in Fig. 11, where the smooth behavior of the three Peierls order parameters a_T, a_L, and Δ are shown as functions of temperature, with $T_{3D} = 50$ K, derived from the resistivity measurements. The critical mean field behavior of the non-Peierls order parameters δ and I_{NP} are shown to vanish at $T_c = 308$ K (Carniero, 1979).

E. Diffuse X-ray Scattering at Room Temperature and the Degree of Partial Oxidation

Owing to the Peierls instability of one-dimensional conductors (Peierls, 1955), the crystal lattices of all the tetracyanoplatinates are distorted in the chain direction, i.e., they all display "modulated structures" (Williams

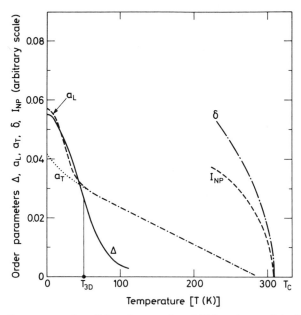

Fig. 11. Parameters describing the two instabilities observed in K(def)TCP. The Peierls instability is shown by the gap Δ from electrical measurements, the two oscillator strengths a_L and a_T derived from Raman scattering and the three dimensional ordering temperature, T_{3D}. The non-Peierls instability is evidenced by δ from electrical measurements, the x-ray intensities I_{NP}, and T_c (see text). [From Williams *et al.* (1982).]

and Schultz, 1979b), with a wave vector $2k_F$ which is related to the degree of partial oxidation (DPO) as

$$DPO = 2[1 - (k_F d_{Pt-Pt}/\pi)], \qquad (4)$$

where d_{Pt-Pt} is the intrachain Pt–Pt distance. This means that all POTCP complexes studied to date have shown diffuse x-ray scattering at room temperature, and this method has proven to be the most accurate for determining the DPO (Williams and Schultz, 1979b). From the previous discussion one may surmise that the DPO has proven to be an important parameter in the chemical and physical characterization of POTCP metals which is a point we will return to later.

Comés *et al.* (1973a,b) were the first to implement the x-ray diffuse scattering technique to measure the $2k_F$ superstructure lines of KCP(Br), and the techniques were later taken up by Schultz *et al.* (1977a), Carneiro *et al.* (1979a), and by Braude *et al.* (1980). Table VIII lists the compounds where $2k_F$ diffuse scattering has been observed and used to determine the DPO. Although it was first thought that the $2k_F$ superstructure in KCP(Br) was commensurate with a DPO = $\frac{1}{3}$, it was soon determined to be incommensurate with DPO = 0.30. From Table VIII it is seen that both commensurate and incommensurate Peierls distortions are found, and this will

TABLE VIII

Values $2k_F$ (in units of $2\pi/d_{Pt-Pt}$) as Determined by X-Ray Diffuse Scattering, Compared to Values from Chemical Analysis and the Intrachain Pt–Pt Distance[a]

Compound	Abbreviation	d_{Pt-Pt}	$2k_F$[b] (chemical)	$2k_F$ (x-ray)
$K_2[Pt(CN)_4]Br_{0.30} \cdot 3H_2O$	KCP(Br)	2.880	1.70	1.67, 1.70
$Rb_2[Pt(CN)_4]Br_{0.23} \cdot xH_2O$	RbCP(Br)	—	1.77	1.75
$[C(NH_2)_3][Pt(CN)_4]Br_{0.23} \cdot xH_2O$	GCP(Br)	2.908	1.77	1.75
$K_{1.75}[Pt(CN)_4] \cdot 1.5H_2O$	K(def)TCP	2.958[d]	1.75	1.76, 1.775
$Rb_{1.75}[Pt(CN)_4] \cdot xH_2O$	Rb(def)TCP	2.94	1.75	1.73
$Cs_{1.75}[Pt(CN)_4] \cdot xH_2O$	Cs(def)TCP	2.88	1.75	1.72
$(NH_4)_2(H_3O)_{0.17}[Pt(CN)_4]Cl_{0.42} \cdot 3H_2O$	ACP(Cl)	2.93	1.58[b]	1.75
$Rb_3(H_3O)_x[Pt(CN)_4](O_3SO \cdot H \cdot OSO_3)_{0.49} \cdot (1-x)H_2O$	RbCP(DSH)	2.826	1.53[c]	1.68
$Cs[Pt(CN)_4](FHF)_{0.39}$	CsCP(FHF)	2.833	1.61	1.60

[a] Taken from Williams et al. (1982).
[b] $2k_F$ (chemical) was determined before (H_3O^+) was inserted in the formula.
[c] This value is for $x = 0$. If $x = 0.12$, then the chemical stoichiometry and the x-ray diffuse scattering are in agreement.
[d] Average of two values, 2.965 and 2.961 Å.

be of importance when we discuss the electronic properties associated with the effects of charge density waves in these materials.

It is of interest to compare the values for $2k_F$ as determined by x rays with those we would obtain from the previous relation when inserting the DPO obtained from quantitative chemical analysis. This is illustrated in Table VIII and usually the disagreement is less than 0.02 units, which is within experimental uncertainty. In two compounds, ACP(Cl) and RbCP(DSH), the disagreement is quite significant however, since the chemically derived DPO values are 0.42 and 0.49, respectively, as opposed to the values of 0.25 and 0.32 from x-ray diffuse scattering. Since many tetracyanoplatinates are crystallized from solutions with $Pt^{(IV)}/Pt^{(II)}$ ratios which correspond to the DPO of the crystal being grown, the DPO's can often be determined by platinum oxidation–reduction titrations, and for both ACP(Cl) and RbCP(DSH) this method yielded values in agreement with the x-ray value for the DPO. This is explained (Carneiro *et al.*, 1979a; Braude *et al.*, 1980) by assuming that additional positive charge is present in the crystal as H_3O^+ to counterbalance part of the anion charge, which is particularly plausible since the crystals in question are grown from acidic solutions. Unfortunately, there is no known way to detect such slight amounts of H_3O^+ directly.

As first pointed out by Williams (1976), the metal–metal bond lengths of POTCP complexes may be derived using the empirical concept of "metallic resonance" which states that

$$d_{Pt-Pt}(\text{Å}) = 2.59 - 0.60 \log_{10} \text{DPO} \qquad (5)$$

This relation was originally proposed by Pauling (1960) in the form $D(n) = D(1) - 0.60 \log_{10} n$, where for metallic systems $D(n)$ is the metallic bond distance for the bond of order n and $D(1)$ is the single-bond distance. The agreement between values derived from this equation and the experimental DPO versus d_{Pt-Pt} is shown in Fig. 12 for the compounds listed in Table VIII. In order to make the significance of this result even clearer, we also indicate in Fig. 12 the corresponding values derived for the case of "no metallic resonance" (Pauling, 1947):

$$d_{Pt-Pt}(\text{Å}) = 2.59 - 0.71 \log_{10} \text{DPO} \qquad (6)$$

which is clearly not obeyed. It is interesting that for ACP(Cl) and RbCP(DSH) the DPOs derived from x-ray diffuse scattering agree well with the observed d_{Pt-Pt}, whereas the chemically derived value gives poor agreement. This is further evidence for the existence of hydronium ions, H_3O^+, in the crystal lattices of these two compounds.

On only one case did the x-ray diffuse scattering lines exhibit a more complicated behavior than is usually the case for tetracyanoplatinates. In

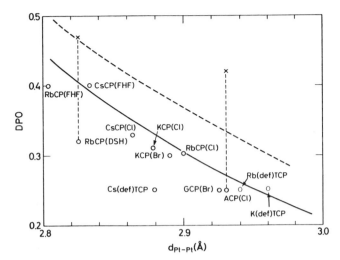

Fig. 12. A plot of DPO versus d_{Pt-Pt}. The full curve is Pauling's formula for metallic resonance and the dashed line is for no metallic resonance (Pauling, 1947). The (○) designation is from x-ray diffuse scattering experiments and (×) is from chemical analyses. [From Williams *et al.* (1982).]

K(def)TCP spots were observed on the uniform $2\mathbf{k}_F$ scattering lines, indicative of a "non-Peierls" superstructure. Upon heating the crystals to 308 K (35°C) the spots disappeared while the lines remained (Braude *et al.*, 1979). At this temperature there is also a rapid change in the electrical conductivity which cannot be ascribed to the Peierls instability, but crystal quality problems have prevented a thorough temperature-dependent x-ray study. Such "non-Peierls" structural instabilities are common in the bisoxalatoplatinates, but such behavior has been observed only in this one tetracyanoplatinate (Comés, 1975).

All POTCP complexes which have been investigated have shown $2\mathbf{k}_F$ diffuse x-ray lines at room temperature (Williams and Schultz, 1979b). Apart from the intrinsic interest from the point of view of the Peierls instability, the diffuse scattering has been important as an aid in understanding the chemical composition and many related structural aspects of POTCP compounds.

F. Temperature Dependence of X-Ray Diffuse Scattering and Elastic Neutron Scattering

In order to further characterize the Peierls instability in KCP(Br), the temperature dependence of the x-ray diffuse scattering has been studied. Comés *et al.* (1973a) found that the Peierls lattice distortion, which was one-dimensional at room temperature, became three-dimensional at liquid nitrogen temperature (77 K). The three-dimensional ordering between chains appears to arise from Coulombic antiferroelastic interactions between chains. Since the x-ray energy is much larger than the energy of a lattice vibration, this technique cannot be used to resolve between the existence of a static or dynamic distortion.

Subsequent neutron scattering studies by Renker *et al.* (1974) have shown that the distortion is both static and dynamic even at room temperature. They also demonstrated that the three-dimensional ordering at low temperatures was incomplete with the result that the diffuse scattering did not develop into spots but always showed a finite width perpendicular to the chains and involved partial ordering at $T_{3D} = 100$ K. Lynn *et al.* (1975) performed a complete analysis of the diffuse elastic lineshapes and established that not only the platinum atoms, but the entire [Pt(CN)$_4$] group participated in the Peierls distortion. Finally, Eagen *et al.* (1975) made the only absolute neutron scattering intensity measurements to date, which prove that the [Pt(CN)$_4$] complexes are sinusoidally modulated with a low-temperature amplitude of $A(0) = 0.0094\ d_{Pt-Pt}$, where $A(T)$ determines the position of the lth ion in a chain

$$x_l = l \cdot d_{Pt-Pt} + A \cos(2\mathbf{k}_F l d_{Pt-Pt} + \phi) \tag{7}$$

During the course of these temperature-dependent studies, the linear thermal expansion coefficients were determined for KCP(Br) and ACP(Cl). The values for KCP are (Carneiro *et al.*, 1976)

$$\alpha_\parallel = 5.1 \times 10^{-5}\ \text{K}^{-1} \tag{8}$$

$$\alpha_\perp = 2.2 \times 10^{-5}\ \text{K}^{-1} \tag{9}$$

where $\parallel$ and $\perp$ denote in-chain and perpendicular, respectively. For a ACP(Cl) (Carneiro *et al.*, 1979a)

$$\alpha_\parallel = 3.9 \times 10^{-5}\ \text{K}^{-1} \tag{10}$$

$$\alpha_\perp = 3.4 \times 10^{-5}\ \text{K}^{-1} \tag{11}$$

Those results show that POTCP compounds have average expansion coefficients similar to lead and in-chain coefficients 3–5 times that of platinum metal. In addition, the chain expands more in the parallel than the perpendicular direction, which is also true for TTF–TCNQ, where $\alpha_\parallel = 0.2 \times 10^{-5}\ \text{deg}^{-1}$ and $\alpha_\perp = 2.6 \times 10^{-5}\ \text{deg}^{-1}$ (Blessing and Coppens, 1974).

The elastic diffuse scattering corresponds to a static lattice distortion with a wavelength $\lambda = 2d_{Pt-Pt}/\text{DPO}$ (6.667 d_{Pt-Pt} for KCP(Br)). It is characterized by a lineshape which depends upon the wave vector $\mathbf{q}$ with components parallel ($\mathbf{q}_\parallel$) and perpendicular ($\mathbf{q}_\perp$) to the metal-atom chain measured from the wave vector $\mathbf{Q} = (\pi/d_\perp, \pi d_\perp, 2\mathbf{k}_F)$:

$$S(\mathbf{Q} + \mathbf{q}) \propto A^2 F_N^2 \frac{1}{1 + (\xi_\parallel \mathbf{q}_\parallel)^2 + (\xi_\perp \mathbf{q}_\perp)^2} \tag{12}$$

where A is the amplitude of distortion, $\xi_\parallel$ and $\xi_\perp$ the correlation lengths along and perpendicular to the chains in a crystal, respectively, and F_N the [Pt(CN)$_4$] structure factor which causes the asymmetry of the line.

Owing to the absence of phase transitions in one-dimensional systems at finite temperatures, particular interest by theorists has been given to the temperature dependence of $\xi_\parallel$ since this parameter describes the critical behavior of the Peierls distorted lattice (and electronic charge-density wave). Based on a calculation by Scalapino et al. (1972), Lee et al. (1973) came to the conclusion that a certain temperature $\xi_\parallel$ would be so large that even a negligible electronic interchain coupling η would trigger a real phase transition. The transition temperature T_{3D} should be about $0.2T_p$, where T_p is the mean field Peierls transition temperature for a one-dimensional lattice.

These conclusions of Lee et al. (1973) are in disagreement with later studies, which concluded that $T_{3D}/T_p \to 0$ as $\eta \to 0$. The more important studies indicate that $\xi_\parallel$ above and below T_p would increase in a way characteristic of one dimension; but as T_{3D} is approached $\xi_\parallel$ diverges as described by a mean field theory that takes the interchain coupling into account. The behaviors of $\xi_\parallel$ and $\xi_\perp$ as a function of temperature do, however, differ between different theories. The fact that the ordering is not infinite at T_{3D} has been ascribed to impurity effects by Sham and Patton (1976) and by Bak (1977). Impurities may cause the lattice to distort quasistatically at finite temperatures and may also explain the finite range of order, but the very long experimental lifetime of the elastic $2\mathbf{k}_F$ distortion is as yet quantitatively unexplained.

In KCP type compounds the ration T_{3D}/T_p is found to be in the range 0.2 to 0.5 from electrical conductivity measurements (vide infra), i.e., above the value suggested by Lee et al. (1974). Later in this chapter the theory of Horowitz et al. (1975) will be used to compute the value of η for a variety of compounds. Regarding the temperature dependence of $\xi_\perp$, the experimental results of Lynn et al. (1975) are shown in Fig. 13 and compared to the results with those derived from theories in which this parameter is given explicitly. In fitting the theoretical expressions to the experimental results, specific parameter values were taken from the independent electrical studies described in the following. Apart from these parameters, the theory of Sham and Patton (1976) contained a number of adjustable

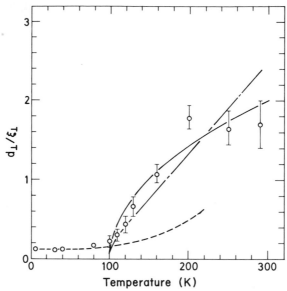

Fig. 13. Temperature of the inverse of the transverse correlation length $\xi_\perp$ of the charge-density wave in KCP(Br). The experimental values are compared to theoretical calculations as discussed in the text: ○, Lynn *et al.*, experiment; ——, Scalapino, Imry, and Pincus, theory; — —, Horowitz, Gutfreund, and Weger, theory; – – –, Sham and Patton, theory. [From Williams *et al.* (1982).]

parameters, whereas that of Scalapino *et al.* (1975) has no adjustable parameters. The two latter theories yield a good fit to the experiments above T_{3D}, as opposed to Sham's and Pattons's, which only provides a good fit below T_{3D}.

This leads to the conclusion that only below T_{3D} is the charge-density wave behavior dominated by impurities, and that the critical behavior of the transverse correlation length $\xi_\perp$ above T_{3D} is dominated by quasi-one-dimensional behavior, since the experimental and theoretical fit of $\xi_\perp$ versus T is good (Scalapino *et al.*, 1975; Horowitz *et al.*, 1975).

Horowitz *et al.* (1975) also found that there is a fixed ratio between $\xi_\parallel$ and $\xi_\perp$:

$$(\xi_\parallel/d_\parallel)/\xi_\perp d_\perp = 1/\eta k_F d_\parallel \qquad (13)$$

where $d_\parallel$ and $d_\perp$ are the intra- and interchain Pt–Pt separations, respectively. Since η, k_F, and $d_\parallel$ can be determined experimentally, the ratio may be estimated to be 50. From this result $d_\parallel$ can then be estimated to range from 450 $d_\parallel$, at low temperature, to approximately 20 $d_\parallel$ at room temperature. These values are beyond presently available resolution and, therefore, $\xi_\parallel$ have not been measured directly by elastic neutron scattering or x rays. However, inelastic neutron scattering does give information about the longitudinal correlation length as discussed in the following.

G. Inelastic Neutron Scattering, Infrared, and Raman Spectroscopy

The unusual lattice dynamical properties of one-dimensional conductors has resulted in intense experimental and theoretical interest. However, owing to the crystal size requirements associated with spectroscopic studies only a few compounds have been examined, i.e., KCP which has been extensively studied (Renker and Comés, 1975; Renker et al., 1974; Lynn et al., 1975; Carneiro et al., 1976; Carneiro, unpublished results; Renker and Comés, unpublished results), K(def)TCP (Carneiro et al., 1976), and ACP(Cl) (Carneiro et al., 1979a).

According to the theory of Lee et al. (1973), the most interesting aspects of the lattice dynamics of a one-dimensional conductor are concentrated in the phase mode which at wave vector $2\mathbf{k}_F$ assumes the value of ω_T and amplitude mode ω_A. The amplitude mode corresponds to oscillations in the amplitude parameter A of the CDW as described above and ω_A has a symmetry which allows detection by Raman and neutron scattering. The phase mode corresponds to oscillation of ϕ in the CDW and was predicted to be observable by infrared (IR) absorption and neutron scattering. Indeed, experiments on KCP(Br) have proven those predictions to be qualitatively true. Figure 14 compares the values for ω_T and ω_A versus temperature as measured by infrared absorption (Brüesch et al., 1975), neutron inelastic scattering (Carneiro et al., 1976), and Raman scattering (Steigmeier et al., 1975). The quantitative disagreement of about 20% between the values of ω_A and ω_T, as obtained from different experiments, is a good example of the difficulty in deriving accurate values for the physical parameters of one-dimensional conductors. These results point out the difficulty in using experimental data to distinguish between the several detailed theories which have been proposed (Horowitz et al., 1975; Barisíc et al., 1973; Dieterich, 1975; Käfer, 1979). Keeping this problem in mind, there is, however, a characteristic feature of the dynamical properties which is well worth pointing out. Despite the fact that mean field theory (Rice and Strässler, 1973) predicts that ω_A should approach zero at temperature T_P, a prediction which was reportedly verified in the early neutron scattering experiments by the observation of a "soft mode" transition at T_{3D}, none of the ω versus T curves of Fig. 14 shows evidence of such soft mode behavior. Since the frequencies plotted in Fig. 14 are all

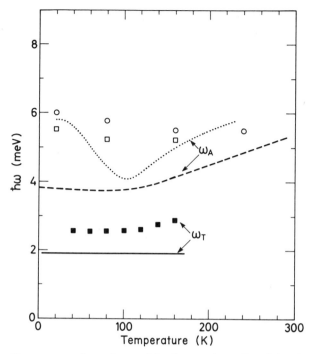

Fig. 14.   Temperature dependence of the frequencies at $2k_F$ of the phase mode ω_T and the amplitude mode ω_A as measured by neutron scattering ($\bigcirc$, $\square$, $\blacksquare$), infrared reflectivity (——), and Raman scattering (– – –). One theoretical calculation ($\cdots$) is shown. [From Williams *et al.* (1982).]

derived from lineshapes obtained at closely spaced temperatures, the evidence indicates that neither ω_A nor ω_T vary strongly with temperature, but rather that ω_A has a slight minimum at $T_{3D} = 100$ K, so that we are left with the unusual situation that correlation length $\xi_\perp$ varies critically, whereas ω_A does not. However, this feature is nicely explained by the theoretical work of Dieterich (1975) who has shown that as fluctuations depress the phase transition from T_P to T_{3D}, and ω_A becomes less "soft" despite the fact that $\xi_\perp$ diverges at T_{3D}. The theoretical result of Dieterich is in semiquantitative agreement with the data as shown in Fig. 14.

Another point which clearly falls within the category of "suggestive," but not strictly "conclusive," is the surprising result of an attempt to derive the longitudinal correlation length $\xi_\parallel$ from the inelastic neutron data. Carneiro *et al.* (1976) derived a wave vector width ΔQ from the intensity of the phason and it is natural to associate $\xi_\parallel$ with $\Delta Q/2$. This offers an opportunity to study ξ versus T. In doing so one finds that $\xi_\parallel$ varies in the same way as $\xi_\perp$ with a temperature-independent ratio $\xi_\parallel/\xi_\perp \gtrsim 20$. We find it interesting that from the previously mentioned work it appears that the theory of Horowitz *et al.* (1975), when combined with the η derived from electrical measurements, predicts this ratio to be 20 and independent of temperature.

Since K(def)TCP is in many respects quite different from KCP, a fact which has not been amply stressed in the literature, one might expect the lattice dynamics to be different as well. In particular, the fact that the DPO in K(def)TCP is commensurate (DPO = 0.25) could cause changes in behavior with respect to KCP which is incommensurate. However, a neutron scattering study of the phonon and amplitude modes gave very similar results for K(def)TCP and KCP and a similar picture has later emerged from Raman experiments of Steigmeier *et al.* (1975) except for the fact that the lower symmetry of K(def)TCP (triclinic) allows the occurrence of anomalies in both the longitudinal and transverse phonons. Since this increases the elastic energy at the expense of the Peierls state, it decreases Δ; but there seems to be no drastic effect of commensurability on the lattice dynamics in K(def)TCP.

Since ACP(Cl) is isostructural with KCP it was not surprising that the acoustic phonons, which were all that could be measured in ACP(Cl) (Carneiro *et al.*, 1979a) were identical to those of KCP. However, it is an astonishingly unifying fact that the unperturbed longitudinal phonon of KCP(Br), K(def)TCP, and ACP(Cl) can all be written as

$$\omega_q^0 = \omega_0 \left| \sin(\tfrac{1}{2} q d_{\text{Pt-Pt}}) \right| \tag{14}$$

with ω_0 independent of compound. However, ω_0 does depend on temperature so that 17.7 meV $> \hbar\omega_0 > 16.7$ meV, when $20 < T < 240$ K. It seems plausible to generalize this and suggest that the important parameter $\omega_{2k_F}^0$ need not be determined experimentally in all cases, but that it may be derived from this empirical relation. As we shall see, $\omega_{2k_F}^0$ derived in this way can be directly correlated with the properties of all the tetracyanoplatinates discussed in Figs. 7 and 10.

Finally, it is worth pointing out that both optical reflection and luminescence studies (Zeller and Brüesch, 1974) offer important information about the electron bands in all the compounds studied. This has proven to be an important simplifying factor in our understanding of POTCP's.

H. Pressure Effects

Pressure studies of the KCP salts have been minimal. Interrante and Bundy (1972) studied KCP at pressures to ~100 kbar. Using x-ray diffraction methods to study the effects of applied pressure they found that the *a* and *c* unit cell dimensions decreased as the pressure increased. The *a* parameter changed by 0.47 Å at 70 kbar, while the c axis (Pt–Pt chain axis) was less pressure sensitive than the chain axis in $Pt(NH_3)_4PtCl_4$ [Magnus Green Salt (MGS)] (Interrante and Bundy, 1971). The ac conductivities increased by a factor of 4 with a maximum occurring at ~25 kbar, and a steady decrease occurring thereafter, presumably due to the onset of steric and repulsive effects. This was probably due to the short d_{Pt-Pt} in the Krogmann salt (2.89 Å) as compared to 3.25 Å in MGS, as well as to a difference in the charge in the metal atoms.

Differences between the behavior under pressure of the cation-deficient and anion-deficient TCP salts were noted by Hara (1975). The anion-deficient compound increased in resistance under pressure by a factor of 100 between 40–100 kbar, after reaching a minimum at ~25 kbar. The cation-deficient compounds showed only a slight drift in resistance upwards with pressure. If one compares d_{Pt-Pt} of these compounds (as follows)

Compound	d_{Pt-Pt} (Å)
$K_2[Pt(CN)_4]Br_{0.3} \cdot 3H_2O$	2.89
$K_{1.75}[Pt(CN)_4] \cdot 1.5H_2O$	2.97, 2.96
$Rb_2[Pt(CN)_4]Cl_{0.30} \cdot 3H_2O$	2.88, 2.92
$Rb_{1.75}[Pt(CN)_4] \cdot 1.5H_2O$	2.94
$Cs_2[Pt(CN)_4]Cl_{0.30}$	2.86
$Cs_{1.75}[Pt(CN)_4] \cdot 1.5H_2O$	>2.88

one observes a larger d_{Pt-Pt} in cation-deficient compounds than in the cation-deficient salts, and these different d_{Pt-Pt} may account for the different behaviors noted. One would anticipate a more dramatic rise in resistance in compounds where the d_{Pt-Pt} is already quite low, and pressure would not tend to lower this distance much further, thus increasing the resistance.

Thielemans *et al.* (1976) observed evidence for a transition from a metal-semiconductor at ~32 kbar pressure in $K_2[Pt(CN)_4] \cdot Br_{0.3} \cdot 3H_2O$, and interpreted the data as an increase in interchain coupling under pressure. They postulated that at $P > 70$ kbar one might suppress a Peierls transition and obtain a superconductor between 1 and 6 K. Thus far, no verification has occurred.

V. SUMMARY AND DISCUSSION

A general discussion of the interrelation between the chemical, structural, and physical properties of partially oxidized tetracyanoplatinate (POTCP) complexes follows. The properties of the 19 compounds listed in Table VI may be summarized as follows.

The challenge to the chemist may be strategically defined as building the $[Pt(CN)_4]^{x-}$ chain structure by embedding it in a crystal structure that also contains the following two or three key features:

(1) A cation which permits formation of a crystalline salt. POTCP's have been prepared with the alkali metals K^+, Rb^+, and Cs^+ as cations but also more complex cations such as ammonium "A" = NH_4^+ and guanidinium "G" = $C(NH_2)_3^+$ have proven successful. Although previously reported, a POTCP salt of Na^+ has not been prepared and at present attempts to produce Li^+ POTCP salts are still underway (Williams and Schultz, unpublished work).

(2) A partial oxidation must occur, either in the form of anion deficiency (deficient in the sense that it does not oxidize Pt^{II} to Pt^{IV}) or in the form of cation deficiency. For this purpose, anions such as the halides Br^-, Cl^-, and to a less extent F^- have been frequently used, but also linear anions have been embedded in the lattice in the form of hydrogen bifluoride $(FHF)^-$, azide $(N_3)^-$, and nonlinear disulphato–hydrogen, "DSH" $[(O_3SO \cdot H \cdot OSO_3)^{3-}]$. Only a few cation deficient salts of the form $M_{1.75}[Pt(CN)_4] \cdot 1.5H_2O$, have been prepared with $M = K^+$, Rb^+, or Cs^+.

(3) The presence of water of crystallization may be important because in many cases the presence of a cation and anion does not suffice in making a crystalline material, and therefore water may be incorporated in the lattice to provide stabilizing hydrogen bonds. In a few cases, namely, ACP(Cl) and RbCP(DSH), water molecules can stabilize the "metallic resonance" by introducing hydronium ions into the lattice which neutralizes any possible charge imbalance.

The variety of structures of the POTCP salts show that these three preparative constraints (among others) still allow the preparation of a wealth of compounds. By modest changes in the chemical constituents it has been possible to modify a *given* structure within certain limits whereas other changes, such as substituting, e.g., a halide anion for $(FHF)^-$, can produce *new* structures with very different room temperature conductivities. Within a given structure type, d_{Pt-Pt} varies monotonically with cation size and (if it is present) with the size of anions; but in going from one structure type to another the variations may not be as uniform.

The physics of ten well-studied compounds out of the 19 POTCP's of Table VI suggests that they are analogues, as demonstrated by their electronic behavior as shown in Figs. 6, 9, and 11, as well as from their common $2\mathbf{k}_F$ Peierls distortion. In order to present this analogy more clearly the present understanding (Underhill *et al.*, 1979–1980; Carneiro, unpublished results) of how the Peierls instability manifests itself in the series of ten POTCP's is shown in Fig. 15.

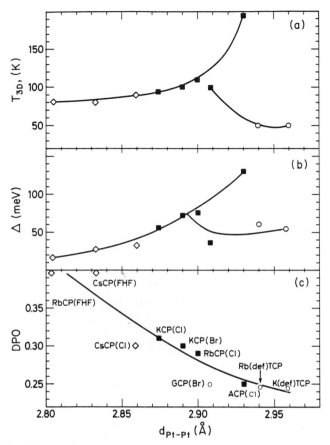

Fig. 15.   Measured values of DPO (a, y axis), Δ (b, y axis), and T_{3D} (c, y axis) versus d_{Pt-Pt}. [From Williams *et al.* (1982).]

First the degree of partial oxidation (DPO) varies regularly with the d_{Pt-Pt} and is related to the Fermi wave vector $\mathbf{k}_F$ (i.e., degree of band filling) by:

$$DPO = 2\{1 - \mathbf{k}_F d_{Pt-Pt}/\pi\}, \tag{15}$$

The band structure, shown in Fig. 16, in the intrachain direction is given by the free electron relation so that the Fermi energy is

$$\epsilon_F = (\hbar \mathbf{k}_F)^2/2m. \tag{16}$$

In this case the Peierls instability will give rise to a band gap 2Δ at low temperatures of

$$\Delta = 8\, \frac{1 - \mathbf{k}_F d_{Pt-Pt}/\pi}{1 + \mathbf{k}_F d_{Pt-Pt}/\pi}\, \epsilon_F e^{-1/\lambda} \tag{17}$$

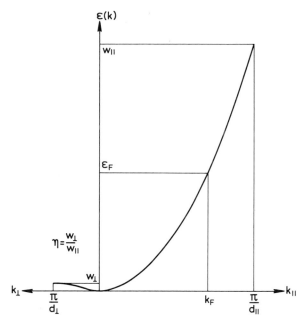

Fig. 16. Schematic band structure of quasi-one-dimensional POTCP's. [From Williams *et al.*, (1982).]

where the dimensionless electron phonon coupling constant is given by

$$\lambda = g^2 (N(\epsilon_F)/\omega^0_{2k_F}) \tag{18}$$

where g is the electron gain per relative displacement of the platinum ions, $N(\epsilon_F)$ the electronic density of states at ϵ_F in the metallic phase, and $\omega^0_{2k_F}$ the bare phonon frequency. The temperature scale T_p of the Peierls instability is related to Δ by

$$T_p = \Delta/(1.77k_B). \tag{19}$$

By performing structural analyses, x-ray diffuse scattering and dc conductivity measurements, d_{Pt-Pt}, k_F, and Δ are obtained, so that λ and T_p can be computed using Eqs. (17) and (19).

Second, the interchain coupling parameter η, defined as the ratio of bandwidth in the inter- and intrachain directions as shown in Fig. 17, determines the three-dimensional ordering temperature (T_{3D}) which, according to the theory of Horowitz *et al.* (1975) becomes

$$T_{3D} = T_p \exp\left(-\frac{2.5}{\eta} \frac{k_B T_p}{\epsilon_F}\right). \tag{20}$$

Since the rapid decrease of $\Delta(T)$ at T_{3D} gives rise to a peak in the logarithmic derivative of the conductivity versus inverse temperature, T_{3D} is

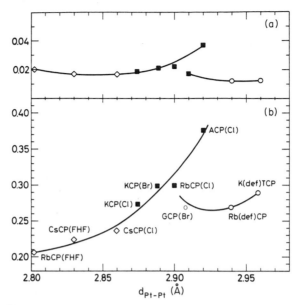

Fig. 17.   Electron phonon coupling constant λ (b) and interchain coupling parameter η (a), as deduced from measurements and shown plotted versus d_{Pt-Pt}. [From Williams *et al.* (1982).]

readily derived from the conductivity, and from Eq. (20) η may then be determined. The measured parameters DPO, Δ, and T_{3D} versus d_{Pt-Pt} are shown in Fig. 15 and the "intrinsic physical" parameters λ and η are shown in Fig. 17.

Two types of behavior emerge from inspection of Figs. 15 and 17. First, the steady increase in the electron–phonon coupling constant λ upon increasing d_{Pt-Pt} going from RbCP(FHF) to ACP(Cl), is understood in terms of the dependence of DPO (and $\mathbf{k}_F$) on d_{Pt-Pt}, since increasing $\mathbf{k}_F$ will give a strongly decreasing $\omega_{2\mathbf{k}_F}^0$ causing λ to rise from Eq. (18), whereas the DPO versus d_{Pt-Pt} relation stabilizes ϵ_F and therefore $g^2N(\epsilon_F)$.

In this series of compounds, the interchain coupling parameter η increases both for short and long d_{Pt-Pt}'s in the presence of more bulky cations or more hydrogen bonding interactions. This provides a nice demonstration of the increase in electron hopping and the effect of increased Coulombic interchain coupling when chains become too close.

The second behavior illustrated in Figs. 15 and 17 indicates that there are exceptions from such regular behavior of the POTCP complexes. In K(def)TCP, the low value of λ could in part be explained from a structural point of view, but of equal importance is the effect of a more complicated hydrogen bonding water network in the lattice, since GCP(Br), although isostructural with ACP(Cl) (space group $P4mm$) has a much smaller λ. The effect of fewer water molecules between distant chains seems to have a clear effect on η, which is very low for the three compounds GCP(Br), Rb(def)TCP, and K(def)TCP.

As is hopefully apparent from the previous discussion, the partially oxidized tetracyanoplatinates are found to be an excellent example of the results that may be obtained when the chemistry and physics of complicated compounds are worked out hand in hand. It seems therefore a challenging task of the future to both characterize all of the compounds in Table VI while at the same time to prepare new derivatives in order to test the theoretical discussions presented here.

VI. PARTIALLY OXIDIZED Bis(Oxalato)platinate SALTS

In 1966 Krogmann found that partially oxidized bis(oxalato)platinate salts had seemingly metallic-like properties in one-dimension, and resembled the POTCP complexes (Krogmann and Dodel, 1966a,b; Krogmann, 1968, 1969). As was the case for the TCP salts, the bis(oxalato)platinate (BOP) salts were discovered over 100 years ago (Döbereiner, 1833; Souchay and Lenfsen, 1958). The lack of progress in the study of these materials has been due to their often incorrect characterization. Lack of crystal size has also slowed down the progress, since this limited the range of techniques that could be used for proper characterizations, as was used for the TCP salts. Nevertheless, some progress has been made and the partially oxidized bis(oxalato)platinate (POBOP) salts bear some resemblance to the POTCP salts.

Partially oxidized bis(oxalato)platinate salts are known with cations of groups IA, IIA, IIB, and IIIB (Söderbaum, 1886; Gmelin, 1940), as well as simple cations derived from the first transition (Söderbaum, 1886; Gmelin, 1940) and lanthanide series (Söderbaum, 1886). Isolated examples exist for cations from other groups, e.g., Pb(II) and Th(IV) salts (Gmelin, 1940). Only cation-deficient salts are known. Table IX tabulates several POBOP salts.

The degree of partial oxidation (DPO) bis(oxalato)platinate salts is tabulated in Table X. These values were determined by x-ray and chemical methods. Figure 18 shows the relationship between DPO and d_{Pt-Pt} for the bis(oxalato)platinates and compares them with those for the TCP salts. Also plotted in the figure are the Pauling empirical relationships which for platinum follow the equation

$$d_{Pt-Pt}(\text{Å}) = 2.59 - \alpha \log_{10} \text{DPO} \qquad (21)$$

TABLE IX

Partially Oxidized Bis(oxalato)platinate Salts, Measured Properties[a]

	Stoichiometry and/or preparation	Electrical conductivity[b]	Diffuse x-ray, $\mathbf{k}_F$	Magnetic	Optical
Monovalent cation salts					
$(H_3O)_{1.6}[Pt(C_2O_4)_2] \cdot 2H_2O$	yes			yes	
$Li_{1.64}[Pt(C_2O_4)_2] \cdot 6H_2O$	yes				
$Na_{1.67}[Pt(C_2O_4)_2] \cdot 4H_2O$	yes				
$K_{1.62}[Pt(C_2O_4)_2] \cdot 2H_2O$	yes[d]	yes[d]			
$K_{1.62}[Pt(C_2O_4)_2] \cdot xH_2O$	yes[d]	yes[d]			
$K_{1.64}[Pt(C_2O_4)_2] \cdot yH_2O$	yes[d]	yes[d]			
$K_{1.64}[Pt(C_2O_4)_2] \cdot zH_2O$	yes[d]	yes[d]			
$K_{1.62}[Pt(C_2O_4)_2] \cdot 2H_2O$ (α-K-OP)	yes	yes	yes		
$K_{1.6}[Pt(C_2O_4)_2] \cdot xH_2O$ (β-K-OP)	yes	yes[d]			
$K_{1.81}[Pt(C_2O_4)_2] \cdot 2H_2O$ (γ-K-OP)	yes	yes	yes		
$K_{1.6}[Pt(C_2O_4)_2] \cdot yH_2O$ (δ-K-OP), (ϵ-K-OP)	yes	yes[d]			
$(NH_4)_{1.64}[Pt(C_2O_4)_2] \cdot H_2O$	yes	yes			
$Rb_{1.51}(H_3O)_{0.17}[Pt(C_2O_4)_2] \cdot 1.3H_2O$	yes	yes	yes		
$Rb_{1.67}[Pt(C_2O_4)_2] \cdot 1.5H_2O$ (α-Rb-OP)	yes	yes	yes		yes
$Rb_{1.5}[Pt(C_2O_4)_2] \cdot H_2O$	yes	yes	yes		
Divalent cation salts					
$Mg_{0.82}[Pt(C_2O_4)_2] \cdot 5.3H_2O$	yes	yes	yes		
$Mg_{0.82}[Pt(C_2O_4)_2] \cdot 4H_2O$	yes				
$Mg_{0.82}[Pt(C_2O_4)_2] \cdot 3.75H_2O$	yes				

Compound				
Ca$_{0.84}$[Pt(C$_2$O$_4$)$_2$]·4H$_2$O	yes			
Sr$_{0.84}$[Pt(C$_2$O$_4$)$_2$]·4H$_2$O	yes			
Ba$_{0.84}$[Pt(C$_2$O$_4$)$_2$]·4H$_2$O	yes			
Mn$_{0.81}$[Pt(C$_2$O$_4$)$_2$]·6H$_2$O	yes	yes		
Fe[Pt(C$_2$O$_4$)$_2$]·6H$_2$O[c]	yes		yes	
Co$_{0.83}$[Pt(C$_2$O$_4$)$_2$]·6H$_2$O	yes	yes	yes	yes
Ni$_{0.84}$[Pt(C$_2$O$_4$)$_2$]·6H$_2$O	yes	yes		
Cu$_{0.84}$[Pt(C$_2$O$_4$)$_2$]·7H$_2$O	yes	yes		
Zn$_{0.81}$[Pt(C$_2$O$_4$)$_2$]·6H$_2$O	yes	yes	yes	yes
Cd[Pt(C$_2$O$_4$)$_2$]·5H$_2$O[c] (olive form)	yes			
Cd[Pt(C$_2$O$_4$)$_2$]·4.5H$_2$O[c] (green form)	yes			
Pb[Pt(C$_2$O$_4$)$_2$]·3H$_2$O[c]	yes			
[Pt(NH$_3$)$_4$][Pt(C$_2$O$_4$)$_2$][c] (blue form)	yes			
[Pt(NH$_3$)$_4$][Pt(C$_2$O$_4$)$_2$][c] (green form)	yes			
Higher-valent cation salts				
Al-OP[c]	yes			
Th[Pt(C$_2$O$_4$)$_2$]$_2$·18H$_2$O	yes			
Y[Pt(C$_2$O$_4$)$_2$]$_2$·12H$_2$O[c]	yes			
Mixed cation salts				
LaNa[Pt(C$_2$O$_4$)$_2$]$_2$·12H$_2$O[c]	yes			
YK[Pt(C$_2$O$_4$)$_2$]$_2$·12H$_2$O[c]	yes			
YNa[Pt(C$_2$O$_4$)$_2$]$_2$·12H$_2$O[c]	yes			

[a] Taken from Underhill et al. (1982).
[b] See Table XIII for details.
[c] Poorly characterized and tentative classification.
[d] Refers to an unspecified phase of K-OP.

TABLE X

Degree of Partial Oxidation (DPO) for Bis(oxalato)platinate Salts[a]

Compound	DPO[b] (chemical)	DPO (from x-ray)
$Mg_{0.82}[Pt(C_2O_4)_2] \cdot 5.3H_2O$	0.36(a)(b)	0.3, 0.31–0.32
$Co_{0.83}[Pt(C_2O_4)_2] \cdot 6H_2O$	0.34(b)	0.30
$Ni_{0.84}[Pt(C_2O_4)_2] \cdot 6H_2O$	0.32(b)	0.30
$Mn_{0.81}[Pt(C_2O_4)_2] \cdot 6H_2O$	0.38(b)	0.28[c]
$Cu_{0.84}[Pt(C_2O_4)_2] \cdot 7H_2O$	0.32(b)	0.28
$Zn_{0.81}[Pt(C_2O_4)_2] \cdot 6H_2O$	0.38(b)	0.32
$Rb_{1.67}[Pt(C_2O_4)_2] \cdot 1.5H_2O$	0.33(b)	0.33
$Rb_{1.51}(H_3O)_{0.17}[Pt(C_2O_4)_2] \cdot 1.3H_2O$		0.32
$K_{1.81}[Pt(C_2O_4)_2] \cdot 2H_2O$ (γ-K-OP)	0.19(b)	0.19
$K_{1.62}[Pt(C_2O_4)_2] \cdot 2H_2O$ (α-K-OP)	0.38(b)	0.36

[a] Taken from Underhill et al. (1982).
[b] (a) DPO by titration and (b) DPO from elemental analysis.
[c] Experimental error bigger than ±0.02.

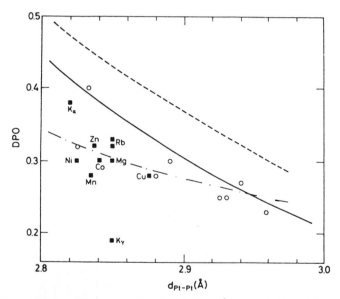

Fig. 18. Relationship between DPO and d_{Pt-Pt} (Å) in partially oxidized bis(oxalato)platinate (■) and tetracyanoplatinate (○) salts. The full curve is Pauling's formula for metallic resonance and the dashed line is for no metallic resonance (Pauling, 1960). [From Underhill et al. (1982).]

where $\alpha = 0.60$ for metallic resonance and $\alpha = 0.71$ for nonmetallic resonance. It can be noted from Fig. 18 that the DPO for the bis(oxalato)platinate salts appear to have a restricted d_{Pt-Pt} value of ~0.3 for the most part.

A. Synthesis

Partially oxidized bis(oxalato)platinate salts have been synthesized by oxidizing bis(oxalato)platinate(II) with a smaller amount of oxidizing agent than is required for complete oxidation to Pt(IV). The following oxidizing agents have been used: $[PtCl_6]^{2-}$ (Krogmann and Dodel, 1966a; Kobayashi et al., 1977), Cl_2, Br_2, and $(Cr_2O_7)^{2-}$ (Krogmann and Dodd, 1966a), H_2O_2 (Thomas et al., 1972), air (Krogmann and Dodel, 1966a; Schultz et al., 1978b), and electrolysis at a Pt anode (Miller, 1976). Other methods of synthesis also exist such as slow diffusion for the K^+ (Kobayashi et al., 1977) and Rb^+ salts (Kobayashi et al., 1978a, 1979), as well as electrolysis of dilute solutions of $K_2[Pt(C_2O_4)_2] \cdot xH_2O$ (Miller, 1976). No one method appears to be universally acceptable because of side reactions. Another problem is that crystalline products consist of several phases (Reis et al., personal communication; Kobayashi et al., 1978b).

B. Structural Studies

Probably the most extensively studied bis(oxalato)platinate salts are those of potassium and rubidium. Table XI tabulates crystal structural data for several salts. Of the five phases of partially oxidized potassium bis(oxalato)platinate (KOP), γ-KOP has been the salt in which detailed crystal studies have been made. The structure (see Fig. 19) has four $[Pt(CN)_2]^{1.81+}$ ions stacked along the b axis and these form a zigzag chain. Kobayashi et al. (1977) found the $d_{Pt-Pt} = 2.837$ and 2.868 Å and Pt(1)–Pt(2)–Pt(3) = 175°. Reis et al. (1976) found these parameters to be 2.833 Å, 2.857 Å, and 177.8°, respectively. There are three nonequivalent platinum atoms in the chain and torsion angles of ~45° are found between $[Pt(C_2O_4)_2]^{1.81-}$ units along the columnar stack. Every other oxalate ligand is either eclipsed or staggered, depending on its position along the platinum chain. A stair-case configuration of the ligand results (Reis et al., 1976). The ligands are nonplanar due to the interactions between the nonbonded oxygen atoms, K^+ ions, and water molecules. Five K^+ sites exist, but only one is fully occupied, whereas all five water sites are crystallographically disordered. The platinum chains are sinusoidally modulated at room temperature. At temperatures <100 K, the modulation wave of γ-KOP has transverse and longitudinal components. The wave amplitude is ~0.17

TABLE XI

Crystal Structure Data for Bis(oxalato)platinate Salts[a]

Compound	Crystal system	a (Å)	b (Å)	c (Å)	α°	β°	γ°	Space group	Intrachain Pt–Pt separation, d_{Pt-Pt} (Å)
Platinum(II) salts									
K$_2$[Pt(C$_2$O$_4$)$_2$]·2H$_2$O	Monoclinic	7.086	14.085	6.630	90	127	90	$P2_1/n$	>8
[Cu(en)$_2$][Pt(C$_2$O$_4$)$_2$] (en = H$_2$NCH$_2$CH$_2$NH$_2$)	Triclinic	6.978[b]	9.418	12.351	73.14	100.59	111.73	$P\bar{1}$	3.554
Ca[Pt(C$_2$O$_4$)$_2$]·3.5H$_2$O	Triclinic	9.33	10.72	6.36	93.7	99.3	115.1		3.18
Partially oxidized salts									
K$_{1.62}$[Pt(C$_2$O$_4$)$_2$]·2H$_2$O (α-K-OP)	Monoclinic	21.178	11.283[b]	17.502	90	92.19	90	Pa	2.82
	Monoclinic	17.48	11.28[b]	21.12	90	92.27	90	$P2/c$	2.82
	Monoclinic	17.58	11.24[b]	21.08	90	92.30	90	$P2_1/c$	2.81
K$_{1.6}$[Pt(C$_2$O$_4$)$_2$]·xH$_2$O (β-K-OP)	Monoclinic	17.637	20.704	8.525[b]	90	90	90		2.84
K$_{1.81}$[Pt(C$_2$O$_4$)$_2$]·2H$_2$O (γ-K-OP)	Triclinic	9.749	11.403	10.694	99.54	115.81	102.32	$P\bar{1}$	2.837 2.868
	Triclinic	9.744	10.700	11.377[b]	80.23	77.97	115.87	$P\bar{1}$	2.833 2.857
K$_{1.6}$[Pt(C$_2$O$_4$)$_2$]·yH$_2$O (ε-K-OP)	Triclinic	10.47	2.83	9.67	101.4	85.4			2.83
	Monoclinic	19.998	17.132[b]	19.547	90	117.355	90	$C2/c$	2.855 (av.)

Compound[a]	Crystal system	a	b	c	α	β	γ	Space group	Pt–Pt distance (Å)
$Rb_{1.67}[Pt(C_2O_4)_2]\cdot1.5H_2O$ (α-Rb-OP)	Triclinic	12.690	17.108	11.357	102.04	115.17	43.58	P1̄	2.717, 2.830, and 3.015 (av.)
$Rb_{1.51}(H_3O)_{0.17}[Pt(C_2O_4)_2]\cdot1.3H_2O$ (β-Rb-OP)	Triclinic	8.998	11.030	17.104[b]	95.73	104.23	110.26	P1 or P1̄	2.85 (av.)
$Rb_{1.5}[Pt(C_2O_4)_2]\cdot2H_2O$ (γ-Rb-OP)	Orthorhombic	11.159	16.596	11.329[b]	90.013	90.023	90.005	P222	2.829 (av.)
Cs-OP	Monoclinic	16.56	18.153	17.065	90	90	90	P2₁	
$Mg_{0.82}[Pt(C_2O_4)_2]\cdot5.3H_2O$ (Mg-OP)	Orthorhombic	16.58	14.27	5.70[b]	90	90	90	Cccm	2.85
$Mg_{0.82}[Pt(C_2O_4)_2]\cdot5.3H_2O$[c] (Mg-OP)	Monoclinic	16.56	14.29	5.72[b]	89.05	90	90	Ccc2	2.86
$Mg_{0.82}[Pt(C_2O_4)_2]\cdot4H_2O$	Triclinic	11.46	9.75	5.72[b]	89.9	93.3	107.0	Cc or C2/c	2.86
$Mg_{0.82}[Pt(C_2O_4)_2]\cdot3.75H_2O$	Orthorhombic	9.71	16.80	2.84[b]	90	90	90		2.84
$Co_{0.83}[Pt(C_2O_4)_2]\cdot6H_2O$ (Co-OP)	Orthorhombic	14.379	16.501	5.682[b]	90	90	90	Cccm	2.841
	Orthorhombic	16.54	14.43	5.70[b]	90	90	90	Pc2m	2.85
$Mn_{0.81}[Pt(C_2O_4)_2]\cdot6H_2O$ (Mn-OP)	Orthorhombic	16.79	14.28	5.67[b]	90	90	90	Cccm	2.835
$Ni_{0.84}[Pt(C_2O_4)_2]\cdot6H_2O$ (Ni-OP)	Orthorhombic	16.40	14.35	5.65[b]	90	90	90	Cccm	2.825
$Zn_{0.81}[Pt(C_2O_4)_2]\cdot6H_2O$ (Zn-OP)	Orthorhombic	16.52	14.36	5.665[b]	90	90	90	Cccm	2.838
$Cu_{0.84}[Pt(C_2O_4)_2]\cdot7H_2O$ (Cu-OP)	Triclinic	5.75[b]	9.95	11.67	107.48	93.50	105.02		2.876

[a] Taken from Underhill et al. (1982).
[b] Denotes Pt atom chain direction.
[c] Exists only below 283 K,

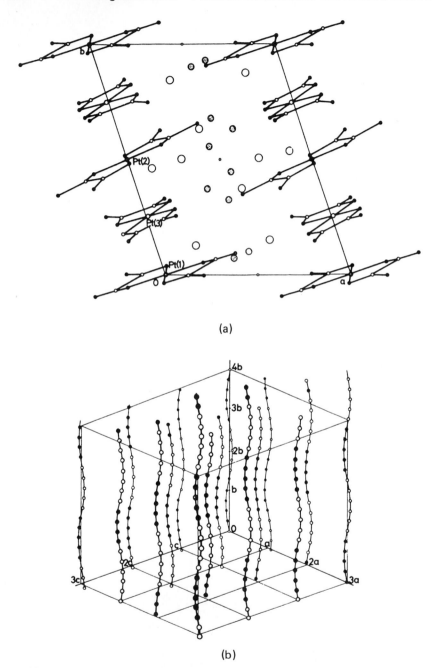

(a)

(b)

Fig. 19.   The crystal structure of γ-KOP (a). The fundamental structure viewed along the c axis. Schematic representation of the sinusoidally modulated lattice(b). The amplitude of the modulation wave is drawn about four times larger than that in the real lattice (Kobayashi *et al.*, 1977). [Underhill *et al.* (1982).]

Å (Kobayashi *et al.,* 1977). This is much larger than that found for
$K_2[Pt(CN)_4]Br_{0.3} \cdot 3H_2O$ (Williams *et al.,* 1982). The modulation wave has
a period $10.5 \, d_{Pt-Pt} = 29.92$ Å equivalent to the Peierls superlattice distortion (30 Å) as given by $2n \, d_{Pt-Pt}/DPO$, where $n = 1$.

The rubidium salts crystallize in at least three phases. The complex
$Rb_{1.69}[Pt(C_2O_4)_2] \cdot 1.5H_2O$ (α-RbOP) has been studied by Kobayashi *et al.*
(1978a, 1979) and possesses a sixfold structure as illustrated in Fig. 20.
The structure features distorted platinum chains along the *b* axis with three
independent Pt–Pt distances of 2.717, 2.830, and 3.015 Å. The spacing of
2.717 Å is the shortest Pt–Pt distance observed in partially oxidized platinum complexes and is shorter than that found in platinum metal (2.775 Å).
The differences in Pt–Pt distances are greater for α-RbOP than in γ-KOP.
In α-RbOP, the ligands are staggered with respect to other oxalate ligands
on adjacent platinum atoms in the chain. The torsion angles are about 46,
55, and 80°, while the alternate ligands are eclipsed or staggered by ~90°.
The oxalate ligands are not quite planar due to the electrostatic interactions
between Rb^+ and water and the nonbonded oxygen atoms on the ligand.
There are five independent rubidium ions and all occupy general positions
in the space group, and are approximately located with the water molecules
in layers parallel to the platinum atom chains. The platinum atom chains

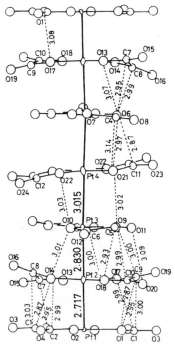

Fig. 20.   Structure of Pt atom chains in α-RbOP (Kobayashi *et al.,* 1979). [From
Underhill *et al.* (1982).]

in α-RbOP are also sinusoidally modulated as in γ-KOP, with similar amplitudes of distortion. The Peierls superstructure lattice in α-RbOP is given by $2n\ d_{Pt-Pt}/DPO$ and is 17.28 Å for $\eta = 1$. This is the same, within experimental error, as the b axis of the unit cell (17.108 Å) representing a sixfold Pt–Pt repeat distance.

C. Non-Peierls Superstructures in Partially Oxidized Bis(Oxalato)platinate Salts

Besides the superstructures which are described by the Peierls insta-bility and which give rise to diffuse x-ray scattering at $2\mathbf{k}_F$, other lattice modulations or ''non-Peierls'' superstructures have been found in the POBOP salts.

The Peierls instability gives rise to two fundamental features:

(1) The rapid change in conductivity observed in the region of the three-dimensional ordering temperature (T_{3D}) separates the high-temper-ature metallic state from the low-temperature semiconducting states.

(2) Above the transition temperature, the $2\mathbf{k}_F$ structural distortion is primarily one dimensional and gives rise to uniform x-ray scattering. Below the transition temperature the $2\mathbf{k}_F$ structural distortion is short range and three-dimensional and gives rise to modulated streaks in the x-ray photographs.

Superstructures which do not follow statements (1) and (2) above are called non-Peierls superstructures. Table XII summarizes the Peierls and non-Peierls superstructures in various partially oxidized BOP salts. A general characteristic of the non-Peierls distortion is that these distortions develop at high temperatures above 0°C.

D. Electrical Conductivity Properties

Table XIII summarizes the electrical conductivity data for partially oxidized BOP salts. The room temperature conductivities in the plat-inum chain directions $\alpha_{\parallel}$ lie in the range of 1–100 (ohm cm)$^{-1}$, and are considerably less than those for partially oxidized TCP salts for similar d_{Pt-Pt}. For example, $Cs_2[Pt(CN)_4](FHF)_{0.39}$ [$d_{Pt-Pt} = 2.833$ Å, $\sigma_{\parallel}$ (300 K) $\simeq 2 \times 10^3$ (ohm cm)$^{-1}$] (Williams $et\ al.$, 1982), compared to $Zn_{0.81}[Pt(C_2O_4)_2]\cdot 6H_2O$ [$d_{Pt-Pt} = 2.838$ Å, $\sigma_{\parallel}$ (300 K) $= 94$ (ohm cm)$^{-1}$] (Jacobsen $et\ al.$, 1980). These results are predicted by band structure calculations of Bullett (1978). The existence at room temperature of su-perstructures in the partially oxidized BOP salts may be an important factor in the lower electrical conductivities in these compounds.

E. Optical Studies

The colors of partially oxidized BOP salts vary from deep red in ZnOP and MgOP to the copper color of KOP and MgOP. These results suggest

TABLE XII

Peierls and Non-Peierls Superstructures in Partially Oxidized Salts[a]

	γ-KOP	α-RbOP	MgOP	CoOP	CuOP	NiOP	ZnOP	K$_{1.75}$[Pt(CN)$_4$]·1.5H$_2$O
Fundamental lattice	Triclinic	Triclinic	Orthorhombic[b] Monoclinic[c]	Orthorhombic	Triclinic	Orthorhombic	Orthorhombic	Triclinic
$d_{\text{Pt-Pt}}$ (Å)[b]	2.837 2.868	2.717 2.830 3.015	2.85[b]	2.841	2.876	2.83	2.83	2.963
$2\mathbf{k}_{\text{F}}$[c]	1.81	1.667	1.70[b]	1.70	1.70	1.70	1.68	1.775
T_{3D} (K)[d]	170	>RT	<100[c]	295	>RT	303	279 ± 5	<60
$q^{\text{NP}c}$	1.81	1.667	0.98–0.99[c]	0.85	0.86	0.85	0.84	1.775
T_c (K)[e]	>RT	>RT	283–285	299–304	>RT	>RT	>RT	308

[a] Taken from Underhill *et al.* (1982).
[b] At room temperature (RT).
[c] In units of $\pi/d_{\text{Pt-Pt}}$.
[d] T_{3D}, Three-dimensional ordering temperature.
[e] T_c, Transition temperature for NP instability.

TABLE XIII

Electrical Conductivity Data for Partially Oxidized Bis(oxalato)platinate Salts[a,b]

Compound	dc studies		35 GHz studies	
	$\sigma_\parallel$(ohm cm)$^{-1}$	E_a (eV)	$\sigma_\parallel$(ohm cm)$^{-1}$	$\sigma_\perp$ (ohm cm)$^{-1}$
α-K-OP	10^2 (max)	0.10 (<190 K)		
γ-K-OP	10 (max)	0.15 (<170 K)		
$K_{1.64}[Pt(C_2O_4)_2] \cdot xH_2O$	(1×10^{-2})–42	0.070–0.086		
$Rb_{1.51}(H_3O)_{0.17}[Pt(C_2O_4)_2] \cdot 1.3H_2O$	5–(18×10^{-3})	0.095 (<180 K)	1.6–(2.6×10^{-1})	1.7×10^{-3}
$Rb_{1.67}[Pt(C_2O_4)_2] \cdot 1.5H_2O$	7×10^{-3}	0.077		
$Mg_{0.82}[Pt(C_2O_4)_2] \cdot 5.3H_2O$	(2×10^{-1})–50	0.005–0.085	9–34	(1.4×10^{-1})–(2.8×10^{-2})
$Co_{0.83}[Pt(C_2O_4)_2]_2 \cdot 6H_2O$	2–25	0.05 (<250 K)	12–38	2–(4×10^{-1})
$Mn_{0.81}[Pt(C_2O_4)_2]_2 \cdot 6H_2O$	10–47	0.05–0.06 (<110 K)		
$Ni_{0.84}[Pt(C_2O_4)_2]_2 \cdot 6H_2O$	2–22	0.053–0.062 (<275 K)		
$Zn_{0.81}[Pt(C_2O_4)_2]_2 \cdot 6H_2O$	29–94	0.053–0.055 (<250 K)		
$Cu_{0.84}[Pt(C_2O_4)_2]_2 \cdot 7H_2O$	(7×10^{-1})–10^c		3.6–5.8	4.0–(5.9×10^{-3})

[a] Taken from Underhill et al. (1982).
[b] Room temperature values.
[c] Reference Watkins et al., unpublished results.

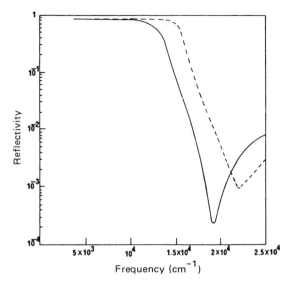

Fig. 21. Reflectivity of single crystals of Zn-OP (solid curve) and $K_2[Pt(CN)_4]Br_{0.3} \cdot 3H_2O$ (dashed curve) at 300 K for light polarized parallel to the metal atom chains (Jacobsen *et al.*, 1980). [From Underhill *et al.* (1982).]

lower plasma frequencies than in $K_2[Pt(CN)_4]Br_{0.3} \cdot 3H_2O$. The reflectivity of ZnOP is highly anisotropic, rising to approximately 80% at ~8000 cm^{-1} (1 eV) for parallel polarized light, and 10% for polarized light perpencicular to the platinum chain direction (see Fig. 21).

F. Magnetic Studies

Two polycrystalline samples of $(H_3O)_{1.6}[Pt(C_2O_4)_2] \cdot nH_2O$ where $n = 2$ or 3, were studied over the temperature range of 1.5–300 K and found to be qualitatively similar in the variation of susceptibility with temperature, parallel to the platinum chains, in $K_2[Pt(CN)_4]Br_{0.3} \cdot H_2O$ (see Fig. 22) (Heitkam *et al.*, 1975). The paramagnetism of both hydrates is characterized by two regions where the slopes of molar magnetic susceptabilities χ_M versus inverse temperature are different. In $K_2[Pt(CN)_4]Br_{0.3} \cdot H_2O$ the break occurs at 40 K but in the $(H_3O)_{1.6}[Pt(C_2O_4)_2] \cdot nH_2O$ the break occurs at 20 ± 3 K.

G. Summary—Comparison with POTCP Salts

Table XIV summarizes a comparison of partially oxidized BOP and TCP salts. As of this writing, only cation-deficient BOP salts have been prepared, while both cation- and anion-deficient TCP salts have been synthesized. Partially oxidized TCP salts occur only with monovalent cations, while the BOP salts are found with mono- or divalent cations and even other cations (e.g., Th^{+4} or Pb^{+2}). Only hydrated BOP salts are

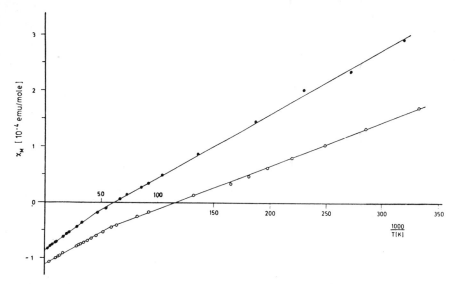

Fig. 22. The molar magnetic susceptibilities, χ_M, of polycrystalline partially oxidized bis(oxalato)platinate acids as functions of the reciprocal temperature: ●, $(H_3O)_{1.6}[Pt(C_2O_4)_2] \cdot mH_2O$ (m is relatively small) and ○, $(H_3O)_{1.6}[Pt(C_2O_4)_2] \cdot nH_{;2}O$ (n is relatively large). In the $\chi_M(1/T)$ plots, n was taken to be 2 and m to be 0 (Heitkamp et al., 1975). [From Underhill et al., (1982).]

TABLE XIV

General Comparison of the Partially Oxidized Bis(oxalato)platinate and Tetracyanoplatinate Salts[a]

	Bis(oxalato)platinate salts	Tetracyanoplatinate salts
Cation deficient	Yes	Yes
monovalent cations	Yes	Yes
divalent cations	Yes	No
Anion deficient	No	Yes
Hydrated	Yes	Yes
Anhydrous	No	Yes
Range of d_{Pt-Pt} (Å)	2.81–2.876	2.789–2.963
Range of DPO	0.19–0.36	0.19–0.40
Range of σ (300 K) (ohm cm)$^{-1}$	$10^2–10^{-2}$	$(2 \times 10^3)–1$

[a] Taken from Underhill et al. (1982).

known, whereas both hydrated and anhydrous TCP salts have been pre-pared. The range of DPO's in BOP salts appears to be from 0.28 to 0.32 with only two exceptions. The TCP salts range from 0.19 to 0.40. The variation of d_{Pt-Pt} in BOP salts is smaller than the variation of TCP salts. For most TCP salts the size, extent of hydration, and hydrogen bonding properties of the cations are directly correlated with rather large variations observed in the d_{Pt-Pt} and ḊPO. In these hydrated salts the cations occupy sites in the upper half and the water molecules are found in the lower half of the unit cell (Williams *et al.*, 1982). For the divalent cation BOP salts, the cations are more symmetrically placed and are surrounded by six water molecules, which may screen and thus reduce the interactions with the platinum chains. There are also stacking differences between the BOP and TCP salts. The torsion angles between the planes of anions are also different, being ~45° in TCP salts, minimizing steric interactions, and varying in BOP salts from 45° to 60° to allow intermolecular backbonding between ligands.

Other differences are noted, e.g., in the POBOP salts the platinum atom chains are often modulated by both three-dimensional non-Peierls super-structures as well as a Peierls distortion. In the TCP salts, with the excep-tion of $K_{1.75}[Pt(CN)_4]1.5H_2O$, the TCP chains are modulated by only the Peierls distortion. The amplitudes of modulation usually are also con-siderably greater in the BOP salts. The superlattices these salts exhibit clearly have an important influence on their electrical properties.

VII.   ONE-DIMENSIONAL METAL DITHIOLATE COMPLEXES

The transition metal complexes of dithiolene [(bis(dicyanoethylene-dithiolato) metal monoanion] have been known since 1962. The planar metal dithiolates of the type $[M(S_2C_2R_2)_2]^{n-}$ occur with a variety of cations (M = Ni, Pd, Pt, Cu, Co, or Fe and R = CN, Me, H, CF_3 or C_4H_4). When R is CN the $(S_2C_2(CN)_2)_2$ entity is abbreviated (mnt). The structure of the monoanion is illustrated in Fig. 23. Earlier studies have concentrated on the salts of bulky cations (Rosseinsky and Malpas, 1979; Miller and Ep-stein, 1976, 1979; Alcácer *et al.*, 1979; Bray *et al.*, 1977; Perez-Abuerne *et*

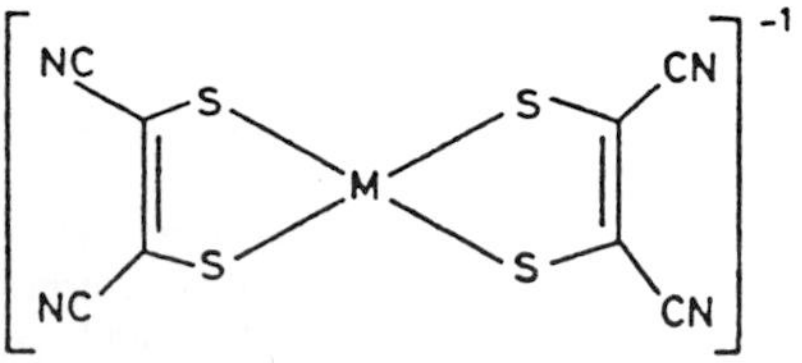

Fig. 23.   Structure of the monoanion of the metal, M = Ni,Pd, bis(dicyano-ethylenedithiolate).

al., 1977). In most instances these complexes demonstrated semiconducting electrical properties. Table XV tabulates electrical conductivities of these salts. The salts containing small cations (as illustrated in Table XV), demonstrated the higher conductivities.

The thiolene ligand is interesting for it has been shown to stabilize unusual formal oxidation states and geometries. When the metal is a transition element, the salts are capable of accommodating an excess electron by reducing Coulomb–Coulomb repulsions by delocalization of the charge over four electronegative cyano groups. The presence of small cations in the lattice of these salts facilitates a short intra-anion distance and provides one-dimensional metallic properties (Underhill and Ahmad, 1981). In fact, Perez-Albuerne *et al.* (1977) reported a salt $Na[M(S_2C_4N_2)_2]1.15H_2O$ where M is Ni or Pd which supports these conclusions. Underhill and Ahmad (1981) reported on single crystals of $[Pt(S_2C_4N_2)_2]^{n-}$ salts in which conduction is through interacting anions and conductivity varies from 30 to 200 $(ohm\ cm)^{-1}$. Figure 24 shows the variation of conductivity of $Li_x[Pt(S_2C_4N_2)_2]\cdot 2H_2O$ (X = ~0.75) with temperature. The temperature dependence is similar to that for $K_2[Pt(CN)_4]Br_{0.3}\cdot 3H_2O$. The conductivity appears to be very sensitive to water of dehydration. The lithium salt has been extensively studied and has been found to have the empirical formula $Li_{0.8}(H_3O)_{0.33}[Pt(S_2C_4N_2)_2]\cdot 1.67H_2O$ (Underhill *et al.*, 1985). The lithium salt is prepared by mixing a solution of $(Et_4N)_2[Pt(mnt)_2]$, in 70% acetone/water, and passing this solution through a protonated ion exchange resin

TABLE XV

Electrical Conductivities of Several Dithiolate Salts

Cation	Anion	σ_{RT} (Compressed pellet)[a] $(ohm\ cm)^{-1}$
NEt_4^+	$[Ni(S_2C_4N_2)_2]^-$	1.0×10^{-8}
NH_4^+ [b] (a)	$[Ni(S_2C_4N_2)_2]^-$	4.0×10^{-4}
(b)		1.0
Na^+ [b] (a)	$Ni(S_2C_4N_2)_2]^-$	4.0×10^{-4}
(b)		2.5×10^{-1}
NEt_4^+	$[Pd(S_2C_4N_2)_2]^-$	2.5×10^{-8}
NH_4^+	$[Pd(S_2C_4N_2)_2]^-$	1.1
Na^+	$[Pd(S_2C_4N_2)_2]^-$	4.0×10^{-1}

[a] Taken in part from Perez-Albuerne *et al.* (1977).

[b] The differences for the NH_4^+ salt a and b for Na were attributed to the state of hydration (a being anhydrous while b is hydrated). It was also determined that two forms of salts could be crystallized, which had different conductivities.

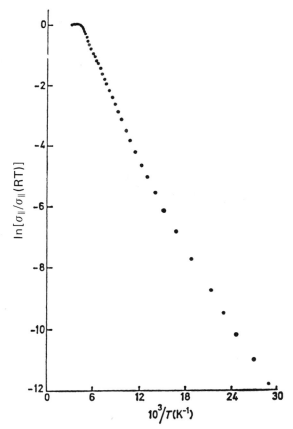

Fig. 24. Variation of conductivity with temperature $\ln[\sigma_{\parallel}/\sigma_{\parallel}(\text{RT})]$ for a single crystal of $\text{Li}_x[\text{Pt}(\text{S}_2\text{C}_4\text{N}_2)_2] \cdot 2\text{H}_2\text{O}$. [From Underhill and Ahmad (1981).]

column (Dowex 20W–50) and then adding an excess of LiCl to the eluted solution. Slow oxidation in air and slow evaporation yields a black solid of small well-shaped crystals and small black platelets. The platelets were found to be $\text{Li}_{0.5}[\text{Pt}(\text{mnt})_2] \cdot 2\text{H}_2\text{O}$ (β-LiPt(mnt)) (Kobayashi *et al.*, 1984; Underhill *et al.*, 1985).

X-ray studies by Kobayashi *et al.* (1982) show that the room temperature structure of LiPt(mnt) consists of stacks of nearly eclipsed Pt(mnt)₂ units along the *c* axis, with *c* = 3.639 Å. The unit cell is triclinic (V_c = 355 Å³, space group $P\bar{1}$). The stacks form sheets along *b* and are separated along *a* by Li⁺ and H₂O. Short chalcogen (S···S) contacts exist between chains within the sheets as well as within the chain, suggesting a relatively two-dimensional network. This is similar to structures observed for the (TMTSF)₂X and (ET)₂ReO₄ conductors. A superstructure below *T* = 215

K, preceded by one-dimensional diffuse scattering above that temperature is indicative of a Peierls transition (Kobayashi *et al.*, unpublished work).

The dc conductivity of LiPt(mnt) was measured by a four-point probe (Jacobsen *et al.*, 1978). At room temperature the conductivity along the *c* axis is 30–200 (ohm cm)$^{-1}$. The temperature dependence is illustrated in Fig. 24, and a $T_{M \to I}$ is indicated at ~200 K. The transition appears, again, to be due to a Peierls instability.

Metal dithiolate complexes of nickel and palladium also exist. For instance, $Li[Ni(mnt)_2] \cdot 2H_2O$ and $Li[Pd(mnt)_2] \cdot 2H_2O$ have been synthesized and room temperature conductivities of these complexes are less than that found for LiPt(mnt) (Underhill *et al.*, 1985).

Hückel MO calculations on LiPt(mnt) have confirmed that the intrastack overlap is predominantly associated with the ligands and with *S* orbitals (Kobayashi *et al.*, 1984), The central metal d_{xy} and d_{yz} orbitals play important roles in the delocalization of the electrons across the whole anion. The overlap of PdS is calculated to be greater than either NiS or PtS (Little, 1983).

The temperature-dependent resistance of the lithium salt has been measured with pressures to 10 kbar, and it has a Peierls transition at 213 K (Carneiro *et al.*, 1985). The authors conclude that the Peierls instability is gradually suppressed by pressure due to an increase in interchain coupling.

LIST OF SYMBOLS AND ABBREVIATIONS

a, b, c	Crystallographic unit cell lengths
ACP	Ammonium cyanoplatinates
ACP(Cl)	$(NH_4)(H_3O)_{0.17}[Pt(CN)_4]Cl_{0.42} \cdot 2.83H_2O$
$A(T)$	Amplitude of lattice distortion
AD	Anion deficient
a_T, a_L	Amplitude mode coupling constants
BOP	Bis(oxalato)platinate
CD	Cation deficient
CDW	Charge-density wave
cm^{-1}	Wave number
$d_{\parallel}, d_{Pt-Pt}$	Intrachain Pt–Pt separation
$d_{\perp}$	Interchain Pt–Pt separation
$D(n)$	Metallic bond distance as a function of bond order
DPO	Degree of partial oxidation
$E_{\perp}$	Perpendicular electric field
$E_{\parallel}$	Parallel electric field
F_N	Structure factor
g	Electron gain per relative displacement
$h, \hbar$	Planck's constants

I	Emission intensity
I	Body centered lattice structural type
I_{NP}	Non-Peierls intensity
K(def)TCP	Potassium deficient tetracyanoplatinate
KOP	Potassium bis(oxalato)platinate
k_B	Boltzmann constant
$\mathbf{k}_F$	Fermi wave vector
M	Metal ion
M	Molar concentration (mol/liter)
$N_{(\epsilon_F)}$	Electronic density of states
NP	Non-Peierls
POBOP	Partially oxidized bis(oxalato)platinate
POTCP	Partially oxidized tetracyanoplatinate
P	Primitive lattice structural type
$\mathbf{q}, \mathbf{Q}$	Wave vector
q^{NP}	Non-Peierls intrachain wave vector
r^+	Cation radius
T	Temperature
T_c	Mean field critical temperature
TCP	Tetracyanoplatinate
T_p	Mean field Peierls transition temperature
T_{3D}	Three-dimensional ordering temperature
XDS	X-ray diffuse scattering
X_M	Molar susceptibility
α	Thermal expansion coefficient
δ	Non-Peierls order parameters
ΔQ	Wave vector width
$\Delta(T)$	Band gap as a function of temperature
ζ	Correlation length
ϵ_F	Fermi energy
η	Interchain coupling constant
λ	Electron-phonon coupling constant
Φ	Phase angle
$\sigma_\parallel$	Parallel dc conductivity
$\sigma_\perp$	Perpendicular dc conductivity
$\omega^0(2\mathbf{k}_F)$	Bare phonon frequency
ω_A, ω_T	Amplitude mode and phase mode frequencies

REFERENCES

Alcácer, L., Novais, H., and Pedroso, F. (1979) in "Molecular Metals" (W. E. Hatfield, ed.), Vol. 1, p. 415, Plenum, New York.

Anderjan, R., Baumann, D., Breer, H., Endres, H., Gitzel, W., Keller, H. J., Lorentz, R., Moroni, W., Megnamissi-Belombé, M., Nothe, D., and Ruppe, H. H. (1974) in "Extended Interactions Between Metal Ions in Transition Metal Complexes" (L. V. Interrante, ed.), ACS Symp. Ser. No. 5, p. 314, Am. Chem. Soc., Washington, D.C.

Andre, J. J., Bieber, A., and Gautier, F. (1976), Ann. Phys. (Paris) 1, 145.

Bak, P. (1977), in "Electron-Phonon Interactions and Phase Transitions" (T. Riste, ed.), pp. 66–87, Plenum, New York.

Barisić, S., Bjelis, A., and Saub, K. (1973), *Solid State Commun.* **13**, 1119.

Blessing, R. H., and Coppens, P. (1974), *Solid State Commun.* **15**, 215.

Blomstrand, C. W. (1971), *J. Prakt. Chem.* **3**, 208.

Bozorth, R. M., and Pauling, L. (1932), *Phys. Rev.* **39**, 537.

Braude, A., Lindegaard-Andersen, A., Carneiro, K. (1979), *Synth. Met.* **1**, 35; Kobayashi, H., Hano, Y., Danno, T., Kobayashi, A., and Sasaki, Y. (1970), *Chem. Lett. Chem. Soc. Jpn.*, p. 198.

Braude, A., Lindegaard-Andersen, A., Carneiro, K., and Peterson, A. S. (1980), *Solid State Commun.* **33** 365.

Bray, J. W., Hart, H. R., Interrante, L. V., Jacobs, I. S., Kasper, J. S., Piacente, P. A., and Watkins, G. D. (1977), *Phys. Rev. B.* **16** (1359).

Brown, R. K., and Williams J. M. (1979), *Inorg. Chem.* **18**, 1922.

Brüesch, P., Strässler, S., and Zeller, H. R. (1975), *Phys. Rev. B.* **12**, 219.

Bullett, D. W. (1978), *Solid State Commun.* **27**, 467.

Carneiro, K. (1979), *in* "Molecular Metals" (W. E. Hatfield, ed.), p. 369, Plenum, New York.

Carneiro, K., unpublished results.

Carneiro, K., Shirane, G., Werner, S. A., and Kaiser, S. (1976), *Phys. Rev. B.* **13**, 4258.

Carneiro, K., Petersen, A. S., Underhill, A. E., Wood, D. J., Watkins, D. M., and Mackenzie, G. A. (1979a), *Phys. Rev. B.* **19**, 6279.

Carneiro, K., Jacobsen, C. S., and Williams, J. M. (1979b), *Solid State Commun.* **31**, 837.

Carneiro, K., Vasquez, J., Underhill, A. C., and Clemensen, P. J. (1985), *Phys. Rev. B.* **31**, 1128.

Comés, R. (1975) *in* "One-Dimensional Conductors." Lecture Notes in Physics (H. G. Schuster, ed.), p. 32, Springer-Verlag, Berlin.

Comés, R., Lambert, M., Launois, H., and Zeller, H. R. (1973a), *Phys. Rev. B.* **8**, 571.

Comés, R., Lambert, M., and Zeller, H. R. (1973b), *Phys. Status Solidi B.* **58**, 578.

Day, P. (1975), *J. Am. Chem. Soc.* **97**, 1588.

Dieterich, W. (1975), *Solid State Commun.* **17**, 445.

Döbereiner, J. W. (1833), *Pogg. Am.* **28**, 180.

Eagen, C. F., Werner, S. A., and Saillant, R. B. (1975), *Phys. Rev. B.* **12**, 2036.

Epstein, A. J., and Miller, J. S. (1979), *Solid State Commun.* **29**, 345.

Gerhardt, V., Pfab, W., Reisinger, J., and Yersin, H. (1979), *J. Lumin.* **18–19**, 375.

Gmelin, X. (1939–1940), "Handbuch der Anorganische Chemie," System No. 68, Platinum, Part C, The Compounds, reprinted 1962, Springer-Verlag, Berlin.

Gmelin, X. (1940), "Handbuch der Anorganische Chemie," Platinum, Part C. Springer-Verlag, Berlin.

Hara, Y., Shirotani, I., and Onodera, A. (1975), *Solid State Commun.* **17**, 827.

Heitkamp, D., Rade, H. S., Keller, M. J., and Rupp, H. H. (1975), *J. Solid State Chem.* **15**, 295.

Holzapfel, W., Yersin, H., Gliemann, G., and Otto, H. H. (1978), *Ber. Bunsenges. Phys. Chem.* **82**, 207.

Horowitz, B., Gutfreund, H., and Weger, M. (1975), *Phys. Rev. B.* **12**, 3174.

Interrante, L. V., and Bundy, F. P. (1971), *Inorg. Chem.* **10**, 1169.

Interrante, L. V., and Bundy, F. P. (1972), *Solid State Commun.* **11**, 1641.

Interrante, L. V., and Messmer, R. P. (1974), *Chem. Phys. Lett.* **26**, 225.

Isci, H., and Mason, W. R. (1975), *Inorg. Chem.* **14**, 907.

Jacobsen, C. S., Mortensen, K., Andersen, J. R., and Bechgaard, K. (1978), *Phys. Rev. B.* **18**, 905.

Jacobsen, C. S., Watkins, D. M., and Underhill, A. E. (1980), *Solid State Commun.* **36**, 477.

Johnson, P. L., Musselman, R. I., and Williams, J. M. (1977), *Acta Crystallogr. Sect. B.* **33**, 3155.

Käfer, K. (1979), *in* "Quasi One-Dimensional Conductors II," Lecture Notes in Physics, Vol. 96, Springer-Verlag, Berlin.

Keefer, K. D., Washecheck, D. M., Enright, N. P., and Williams, J. M. (1976), *Inorg. Chem.* **15**, 2446; Reis, A. M., Peterson, S. W., Washecheck, D. M., and Miller, J. S. (1976), *J. Am. Chem. Soc.* **98**, 234.

Knop, W. (1842), *Justus Liebigs Ann. Chem.* **43**, 111.

Knop, W., and Schnedermann, J. (1845), *J. Prakt. Chem.* **37**, 461.

Kobayashi, H., Shirotani, I., Kobayashi, A., and Sasaki, Y. (1977), *Solid State Commun.* **23**, 409.

Kobayashi, A., Sasaki, Y., and Kobayashi, H. (1978a), *Chem. Lett. Chem. Soc. Jpn.,* p. 1167.

Kobayashi, A., Sasaki, Y., Shirotani, I., and Kobayashi, H. (1978b), *Solid State Commun.* **26**, 653.

Kobayashi, A., Sasaki, Y., and Kobayashi, H. (1979), *Bull. Chem. Soc. Jpn.* **52**, 3682.

Kobayashi, A., Mori, T., Sasaki, Y., Kobayashi, A., Underhill, A. E., and Ahmad, M. M. (1982), *J. Chem. Soc. Chem. Commun.,* p. 390.

Kobayashi, A., Mori, T., Sasaki, Y., Kobayashi, H., Ahmad, M. M., and Underhill, A. E. (1984), *Bull. Chem. Soc. Jpn.* **57,** 3262.

Kobayashi, A., Sasaki, Y., Kobayashi, H., Underhill, A. E., and Ahmed, A. A. (1984), *Chem Lett Chem Soc. Jpn.,* p. 305.

Kobayashi, A., Underhill, A. E., Sasaki, Y., Kobayashi, H., and Ahmad, M. M., unpublished work.

Koch, T. R., Johnson, P. L., and Williams, J. M. (1977a), *Inorg. Chem.* **16**, 640.

Koch, T. R., Johnson, P. L., Washecheck, D. M., Cornish, T. L., and Williams, J. M. (1977b), *Acta Crystallogr. Sect. B.* **33**, 3248.

Krogmann, K., and Dodel, P. (1966a), *Chem. Ber.* **99**, 3402.

Krogmann, K., and Dodel, P. (1966b), *Chem. Ber.* **99**, 3408.

Krogmann, K. (1968), *Z. Anorg. Allg. Chem.* **358**, 97.

Krogmann, K. (1969), *Angew. Chem. Int. Ed. Engl.* **8**, 35.

Krogmann, K., and Hausen, H. D. (1968), *Z. Anorg. Chem.* **358**, 167.

Krogmann, K., and Stephan, D. (1968), *Z. Anorg. Chem.* **362**, 290.

Lee, P. A., Rice, T. M., and Anderson, P. N. (1973), *Phys. Rev. Lett.* **31**, 462.

Lee, P. A., Rice, T. M., and Anderson, P. N. (1974), *Solid State Commun.* **14**, 703.

Levy, L. A. (1912), *J. Chem. Soc.,* p. 1081.

Little, W. A. (1983), *J. Phys. (Orsay, Fr.)* **44**, 819.

Lynn, J. W., Iizumi, M., Shirane, G., Werner, S. A., and Saillant, R. B. (1975), *Phys. Rev. B.* **12**, 1154.

Maffly, R. L., Johnson, P. L., and Williams, J. M. (1977), *Acta Crystallogra. Sect. B.* **33**, 884.

Marsh, D. G., and Miller, J. S. (1976), *Inorg. Chem.* **15**, 720.

Mason, W. R. (1972), *Coord. Chem. Rev.* **7**, 241.

Mason, W. R., and Gray, H. B. (1968), *J. Am. Chem. Soc.* **90**, 5721.

McClure, D. S. (1959), *Solid State Phys.* **9**, 399.

Miller, J. S. (1976), *Science* **194**, 189.

Miller, J. S. (1979), *in* "Inorganic Syntheses" (D. F. Shriver, ed.), Vol. 19, pp. 13–18, Wiley, New York.

Miller, J. S., and Epstein, A. (1976), *Prog. Inorg. Chem.* **20**, 93.

Miller, J. S., and Epstein, A. (1979), *J. Coord. Chem.* **8**, 191.

Minot, M. J., and Perlstein, J. H. (1971), *Phys. Rev. Lett.* **26**, 371.

Moncuit, C., and Poulet, H. (1962), *J. Phys. Radium* **23**, 353.

Moreau-Colin, M. L. (1965), *Bull. Soc. R. Sci. Leige* **34**, 778.

Moreau-Colin, M. L. (1972), *Struct. Bonding (Berlin)* **10**, 167.

Musselman, R., and Williams, J. M. (1977), *J. Chem. Soc. Chem. Commun.* p. 186.

Needham, G. F., Johnson, P. L., Cornish, T. L., and Williams, J. M. (1977), *Acta. Crystallogr. Sect. B.* **33**, 887.

O'Neil, J. H., Underhill, A. E., and Toombs, G. A. (1979), *Solid State Commun.* **29**, 557.

Osso, R., and Rund, R. V. (1978), *J. Coord. Chem.* **8**, 169.

Otto, H. H., Schulz, H., Thiemann, K. H., Yersin, H., and Gliemann, G. Z. (1977), *Z. Naturforsch.* **32b**, 127.

Pauling, L. (1947), *J. Am. Chem. Soc.* **69**, 542.

Pauling, L. (1960), "The Nature of the Chemical Bond and Structure of Molecules and Crystals," pp. 398–404, Cornell Univ. Press, Ithaca, New York.

Peierls, R. E. (1955), "Quantum Theory of Solids," p. 108, Oxford Univ. Press, London and New York.

Perez-Albuerne, E. A., Isett, L. C., and Haller, R. K. (1977), *J. Chem. Soc. Chem. Commun.* p. 417.

Phillips, T. E., and Hoffman, B. M. (1977), *J. Am. Chem. Soc.* **99**, 7734.

Phillips, T. E., Scaringe, R. P., Hoffman, B. M., and Ibers, J. A. (1980), *J. Am. Chem. Soc.* **102**, 3435.

Piepho, S. B., Shatz, P. N., and McCaffery, A. J. (1969), *J. Am. Chem. Soc.* **91**, 5994.

Reis, A. H., Peterson, S. W., and Lin, S. C. (1976), *J. Am. Chem. Soc.* **98**, 739.

Reis, A. H., Peterson, S. W., and Lin. S. C., personal communication.

Renker, V., and Comés, R. (1975) *in* "Low-Dimensional Cooperative Phenomena," (H. J. Keller, ed.), p. 235, Plenum, New York.

Renker, B., and Comés, R., unpublished results privately communicated to Käfer, K. (1979), *in* "Quasi One-Dimensional Conductors II," Lecture Notes in Physics (H. G. Schuster, ed.), Vol. 96, Springer-Verlag, Berlin.

Renker, B., Pintschovious, L., Gläser, W., Reitschel, H., Comés, R., Liebert, L., and Drexel, W. (1974), *Phys. Rev. Lett.* **32**, 836.

Renker, B., Pintschovious, L., Gläser, W., Reitschel, H., and Comés, R. (1975), *in* "Quasi-One-Dimensinal Conductors" (H. G. Schuster, ed.), Lecture Notes in Physics, Vol. 34, p. 53, Springer-Verlag, Berlin.

Rice, M. J., and Strässler, S. (1973), *Solid State Commun.* **13**, 125.

Rosseinsky, D. R., and Malpas, R. E. (1979), *J. Chem. Soc., Dalton Trans.*, p. 740.

Saillant, R. B., and Jaklevic, R. C. (1974), *in* "Extended Interactions between Metal Ions in Transition Metal Complexes" (L. V. Interrante, ed.), p. 376, ACS Symp. Ser. No. 5, Am. Chem. Soc., Washington, D. C.

Scalapino, D. J., Sears, M., and Ferrell, R. A. (1972), *Phys. Rev. B.* **6**, 3409.

Scalapino, D. J., Imry, Y., and Pincus, P. (1975), *Phys. Rev. B.* **11**, 2042.

Schindler, J. W., Fukuda, R., and Adamson, A. W. (1979), *in* "Abstracts on Papers," Am Chem. Soc. Chem. Soc. Jpn. Joint Meet. Honolulu, Hawaii, INOR 447, Am. Chem. Soc., Washington, D. C.

Schultz, A. J., Stucky, G. D., Williams, J. M., Koch, T. R., and Maffly, R. L. (1977a), *Solid State Commun.* **21**, p. 197.

Schultz, A. J., Coffee, C. C., Lee, G. C., and Williams, J. M. (1977b), *Inorg. Chem.* **16**, 2129.

Schultz, A. J., Gerrity, D. P., and Williams, J. M. (1978a), *Acta Crystallogr. Sect. B.* **34**, 1673.

Schultz, A. J., Underhill, A. E., and Williams, J. M. (1978b), *Inorg. Chem.* **17**, 1313.

Sham, L. J., and Patton, B. R. (1976), *Phys. Rev. Lett.* **36**, 733.

Söderbaum, H. G. (1886), *Bull. Soc. Chim. Fr.* **45**, 188.

Soos, Z. G., and Klein, D. J. (1975) *in* "Molecular Association" (R. Foster, ed.), Vol. 1, pp. 1–119, Academic Press, New York.

Souchay, A., and Lenfsen, E. (1858), *Liebigs Ann. Chem.* **105**, 256.

Steigmeier, E. F., Loudon, R., Harbeke, G., Anderset, T. H., and Schieber, G. (1975), *Solid State Commun.* **17**, 1447.

Steigmeier, E. F., Baeriswyl, D., Auderset, H., and Williams, J. M. (1979) *in* "Quasi One-Dimensional Conductors II" (S. Barisić *et al.*, eds.), p. 229, Springer-Verlag, Berlin.

Stock, M., and Yersin, H. (1976), *Chem. Phys. Lett.* **40**, 423.

Stock, M., and Yersin, H. (1978), *Solid State Commun.* **27**, 1305.

Terry, H. (1928), *J. Chem. Soc.* **202**, 61.

Thielemans, M., Deltour, R., Jérome, D., and Cooper, J. R. (1976), *Solid State Commun.* **19**, 21.

Thomas, T. W., Che-Hsiung, Hsu, Labes, M. M., Gomm, P. S., Underhill, A. E., and Watkins, D. M. (1972), *J. Chem. Soc. A.,* p. 2050.

Torrance, J. B. (1977) *in* "Chemistry and Physics on One-Dimensional Metals" (H. J. Keller, ed.), pp. 137–166. Plenum, New York.

Underhill, A. E. (1974), *in* "Low-Dimensional Cooperative Phenomena and the Possiblity of a High Temperature Superconductor," NATO Advanced Study Institute, Starnberg, Germany, Sept. 1974 (H. J. Keller, ed.), Plenum, New York.

Underhill, A. E., and Ahmad, M. M. (1981), *J. Chem. Soc. Chem. Commun.,* p. 67.

Underhill, A. E., Watkins, D. M., and Wood, D. J. (1976), *J. Chem. Soc. Chem. Commun.,* p. 805.

Underhill, A. E., Wood, D. J., and Carneiro, K. (1979–1980), *Synth. Met.* **1**, 395.

Underhill, A. E., Watkins, D. M., Williams, J. M., and Carneiro, K. (1982), *in* "Extended Linear Chain Compounds" (J. S. Miller, ed.), Vol. 1, pp. 119–156, Plenum, New York.

Underhill, A. E., Ahmad, M. M., Turner, D. J., Clemenson, P. I., Carneiro, K., Yueqiuan, S., and Mortensen, K. (1985), *Mol. Cryst. Liq. Cryst.* **120**, 369.

Washecheck, D. M., and Williams, J. M., unpublished work.

Watkins, D. M., Wood, D. J., Underhill, A. E., Lundegaard- Anderson, A., Rindorf, G., and Braude, A., unpublished work.

Williams, J. M. (1976), *Inorg Nucl. Chem. Lett.* **12**, 651.

Williams, J. M. (1979), *in* "Inorganic Synthesis" (D. F. Shriver, ed.), Vol. 19, pp. 1–13, Wiley, New York.

Williams, J. M. (1980), ibid, pp. 13–18.

Williams, J. M., and Schultz, A. J. (1978), *Ann., N. Y. Acad. Sci.* **313**, 509.

Williams, J. M., and Schultz, A. J. (1979a), *in* "Molecular Metals" (W. E. Hatfield, ed.), Vol. III, pp. 337, 368, Plenum, New York.

Williams, J. M., and Schultz, A. J. (1979b), *in* "Modulated Structures" (J. M. Cowley, J. B. Cohen, M. B. Salamon, and B. J. Wiensch, eds.), pp. 187–192, Am. Inst. Phys., New York.

Williams, J. M., and Schultz, A. J., unpublished work.

Williams, J. M., Keefer, K. D., Washecheck, D. M., and Enright, N. P. (1976), *Inorg. Chem.* **15**, 2446.

Williams, J. M., Johnson, P. L., Schultz, A. J., and Coffey, C. (1978), *Inorg. Chem.* **17**, 834.

Williams, J. M., Schultz, A. J., Underhill, A. E., and Carneiro, K. (1982), *in* "Extended Linear Chain Compounds" (J. S. Miller, ed.), Vol. I, pp. 73–156, Plenum, New York.

Wilm, T. (1888), *Chem. Ber.* **21**, 1436.

Wood, D. J., Underhill, A. E., and Williams, J. M. (1979a), *Solid State Commun.* **31**, 219.
Wood, D. J., Underhill, A. E., Schultz, A. J., and Williams, J. M. (1979b), *Solid State Commun.* **30**, 501.
Yersin, H. (1976), *Ber. Bunsenges. Phys. Chem.* **80**, 1237.
Yersin, H. (1978), *J. Chem. Phys.* **68**, 4707.
Yersin, H., and Gliemann, G. (1975a), *Ber. Bunsenges. Phys. Chem.* **79**, 1050.
Yersin, H., and Gliemann, G. (1975b), *Z. Naturforsch.* **306**, 183.
Yersin, H., and Gliemann, G. (1978), *Ann. N. Y. Acad. Sci.* **313**, 539.
Yersin, H., Gliemann, G., and Rössler, U. (1977), *Solid State Commun.* **21**, 915.
Yersin, H., Ammon, W. Y., Stock, M., and Gliemann, G. (1979), *J. Lumin.* **18** (19), 774.
Zeller, H. R., and Beck, A. S. (1974), *J. Phys. Chem. Solids* **35**, 771.
Zeller, H. R., and Brüesch, P. (1974), *Phys. Status Solidi* **65** (6), 537.

Review Articles

Brüesch, P. (1975), Optical properties of the one-dimensional Pt-complex compounds, *in* "One-Dimensional Conductors" (H. G. Schuster, ed.), Lecture Notes on Physics No. 34, pp. 194–243, Springer-Verlag, Berlin.
Ferraro, J. R. (1982), Structural conductivity and Spectroscopic aspects of one-dimensional transition metal complexes, *Coord. Chem. Rev.* **43**, 205–232.
Krogmann, K. (1975), Chemical aspects of planar complex one-dimensional metals, *in* "One-Dimensional Conductors" (H. G. Schuster, ed.), Lecture Notes on Physics No. 34, pp. 12–19, Springer-Verlag, Berlin.
Miller, J. S., and Epstein, A. J. (1976), One-dimensional inorganic complexes, *in* "Progress in Inorganic Chemistry" (S. J. Lippard, ed.), pp. 1–151, Wiley, New York.
Underhill, A. E., Watkins, D. M., Williams, J. M., and Carneiro, K. (1982), Linear chain bis(oxalato)platinate salts, *in* "Extended Linear Chain Compounds" (J. S. Miller, ed.), Vol. 1, pp. 119–156, Plenum, New York.
Williams, J. M. (1983), One-dimensional inorganic platinum—chain electrical conductors, *in* "Advances in Inorganic Chemistry and Radiochemistry" (H. J. Emeléus and A. G. Sharpe, eds.), Vol. 26, pp. 235–268, Academic Press, New York.
Williams, J. M., and Schultz, A. J. (1979), One-dimensional partially oxidized tetracyanoplatinate metals: new results and summary, *in* "Molecular Metals" (W. E. Hatfield, ed.), pp. 337–368, Plenum, New York.

Books

Hatfield, W. E., ed. (1979), "Molecular Metals," pp. 1–555, Plenum, New York.
Miller, J. S., ed. (1982), "Extended Linear Chain Compounds," Vol. 1, pp. 1–481, Plenum, New York.
Miller, J. S., and Epstein, A. J., eds. (1978), "Synthesis and Properties of Low-Dimensional Materials," pp. 1–828, N. Y. Acad. Sci., New York.

5 TRANSITION ELEMENT— MACROCYCLIC LIGAND COMPLEXES

I. INTRODUCTION

Since the mid 1970s, interest in another class of conductive molecular solids containing stacks of metallomacrocyclic M(mac) molecules (M for transition metal and mac for macrocyclic ligands) has developed. These materials become conductive when oxidized (by I_2 and Br_2), and produce lattices which contain one-dimensional arrays of planar donor molecules M(mac) with fractionally occupied electronic-valence shells. In effect a partial oxidation state of the metal cation radical ensues and chains of polyhalide ions parallel to the M(mac) stacks occur in the structure. The types of macrocyclic (mac) molecules generally encountered are shown in Fig. 1. A variety of transition metals are found to take part in these complexes including Fe(II), Co(II), Ni(II), Cu(II), Zn(II), and Pt(II). Notable about this series is the occurrence of first row transition metals not having a d^8 electronic configuration. For comparison, the POTCP complexes discussed in Chapter 4, only Pt (d^8) transition metal formed these types of complexes.

The general oxidation reaction in these complexes can be considered to follow the equation:

$$M(mac) + 1/_2I_2 \rightarrow M(mac)I \qquad (1)$$

If the complex contained the iodine as I^- it would certainly necessitate an oxidation of the divalent metal to a trivalent state. Data from resonance Raman, Mössbauer (iodine), x-ray diffuse scattering, electron spin resonance, magnetic susceptibility, and charge transport measurements indi-

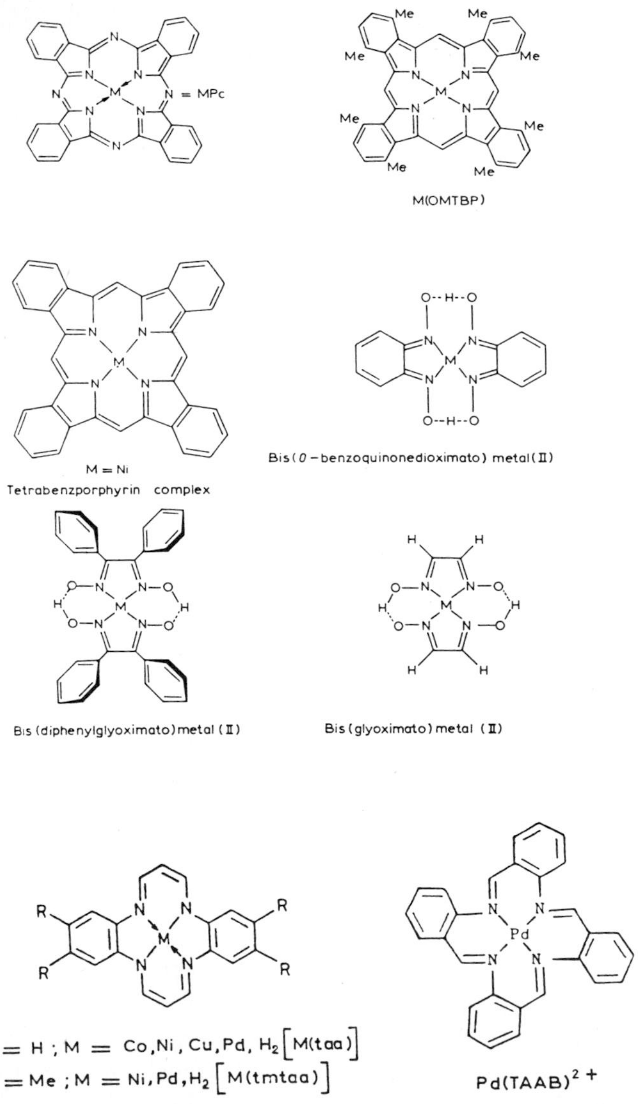

M(OMTBP)

M = Ni
Tetrabenzporphyrin complex

Bis(0-benzoquinonedioximato) metal(II)

Bis (diphenylglyoximato)metal (II)

Bis (glyoximato) metal (II)

(1) R $=$ H ; M $=$ Co,Ni, Cu,Pd, H₂$\left[M(taa)\right]$

(2) R $=$ Me ; M $=$ Ni,Pd,H₂$\left[M(tmtaa)\right]$

Pd(TAAB)$^{2+}$

Fig. 1. Structures (schematic) of various organometallic complexes. [From Ferraro (1982).]

cates that the iodine- or bromine-containing moiety is present as a polyhalide specie (e.g., I_3^-, I_5^-, Br_3^-, and Br_5^-). As a consequence, in these materials the metal retains its formal M^{2+} oxidation state and plays a secondary role in the conduction process, while the charge carriers are associated with delocalized π orbitals on the macrocycle. Electrons are thus removed from the ligand π system and not from the metal. In the complex M(mac)I where iodine is present as I_3^-, the M(mac) cation can be considered to have a formal charge of 2.33^+ with the metal oxidation state remaining divalent. The formulation can be written as $[M(mac)]^{+0.33}(I_3^-)_{0.33}$. If the iodine is present as the I_5^- anion the formulation may be $[M(mac)]^{+0.2}(I_5^-)_{0.2}$. Single crystal electron spin resonance studies of $[Ni(mac)]^{+0.33}(I_3^-)_{0.33}$ reveal that the iodine oxidation occurs at the ligand yielding π radical cations. The additional charge on the M(mac) cation (fraction greater than $2+$) is the partial oxidation of the cation or the degree of charge transfer in the complex and corresponds to the concept formulated by Torrance in Chapter 2 for the charge transfer metals therein described (see Torrance, 1977, 1979, 1985).

The high capability of iodine and halogens, in general, to accept electronic charge from donor molecules is related to their strong oxidizing power, i.e., $F_2 > Cl_2 > Br_2 > I_2$. The less powerful oxidizing halogens, iodine and bromine, are of greater value for the reaction illustrated in Eq. (1) due to the ease with which they form polyhalide chains. The formation of polyhalide anions in these compounds can be depicted as

$$X_2 + 2\bar{e} \rightleftarrows 2X^- \tag{2}$$

$$X^- + X_2 \rightleftarrows X_3^-. \tag{3}$$

As previously mentioned, nonintegral oxidation states are obtained for the M(mac) cation as was found for the POTCP complexes (Chapter 4) and appears to be a requirement for these materials to exhibit metallic conductivity.

II. DIPHENYLGLYOXIMATES

Foust and Soderberg (1967) showed that halogen-oxidized $M(dpg)_2X$ compounds, where dpg is for diphenylglyoxime, form tetragonal crystal structures in which the square coplanar $M(dpg)_2$ units form columnar stacks with equivalent M–M distances. The halogen atoms were situated in channels running parallel to the metal chains and were surrounded by the phenyl groups of the ligand. Keller and Siebold (1971) found that the compound formulated as $Pd(dpg)_2I$ contained I_3^- anions indicating an average nonintegral oxidation state for the $Pd(dpg)_2$ cation of 0.33^+. As we shall see later the diphenylglyoximates exhibit increases in conductivity of about six orders of magnitude when the starting material $M(dpg)_2$ is oxidized by I_2 or Br_2, but, at best, these systems are semiconductors.

A. Synthesis

The slow cooling of hot o-dichlorobenzene solutions of $Pd(dpg)_2$ with iodine causes lustrous needlelike crystals to form, which are black, diamagnetic, and which have the formula of $Pd(dpg)_2I$ (Endres et al., 1975).

Similar synthetic procedures with $Ni(dpg)_2$ have been used, and lustrous golden-olive crystals have been obtained. Formulas conforming to $Pd(dpg)_2I$ and $Ni(dpg)_2I$ were obtained (Edelmann, 1950; Underhill *et al.*, 1973; Miller and Griffiths, 1977; Mehne and Wayland, 1975; Keller and Siebold, 1971; Foust and Soderberg, 1967). With bromine, lustrous red–brown needles of the brominated $M(dpg)_2$ were obtained (Kalina *et al.*, 1980; Marks and Kalina, 1982). Elemental analyses corresponded to the formulas $Ni(dpg)_2Br_{1.0}$ and $Pd(dpg)_2Br_{1.1}$.

B. Structural Data

Table I tabulates the structural data for partially oxidized $M(dpg)_2$ complexes. The iodides and bromides of $Ni(dpg)_2$ and $Pd(dpg)_2$ are iso-structural (tetragonal space group $P4/ncc$). The crystal structure of $Ni(dpg)_2I$, which typifies this group of materials, appears in Fig. 2. Stacks of $Ni(dpg)_2$ units are staggered by 90°, and parallel chains of iodine atoms occur along the crystallographic c axis. Slight distortions from planarity are observed. A decrease in d_{M-M} stacking distance ($c/2$) is observed in the tetragonal $M(dpg)_2X$ materials from d_{M-M} (3.223 Å) to the d_{M-M} (3.547 Å) in the nonoxidized $M(dpg)_2$ complexes (Cowie *et al.*, 1979; Gleizes *et al.*, 1975; Mehne and Wayland, 1975). It should be noted that the form of the iodine specie in $Ni(dpg)_2I$ could not be determined from the x-ray scattering data due to the iodine disorder which occurs along the c axis (Cowie *et al.*, 1979; Gleizes *et al.*, 1975). The diffuse scattering pattern was compatible with I_5^- (Cowie *et al.*, 1979) and was later confirmed by resonance Raman and ^{129}I Mössbauer data (Cowie *et al.*, 1979). Figure 3 shows the Raman data for $Ni(dpg)_2I$ and $Pd(dpg)_2I$ in which an intense band is observed at 161 cm^{-1} and a weak band at 107 cm^{-1} along with overtones and combinations due to the resonance effect. The spectra are typical of I_5^- moieties (Marks and Kalina, 1982). Figure 4 shows the ^{129}I

TABLE I

Structural Data for Partially Oxidized $M(dpg)_2$ Complexes[a]

Compound	Crystal class	Space group	Unit cell (Å)
$Ni(dpg)_2$	Orthorhombic	—	—
$Ni(dgp)_2I$	Tetragonal	$P4/ncc$	$a = 19.887(4)$, $c = 6.542(2)$
$Ni(dpg)_2Br$	Tetragonal	$P4/ncc$	$a = 19.51(6)$, $c = 6.72(2)$
$Pd(dpg)_2$	Orthorhombic	—	—
$Pd(dpg)_2I$	Tetragonal	$P4/ncc$	$a = 20.17(6)$, $c = 6.52(2)$
$Pd(dpg)_2Br_x$	Tetragonal	$P4/ncc$	$a = 19.78(6)$, $c = 6.54(2)$

[a] From Marks and Kalina (1982).

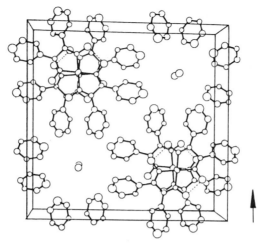

Fig. 2. Perspective view down the c axis of the unit cell of Ni(dpg)$_2$I or [Ni(dpg)$_2$](I$_5$)$_{0.2}$. Dotted lines show the diphenylglyoximate O—H—O hydrogen bonds. [From Ferraro (1982).]

Mössbauer spectra Ni(dpg)$_2$I and Pd(dpg)$_2$I and compares it to that for (trimesic acid·H$_2$O)$_{10}$·HI$_5$. The latter compounds contain I$_5^-$ moieties, and the similarities with the dpg compounds are obvious. Similar methods of investigation were used for the study of M(dpg)$_2$Br$_x$ materials and it was concluded that Br$_5^-$ species also occurred in these compounds (Kalina *et al.*, 1980; Marks and Kalina, 1982). The M(dpg)$_2$I materials can then be formulated as [M(dpg)$_2^{+0.2}$](I$_5^-$)$_{0.2}$. The M(dpg)$_2$Br compounds correspond to [Ni(dpg)$_2^{+0.2}$](Br$_5^-$)$_{0.2}$ and [Pd(dpg)$_2^{+0.22}$](Br$_5^-$)$_{0.22}$.

C. Electrical Conductivities

The formulations shown in the previous section indicate that these materials bear a resemblance to the POTCP conductors discussed in Chapter 4. Both classes of materials exhibit partial oxidation states and incomplete charge transfer, an important requisite for high conductivity in many quasi-one-dimensional materials. This is illustrated by the rise in conductivities as one oxidizes the M(dpg)$_2$ units (see Table II). It should be emphasized that whereas the POTCP salts indicate that the residual fractional charge resides on the platinum atom, in oxidized M(dpg)$_2$ compounds the residual fraction charge resides on the M(dpg)$_2$ units. Figures 5 and 6 show the electrical conductivity of M(dpg)$_2$I compounds as a function of temperature, and a linear dependence of ln σ versus $1/T$ was observed.

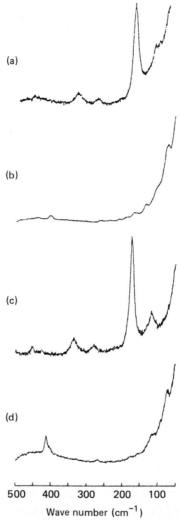

Fig. 3.   Raman spectra (5145 Å excitation, polycrystalline samples) of (a) Pd(dpg)$_2$I; (b) Pd(dpg)$_2$; (c) Ni(dpg)$_2$I; and (d) Ni(dgp)$_2$. [From Marks and Kalina (1982).]

D.   Pressure Effects

Onodera *et al.* (1979) studied platinum(II) glyoximates under applied pressure. The electrical resistivity of polycrystalline Pt(dmg)$_2$ at ambient pressure is greater than 10^5 ohm cm. The resistivity decreases significantly up to a pressure of 40 kbar, at which point a minimum occurs. The resistivity at the minimum is 1×10^{-1} ohm cm. The behavior was found to be reversible with some hysteresis.

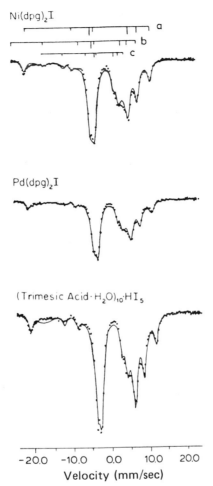

Fig. 4. ^{129}I Mössbauer spectra of several (dpg) complexes at 4 K. The solid lines represent the best computer fit to data points. Stick figures representing contributing transitions are shown for the Ni(dpg)$_2$I spectrum. ZnTe is used for the reference of isomer shifts. [From Marks and Kalina (1982).]

III.   Bis(BENZOQUINONE)DIOXIMATES

The bis(benzoquinone)dioximates, where benzoquinonedioxine (bqd) are closely related to the diphenylglyoximates (Brown *et al.*, 1979) (see Fig. 1). Early work was done by Endres *et al.* (1974, 1975), who established the existence of a quasi-one-dimensional material of stoichiometry M(bqd)$_2$I$_{0.5}$ (M = Ni or Pd). In some cases the compounds incorporate solvent molecules.

TABLE II

Single-Crystal (*c*-Axis) Electrical Conductivities for Bis(diphenyl)glyoximates and Bis(benzoquinone)dioximates[a]

Compound	dc Conductivity at 300 K (ohm cm)$^{-1}$
$Ni(bqd)_2$	$<9 \times 10^{-9}$
$Ni(bqd)_2I_{0.52} \cdot S^b$	$(1.8-11) \times 10^{-6}$
$Pd(bqd)_2I_{0.50} \cdot S^c$	$(7.8-810) \times 10^{-5}$
$Ni(dpg)_2$	$<8 \times 10^{-9}$
$Ni(dpg)_2I$	$(2.3-11) \times 10^{-3}$
$Ni(dpg)Br_{1.0}$	1.8×10^{-3}
$Pd(dpg)_2$	$<8 \times 10^{-9}$
$Pd(dpg)_2I$	$(7.7-47) \times 10^{-5}$
$Pd(dpg)_2Br_{1.1}$	$(8.9-13) \times 10^{-5}$

[a] Taken in part from Brown *et al.* (1979), Kalina *et al.* (1980), Marks and Kalina (1982), and Underhill *et al.* (1973).
[b] S is 0.32 toluene.
[c] S is 0.52 *o*-C$_6$H$_4$Cl$_2$.

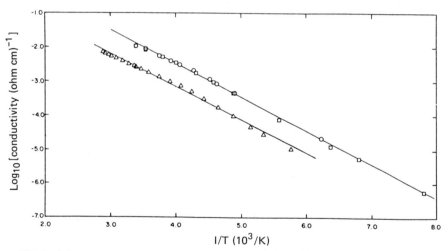

Fig. 5. Electrical conductivity (dc) in the crystallographic *c* direction of representative Ni(dpg)$_2$I crystals as a function of temperature. [From Brown *et al.*, 1979).]

A. Synthesis

Syntheses of the bis(benzoquinone)dioximates were accomplished by methods similar to those given in the discussion for the dpg complexes. Upon oxidation with iodine, golden–black, needlelike crystals of the iodine complexes were isolated from an aromatic solvent (*o*-C$_6$H$_4$Cl$_2$).

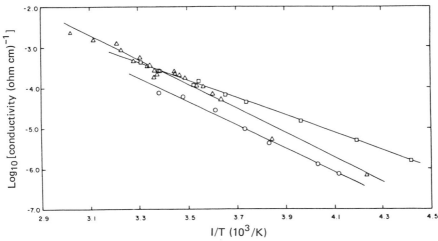

Fig. 6.   Electrical conductivity (dc) in the crystallographic c direction of representative Pd(dpg)$_2$I crystals as a function of temperature. [From Brown *et al.* (1979).]

B.   Structural Data

The compound Pd(bqd)$_2$I$_{0.5}$·0.52 (o-C$_6$H$_4$Cl$_2$) is found to belong to the tetragonal space group D_{4h}^2-$P4/mcc$ with $Z = 4$ and unit cell dimensions $a = 16.048(7)$ and $c = 6.367(3)$ Å (Fig. 7). The crystal structures of metal bis(benzoquinone)dioximates are illustrated in Fig. 8, and consist of stacked Pd(bqd)$_2$ units, each staggered by 65° with respect to its neighbors. Disordered chains of iodine atoms extend in the c direction. The solvent molecules are disordered throughout tunnels which also extend in a direction parallel to c. The Pd–Pd distance is 3.184(3) Å (Brown *et al.*, 1979). The Pd(bqd)$_2$ units are planar. Resonance Raman studies reveal that the iodine is present as I$_3^-$ with a strong band at 107 cm^{-1} (Fig. 9). The nickel compound Ni(bqd)$_2$I$_{0.018}$ crystallizes having the D_{2h}^{26}–Ibam space group with $Z = 4$ with unit cell dimensions of $a = 16.438(2)$, $b = 14.759(4)$, and $c = 6.360(2)$ Å. The planar Ni(bqd)$_2$ units are stacked along the c axis and each molecule is staggered by 68° with respect to its nearest neighbors. The Ni–Ni distance is 3.180(2) Å (Brown *et al.*, 1979) and these distances are shorter than they are in the parent compound Ni(bqd)$_2$. Resonance Raman data has shown the existence of the I$_3^-$ moiety in the Ni compound as well (Brown *et al.*, 1979; Marks, 1976). The compounds can be formulated as [M(bqd)$_2^{+0.17}$](I$_3^-$)$_{0.17}$$n$S similar to the M(dpg)$_2$I materials with a partial oxidation state of +0.17, where S is solvent and n is moles of solvent.

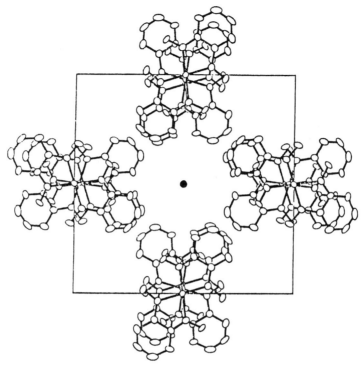

Fig. 7.   View of the unit cell of Pd(bqd)$_2$I$_{0.5}$ · 0.52(o-dichlorobenzene) along the stacking direction. The a and b axes are in the plane of the page and the c axis is toward the reader. The vibrational ellipsoids are drawn at the 50% level; the dark circle is an iodine atom. [From Marks and Kalina (1982).]

C. Electrical Conductivities

Table II lists the electrical conductivities of several bqd compounds and makes comparison with M(dpg)$_2$I compounds. It may be observed that the bqd compounds are poorer conductors than the dpg compounds, despite the shorter d_{M-M} distances in the former compounds. The conductivity behavior versus temperature is typically metallic and shows a linear dependence of ln σ versus $1/T$. See Figs. 10 and 11.

IV.   OXDIZED DIHYDRODIBENZO[b,i][1,4,8,11]TETRA-AZACYCLOTETRADECINE COMPLEXES

The metallic-like properties of doped metallomacrocycles, when the macrocycles were conjugated, has generated considerable interest in other conjugated macrocyclic π-electron systems. Another new class of com-

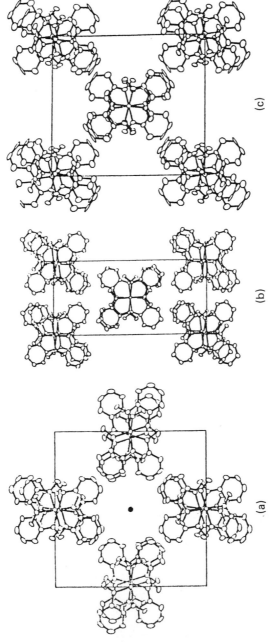

(a)

(b)

(c)

Fig. 8. Comparison of metal bis(benzoquinone)dioximate crystal structures viewed along the stacking direction. (a) Structure for $Ni(bqd)_2I_{0.5}$ or $Pd(bqd)_2I_{0.5}$. (b) Structure for $Pd(bqd)_2$. (c) Structure for $Ni(bqd)_2I_{0.018}$. [From Marks and Kalina (1982).]

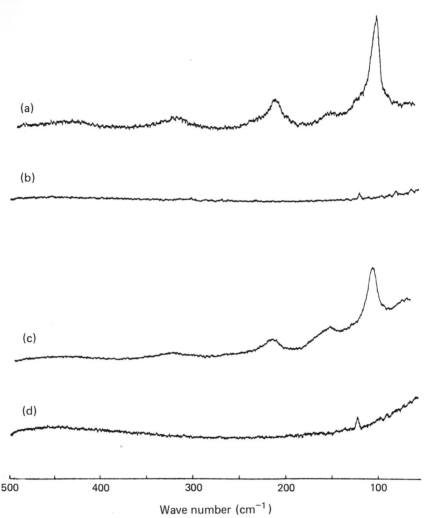

Fig. 9. Resonance Raman spectra (5145 Å excitation, polycrystalline samples) of (a) Pd(bqd)$_2$I$_{0.5}$·0.52 (*o*-dichlorobenzene); (b) Pd(bqd)$_2$; (c) Ni(bqd)$_2$I$_{0.52}$·0.32 (toluene); and (d) Ni(bqd)$_2$. Weak transitions in (b) and (d) at 117 and 77 cm^{-1} result from laser plasma emission. [From Marks and Kalina (1982).]

pounds consists of metal complexes of dihydrodibenzo[*b*,*i*][1,4,8,11]tetra-azacyclotetradecine-M(TAA) (Lin *et al.*, 1980). Methyl group substitution in TAA produces TMTAA (see Fig. 1).

Redox properties of M(TAA) and M(TMTAA) are similar to those encountered in the metallophthalocyanines (see Section V) (Dabrowiak *et al.*, 1979). Table III tabulates electrical conductivities and the activation

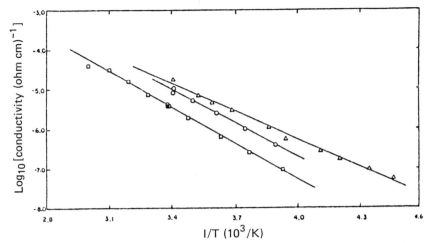

Fig. 10.   Electrical conductivity (dc) of representative Ni(bqd)$_2$I$_{0.52}$·0.32 (toluene) crystals as a function of temperature. Data are measured in the crystallographic *c* direction. [From Brown *et al.* (1979).]

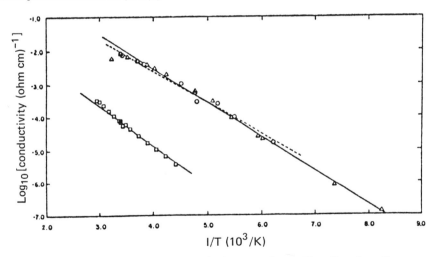

Fig. 11.   Electrical conductivity (dc) in the crystallographic *c* direction of representative Pd(bqd)$_2$I$_{0.50}$·0.52 (*o*-dichlorobenzene) crystals as a function of temperature. [From Brown *et al.* (1979).]

energy E_a for these complexes. Resonance Raman spectroscopy showed that the polyiodide species is present in (MTAA)I$_x$ and (MTMTTA)I$_x$ for $x = {\leq}3$ as I$_3^-$ (see Fig. 12) and for $x \gtrsim 3$ as I$_5^-$ (McClure *et al.*, 1980; Lin *et al.*, 1980). This implies that partial oxidation has occurred and the complexes can be formulated as Ni(TAA)$^{+0.6}$(I$_3^-$)$_{0.6}$ and

TABLE III

Electrical Conductivities for (MTAA)I_x and (MTMTAA)I_x Materials[a]

Compound	Room temperature conductivity[b] (ohm cm)$^{-1}$	E_a (eV)
NiTAA[c]	1.8×10^{-15}	2.25
(NiTAA)$I_{0.8}$	2.1×10^{-1}	0.093
(NiTAA)$I_{1.0}$	4.5×10^{-1}	0.051
(NiTAA)$I_{2.6}$	8.1×10^{-2}	
(NiTAA)$I_{7.0}$	2.3×10^{-2}	
NiTMTAA[c]	4.0×10^{-14}	1.36
(NiTMTAA)$I_{1.7}$	1.4×10^{-2}	
(NiTMTAA)$I_{2.9}$	3.8×10^{-2}	0.093
CoTAA[c]	7.9×10^{-13}	1.33
(CoTAA)$I_{1.9}$	1.1×10^{-2}	
CuTAA	1.8×10^{-10}	0.89
(CuTAA)$I_{1.8}$	1.1×10^{-3}	
(PdTAA)$I_{0.4}$	2.0×10^{-2}	
(PdTAA)$I_{0.8}$	1.3×10^{-1}	0.080
(PdTMTAA)$I_{0.4}$	9.3×10^{-4}	0.154
(H$_2$TAA)$I_{1.4}$	2.4×10^{-5}	
(H$_2$TMTAA)$I_{1.6}$	3.8×10^{-6}	
(NiTMTAA)$I_{2.7}$	$<10^{-7}$	
(NiTMTAA)$I_{4.9}$	2.9×10^{-6}	
(NiPc)$I_{1.0}$[d]	7.0×10^{-1}	0.036

[a] Taken from Lin *et al.* (1980) and Marks and Kalina (1982).
[b] Pressed-pellet samples at room temperature.
[c] Müller and Wöhrle (1975).
[d] Schramm *et al.* (1978).

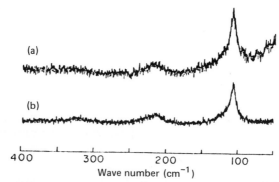

Fig. 12.   Resonance Raman spectra ($\nu_0 = 5145$ Å) for crystalline samples of (a) (NiTAA)$I_{1.80}$ and (b) (NiTMTAA)$I_{2.44}$. [From Marks and Kalina (1982).]

$Ni(TMTAA)^{+0.8}(I_3^-)_{0.8}$ for $(NiTAA)I_{1.8}$ and $(NiTMTAA)I_{2.44}$, respectively. The $M(TAA)I_x$ complexes appear to be more conductive (see Table III) than the $M(TMTAA)I_x$ complexes and this may be related to the buckling of the TMTAA ligand which decreases delocalization and prevents close stacking (Tsutsui *et al.*, 1979; Goedken *et al.*, 1976). Replacement of metal with H_2 in both types of complexes causes large losses in conductivity, a behavior different than that found for the metallophthalocyanines. Figure 13 shows the conductivities of various metal complexes of TAA and TMTAA as a function of temperature.

V. PHTHALOCYANINES

The success of producing M(mac) complexes, which when partially oxidized became electrical conductors, created interest in synthesizing additional highly conjugated, but more planar π-electron systems. This research led to the metallophthalocyanines. These metal complexes have accessible redox states and demonstrate increased planar flexibility over the previously discussed M(mac) conductors. Oxidation of metallophthalocyanines M(Pc) or the metal-free macrocycle (H_2Pc) with iodine resulted in the formation of a new series of molecular conductors (Peterson *et al.*, 1977; Marks, 1976; Schramm *et al.*, 1978; Marks and Kalina, 1982; Marks, 1985). It was observed that the ratios of $I/(M(Pc))$ could be varied, and were dependent of the experimental conditions. The oxidation reaction can be depicted as

$$M(Pc) + (x/2)I_2 \rightleftarrows M(Pc)I_x, \qquad M = Fe, Co, Ni, Cu, Zn, Pt, H_2$$

The iodination reaction is reversible and heating the materials in vacuum results in loss of iodine. Resonance Raman spectra (Fig. 14) and Mössbauer studies indicate that these substances are partially oxidized.

A. Synthesis

Finely powdered Ni(Pc) was stirred in a closed vial with various amounts of iodine in chlorobenzene for 48 hours. The iodinated Ni(Pc) was collected by centrifugation, washed with hexane, and dried with dry nitrogen. This method produced polycrystalline material. Single crystals of $Ni(Pc)I_{1.0}$ were grown in an H-tube by diffusing together solutions of iodine and Ni(Pc) in 1,2,4-trichlorobenzene or 1-chloronaphthalene (Schramm *et al.*, 1980).

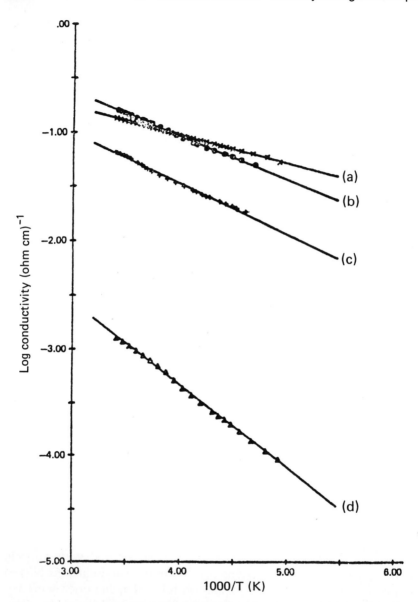

Fig. 13. Four-probe, variable-temperature electrical conductivity data for compressed polycrystalline samples of the partially oxidized TAA and TMTAA complexes. (a) $(NiTAA)I_{1.0}$, (b) $(PdTAA)I_{0.8}$, (c) $(NiTMTAA)I_{2.9}$, and (d) $(PdTMTAA)I_{0.4}$. [From Marks and Kalina (1982).]

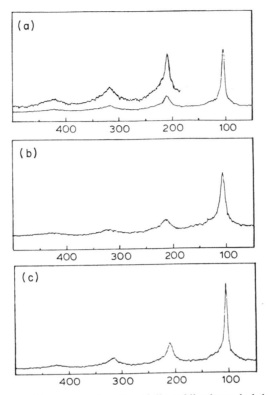

Fig. 14. Resonance Raman spectra of partially oxidized metal phthalocyanine materials with 5145 Å excitation; (a) $FePcI_{1.9}$, (b) $CoPcI_{1.0}$, and (c) $NiPcI_{1.0}$. [From Marks and Kalina (1982).]

B. Structural Data

Figure 15 shows the crystal structure of $Ni(Pc)I$ or $Ni(Pc)(I_3)_{0.33}$. The figure also shows the arrangement of the stack of donor molecules and the disordered polyiodide chains. The $Ni(Pc)$ units are staggered with respect to neighboring units by 39.5° and the interplanar distances are 3.244 Å. The iodine is present as I_3^- as determined by analysis of x-ray diffuse scattering data and resonance Raman measurements (see Fig. 14). Thus, $Ni(Pc)I$ may be represented as $(NiPc)^{+0.33}(I_3^-)_{0.33}$, where the positive charge resides on the cation. For the complexes $M(Pc)I_x$, I_3^- appears to be the chief polyiodide moiety where $X = \leq 3$ (Scaringe et al., 1980; Peterson et al., 1977; Marks, 1976). At higher levels of dopant, I_5^- and other coordinated iodine species have been detected.

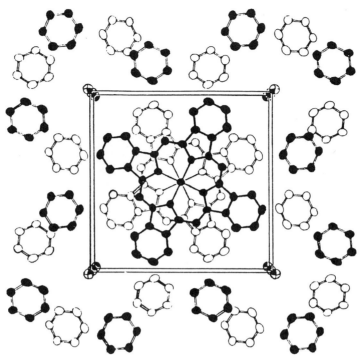

Fig. 15.   View down the c axis on the unit cell of $Ni(Pc)(I_3)_{0.33}$. [From Marks and Kalina (1982).]

C.   Electrical Conductivities

Table IV tabulates the electrical conductivities for the oxidized and unoxidized M(Pc) complexes. We are able to observe an increase in the conductivities over those observed for the dioximates and benzoquin-onedioximates. The temperature dependence of the conductivity is thermally activated (semiconduction) and is comparable to those of other known molecular metals of this type (Scaringe *et al.*, 1980; Peterson *et al.*, 1977; Heeger, 1979). From Table IV one observes that replacement of the metal in M(Pc) complexes with H_2 causes only slight conductivity changes. Figure 16 demonstrates the electrical conductivity of Ni(Pc)I in the c axis direction as a function of temperature. At 60 K, the conductivity goes through an abrupt transition to semiconducting behavior. Figure 17 illustrates variable temperature conductivities of several iodinated metallo-phthalocyanines. Table IV also shows that the activation energy decreases drastically upon oxidation ($\sim$50-fold). The data in Table IV can be fit to Eq. (4), in which E_a is the activation energy

$$\sigma = \sigma_\infty e^{-E_a/kT} \tag{4}$$

Values of E_a are obtained from a least-squares fit of the data.

TABLE IV

Conductivities for Various Phthalocyanine Complexes[a,b]

Material	x	Room temperature conductivity $(\text{ohm cm})^{-1}$	E_a (eV)
(NiPc)I_x	0.00[c]	1×10^{-11}	1.6
	0.56	0.7	0.024
	1.0	0.7	0.036
	1.7	0.8	0.021
(CoPc)I_x	0.00[c]	2×10^{-11}	1.6
	0.60	0.1	0.065
	1.0	0.06	0.082
(FePc)I_x	0.00[c]	2×10^{-10}	
	1.93	4×10^{-3}	0.127
	2.74	2×10^{-3}	0.070
	3.85	1×10^{-4}	0.254
(CuPc)I_x	1.71	4.2	0.021
(PtPc)I_x	0.93	2.4	0.016
(PcH$_2$)I_x	2.20	2.3	0.040

[a] Taken from Marks and Kalina (1982).

[b] Measurements made on pressed pellets by four-probe van der Pauw method. Single-crystal conductivity for (NiPc)$I_{1.0}$ is several orders of magnitude higher.

[c] From Gutmann, P., and Lyons, L. E. (1967), "Organic Semiconductors," p. 718, Wiley, New York.

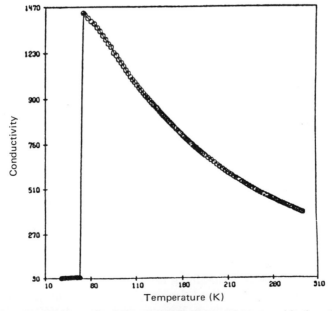

Fig. 16. Electrical conductivity of a Ni(Pc)I crystal measured in the c direction as a function of temperature. [From Marks and Kalina (1982).]

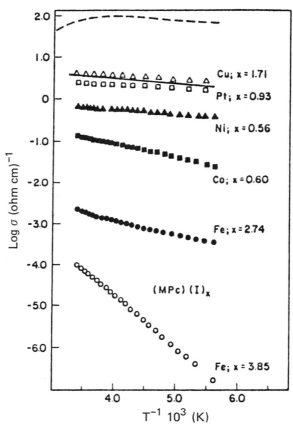

Fig. 17.   Variable-temperature conductivity data for iodinated phthalocyanines. The solid line is for compressed polycrystalline samples and the dashed line for single crystals of (quinolinium)(TCNQ)$_2$. [From Marks and Kalina (1982).]

D.   Phthalocyanine Derivatives

Numerous derivatives of phthalocyanines have also been investigated. Metal complexes have been prepared and oxidized, and conductive properties determined. Substitution in the 3 or 4 position of phthalocyanine with a methyl substitutent gives the (3-Me) and (4-Me) phthalocyanine derivatives, respectively (Stojakovic, 1977; Kalina *et al.*, unpublished data). The metal complexes are depicted in Fig. 18. When these materials are oxidized by iodine they exhibit metallic conductivities as illustrated in Table V. The [M(3-Me)$_4$Pc]I$_x$ compounds show the highest conductivities. This may be related to the fact that the methyl group extends less into the surrounding lattice, thus maintaining the requisites of close stacking and delocalization of electrons. The I$_5^-$ moiety is observed in these materials even at low iodine concentrations. When compared with the unsubstituted

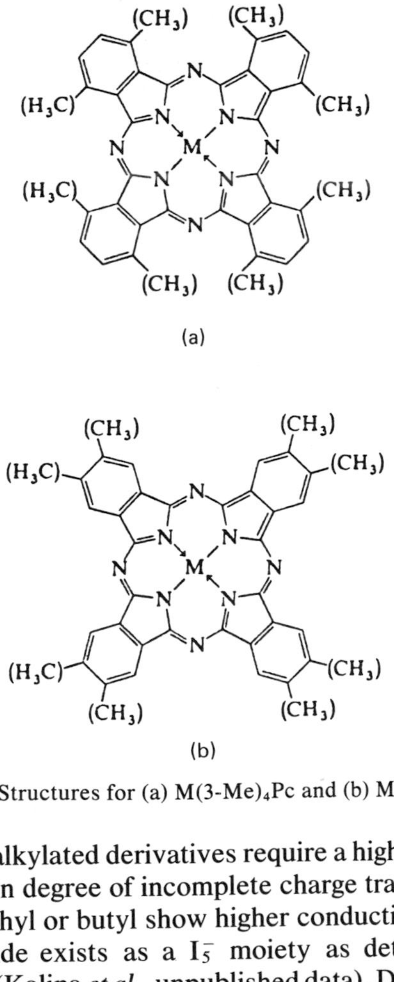

(a)

(b)

Fig. 18. Structures for (a) M(3-Me)$_4$Pc and (b) M(4-Me)$_4$Pc.

Pc complexes, the alkylated derivatives require a higher iodine concentration to reach a given degree of incomplete charge transfer. Complexes of 4,5-substituted methyl or butyl show higher conductivities (see Table V). Again the polyiodide exists as a I$_5^-$ moiety as determined by Raman scattering (Fig. 19) (Kalina *et al.*, unpublished data). Derivatives with fused benzo groups (MNc) are illustrated in Fig. 20 (Kalina *et al.*, 1980; Cuellar and Marks, 1981). These materials show electrical conductivities comparable to M(Pc)I$_x$. The polyiodide is in the form of I$_5^-$ for (MNc)I$_x$ materials for $x \leq 1$.

E. Pressure Effects

No pressure measurements on the partially oxidized metallophthalocyanines have been reported. However, Onodera *et al.* (1975) has examined several M(Pc) complexes under very high pressures (see Table VI). Vanadyl phthalocyanine showed the lowest resistivity at 290 kbar. Each M(Pc) compound exhibited a resistivity minimum.

TABLE V

Conductivities for Several Ring-Substituted Phthalocyanines[a]

Material	x	Room temperature conductivity $(ohm \ cm)^{-1}$
[Cu(4-Me)$_4$Pc]I$_x$	0.00	$<10^{-7}$
	1.24	8.3×10^{-4}
	4.15	5.4×10^{-3}
[Cu(3-Me)$_4$Pc]I$_x$	0.00	$<10^{-7}$
	1.22	3.0×10^{-2}
	4.85	2.0×10^{-1}
[Ni(4-Me)$_4$Pc]I$_x$	0.00	$<10^{-7}$
	1.24	1.4×10^{-4}
	1.91	1.1×10^{-3}
	2.55	1.0×10^{-3}
	2.86	1.6×10^{-3}
[Ni(4,5-Me$_2$)$_4$Pc]I$_x$	0.00	$<10^{-7}$
	0.52	2.3×10^{-3}
	1.06	1.4×10^{-2}
[H$_2$(4,5-Bu$_2$)$_4$Pc]I$_x$	0.00	$<10^{-7}$
	0.19	2.0×10^{-2}
(NiNc)I$_x$	0.00	$<10^{-7}$
	1.03	1.6×10^{-1}
	2.08	1.6×10^{-1}
(CuNc)I$_x$	0.00	$<10^{-7}$
	0.94	3.4×10^{-2}
	1.84	3.1×10^{-2}

[a] Four-probe van der Pauw measurements of pressed pellets. Taken from Marks and Kalina (1982).

VI. BENZOPHORPHYRINATES

In keeping with the apparent criteria that to be conductive, M(mac) materials require some degree of planarity of the metal and the organic ligand in addition to the ability to adopt a nonintegral oxidation state, the phorphyrin and substituted phorphyrin metal complexes were eventually synthesized. Porphyrins closely resemble the phthalocyanines in a molecular and electronic sense and thus appeared as promising materials for new molecular conductors.

These materials are prepared by reacting ML complexes with iodine [L = tetrabenzophorphyrin(TBP),1,4,5,8,9,12,13,16-octamethyltetrabenzoporphyrin (OMTBP)] and 1,2,3,4,5,6,7,8-octaethylphorphyrin (OEP) and, again, partially oxidized complexes result (Phillips and Hoffman,

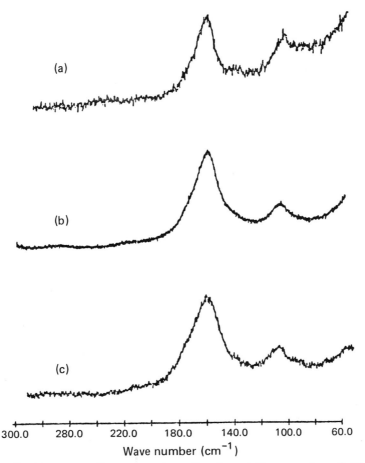

Fig. 19. Resonance Raman spectra of polycrystalline samples of (a) [Ni(4-Me)$_4$Pc]I$_{1.24}$; (b) [Cu(3-Me)$_4$Pc]I$_{1.56}$; and (c) [Cu(3-Me)$_4$Pc]I$_{3.41}$, $\nu_0 = 5145$ Å. [From Marks and Kalina (1982).]

1977; Phillips *et al.*, 1980; Martinsen *et al.*, 1982; Hoffman *et al.*, 1979). Figures 1 and 21 depict the structures of these metallomacrocycles.

Complexes with the stoichiometry of MLI$_x$ are generally formed. In the case of the NiOMTBP complex, two forms are obtained on iodination: (NiOMTBP)I$_{1.08}$ and (NiOMTBP)I$_{2.91}$.

Crystalline Ni(TBP)I$_{1.0}$ is found to have the D_{4h}^2-P4/mcc space group with $Z = 2$ in a unit cell of dimensions of $a = 14.081(25)$ Å and $c = 6.43(11)$ Å at 113 K. The crystal structure consists of planar Ni(TBP) units stacked metal over metal with a Ni–Ni spacing of 3.217(5) Å at 113 K and

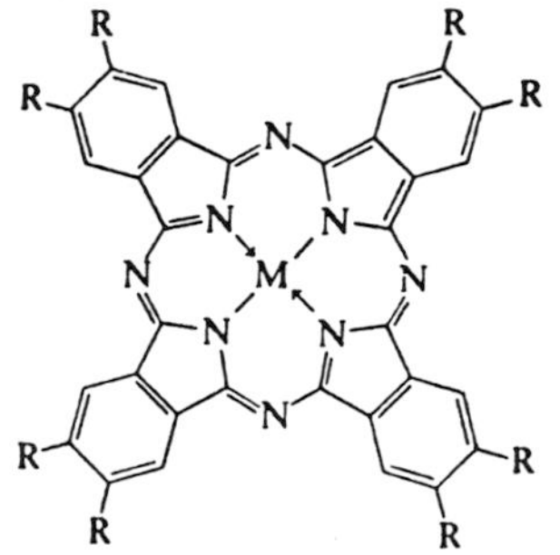

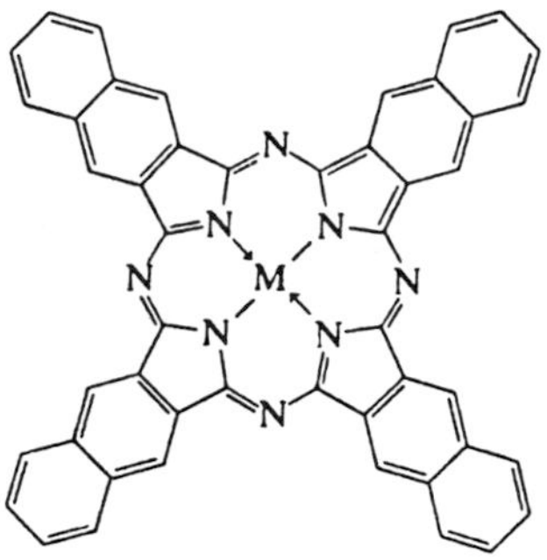

Fig. 20.   Structures of fused benzo-group substituted MNc complexes.

TABLE VI

Pressure Effects on the Conductivity of Several M(Pc) Complexes[a]

M(Pc)	Resistivity (ohm cm)	Pressure (kbar)
VOPc	1×10^{-1}	290
ThPc	2×10^{-1}	330
SnPc$_2$	2×10	360
PbPc	2	—
CuPc	4	620
ZnPc	1×10	640
H$_2$Pc	4×10	570
CuPcCl$_{16}$	8×10^{15}	520

[a] Taken from Onodera *et al.* (1975).

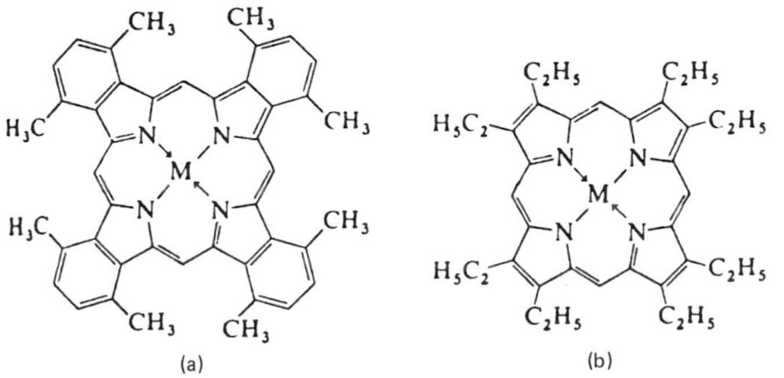

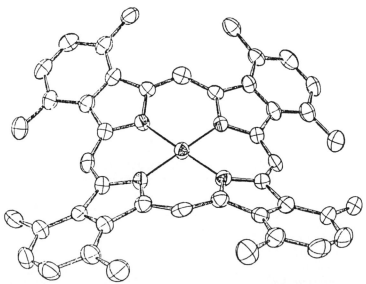

Fig. 21. Structures of (a) metal octamethyltetrabenzophorphyrin [M(OMTBP)] and (b) metal octaethylphorphyrin [M(OEP)].

with the units staggered by 41°. Chains of disordered I_3^- ions occur parallel to these stacks.

The crystal structure of $(NiOMTBP)I_{1.08}$ is illustrated in Fig. 22 (Phillips *et al.*, 1980). The packing is like that found for Ni(Pc)I. The complex crystallizes having the space group D_{4h}^{11}-$P4_2/nbc$ of the tetragonal system with 4 formula weights per unit cell. The structure consists of ruffled macrocycles stacked with relatively large intermolecular spacings along the *c* axis ($d_{Ni-Ni} = 3.778(5)$ Å); Ni(Pc)I having d_{Ni-Ni} distances of 3.244(2)

Fig. 22. Molecular structure of the metallomacrocycle in $(NiOMTBP)I_{1.08}$. [From Marks and Kalina (1982).]

Å. Chains of extensively disordered linear I_3^- units occur parallel to the stacking axis between adjacent macrocycle columns. Methyl substitution of tetrabenzophorphyrin (TBP) increases the d_{Ni-Ni} spacing from 3.22 to 3.77 Å. The difference in metal–metal spacing affects the conductivities of these materials. The TBP complex shows a room temperature electrical conductivity of ~350 (ohm cm)$^{-1}$, while the OMTBP complex has a conductivity of ~10 (ohm cm)$^{-1}$ (Hoffman et al., 1979) (see Table VII).

The Raman spectra shows that the polyiodides in the phorphyrinates are present as I_3^- moieties, which implies partial oxidation and that partial charge transfer has occurred. Electron spin resonance studies show that the partial oxidation is predominantly ligand centered, as found for previously discussed metallomacrocycles. The complex (NiOMTP)$I_{1.08}$ can be depicted as (NiOMTBP)$^{+0.36}(I_3^-)_{0.36}$, (NiTBP)$I_{1.0}$ as (NiTBP)$_{0.33}(I_3^-)_{0.33}$, and (NiOMTBP)$I_{2.91}$ as (NiOMTBP)$^{+0.97}(I_3^-)_{0.97}$.

Iodination of the MOEP complexes where M is Ni, Cu, or H_2 produces complexes of (MOEP)I_x (Wright et al., 1979/1980). Resonance Raman studies show the presence of I_5^- ions again suggesting that partial oxidation has occurred. The complexes (MOEP)I_x can be written as (MOEP)$^{+x/5}(I_5^-)_{x/5}$ and are tabulated in Table VII along with those of TBP and OMTBP complexes.

It may be observed that the conductivities follow a trend of TBP > OMTBP > OEP. Replacement of metal with H_2 causes a lowering of conductivity in the OEP complexes—results that are different from those found for the metallophthalocyanines. The phorphyrins demonstrate lower conductivities than (NiPc)I. Figure 23 illustrates the conductivity of (NiOMTBP)$I_{1.08}$ and (NiOMTBP)$I_{2.91}$ as a function of temperature. A short region of metal-like behavior is observed in both materials near room temperature but a transition to a less conductive state occurs as the temperature is lowered.

TABLE VII

Conductivities of Several Metal Phorphyrinates

Compound	Conductivity (ohm cm)$^{-1}$
(NiOEP)$I_{5.7}$	2.8×10^{-2}
(CuOEP)$I_{3.5}$	6.0×10^{-3}
(H$_2$OEP)$I_{5.8}$	3×10^{-7}
(NiTBP)$I_{1.0}$	3.30×10^2
(NiOMTBP)$I_{0.97}$	~10

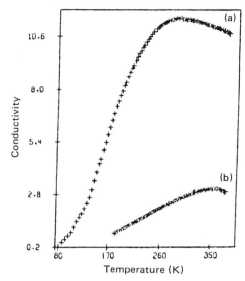

Fig. 23. Variable-temperature needle-axis electrical conductivity data in (ohm cm)$^{-1}$ for samples of (a) $(NiOMTBP)I_{1.08}$ and (b) $(NiOMTBP)I_{2.91}$. [From Marks and Kalina (1982).]

VII. PARTIALLY OXIDIZED FACE-TO-FACE LINKED METALLOMACROCYCLES.

In the current collection of electrical conductors discussed in this chapter, many shortcomings in their properties are observed. For example, they have low water, air, and thermal stability. Additionally, many of these substances have low mechanical stability, problems in processibility, and low solubility. However, most of the molecular–macromolecular phthalocyanine (Pc) compounds have high environmental and thermal stabilities and are soluble in strong acids with no decomposition (Marks, 1985; Marks and Kalina, 1982; Dirk *et al.*, 1983).

Recently, some of the Pc complexes have been modified in an attempt to control molecular stacking as well as the other aforementioned factors associated with the metallomacrocycles. Several oligomers have been prepared and partially oxidized, and these metallomacrocycles are covalently linked forming O–M–O chains. These materials are synthesized by preparing the $M(Pc)Cl_2$ complexes (M = Si, Ge, Sn), which are then hydrolyzed in pyridine (Davison and Wynne, 1978) to $M(Pc)(OH)_2$. Polymerization takes place at 300–400°C (Meyer and Wöhrle, 1974; Meyer *et al.*, 1975; Joyner and Kenney, 1960). Linear high polymers are produced according to the reactions

$$M(Pc)Cl_2 \xrightarrow[\text{Pyridine}]{H_2O} M(Pc)(OH)_2 \xrightarrow[\substack{-H_2O \\ 300–400°C}]{} [M(Pc)O]_n \xrightarrow{I_2} [(MPc)OI_x]_n$$

electrically conductive polymer

The iodination can be accomplished by a variety of chemical or electrochemical methods. Polymers that are covalently linked and have nonintegral valences and chain structures are produced. For example, if M is Si or Ge the process is

$$[M(Pc)O]_n + 0.55nX_2 \rightarrow \{[M(Pc)O]X_{1.1}\}_n \qquad [X = Br, I]$$

$$[M(Pc)O]_n + 0.35nNO^+Y^- \rightarrow \{[M(Pc)O]Y_{0.35}\}_n + 0.35nNO(g)$$
$$[Y = BF_4^-, PF_6^-, SbF_6^-]$$

$$[M(Pc)O]_n + nyQ \rightarrow \{[M(Pc)O]Qy\}_n$$
$$[Q = TCNQ \text{ or } DDQ = (C_6Cl_2(CN)_2O_2)]$$

Halogen doping of planar, conjugated metallomacrocycles is effective in synthesizing conductive materials. However, the process depends on the solid state structures which are completely dependent on intermolecular forces that are unpredictable and uncontrollable. These forces dictate whether the stacks which form are segregated or mixed, the relative orientation of donors with respect to the acceptors in the molecule, and of the units within a stack, as well as stacking distances (Schoch *et al.*, 1979). The M(Pc) stacking is stabilized in these polymers with $n = 50–200$. The Pc–Pc interplanar separations increase with an increasing ionic radius of M: 3.33(2) Å, M = Si; 3.52(2) Å, M = Ge; and 3.82(2) Å, M = Sn. The halogen-doped materials are stable in air for years and thermally stable up to 573 K and may be recrystallized from sulfuric acid. However, the halogen-doping process is inhomogeneous, and therefore, the dopant is not uniformly distributed throughout the lattice and a wide range of stoichiometries is formed. More importantly, the original polymers, which are insulators, become electrical conductors upon doping. Table VIII tabulates electrical conductivites for several such doped polymers. Incremental doping with iodine causes both $[Si(Pc)O]_n$ and $[Ge(Pc)O]_n$ to exhibit dramatic increases in electrical conductivity from undoped materials, as demonstrated in Fig. 24. The temperature dependence of the electrical conductivity for $\{[Si(Pc)O]Iy\}_n$, shown in Fig. 25, is seen to be thermally activated (semiconducting behavior). The room temperature conductivities parallel to interplanar ring–ring spacings follow the trend $\sigma_{Ni} > \sigma_{Si} \gtrsim \sigma_{Ge} > \sigma_{Sn}$.

Raman spectra have demonstrated that the predominant band, at 108

TABLE VIII

Pressed Powder Electrical Conductivity Data for $\{[M(Pc)O]X_y\}_n$
Doped Polymers at Room Temperature[a]

Compound	σ_{RT} (ohm cm)$^{-1,b}$
$[Si(Pc)O]_n$	5.5×10^{-6}
$\{[Si(Pc)O]I_{0.12}\}_n$	1.4×10^{-3}
$\{[Si(Pc)O]I_{0.23}\}_n$	1.7×10^{-2}
$\{[Si(Pc)O]I_{0.31}\}_n$	7.6×10^{-2}
$\{[Si(Pc)O]I_{0.71}\}_n$	3.2×10^{-1}
$\{[Si(Pc)O]I_{1.12}\}_n$	5.8×10^{-1}
$\{[Si(Pc)O]I_{1.13}\}_n$	6.7×10^{-1}
$\{[Si(Pc)O]I_{1.55}\}_n$	1.4
$\{[Si(Pc)O]Br_{1.12}\}_n$	9.6×10^{-1}
$\{[Si(Pc)O]Br_{1.8}\}_n$	7.2×10^{-1}
$[Ge(Pc)O]_n$	2.2×10^{-10}
$\{[Ge(Pc)O]I_{0.14}\}_n$	2.5×10^{-6}
$\{[Ge(Pc)O]I_{0.31}\}_n$	1.6×10^{-3}
$\{[Ge(Pc)O]I_{0.56}\}_n$	1.9×10^{-2}
$\{[Ge(Pc)O]I_{1.08}\}_n$	1.1×10^{-1}
$[Sn(Pc)O]_n$	1.2×10^{-9}
$\{[Sn(Pc)O]I_{0.14}\}_n$	5.2×10^{-9}
$\{[Sn(Pc)O]I_{0.32}\}_n$	5.2×10^{-9}
$\{[Sn(Pc)O]I_{0.60}\}_n$	2.2×10^{-8}
$\{[Sn(Pc)O]I_{1.11}\}_n$	2.2×10^{-8}
$\{[Sn(Pc)O]I_{1.76}\}_n$	6.5×10^{-7}

[a] Taken from Dirk et al. (1983).
[b] Measured by four-probe techniques.

cm^{-1}, is characteristic of that of an I_3^- moiety (Teitelbaum et al., 1979, 1980; Schoch et al., 1979). Therefore, the formal degree of charge transfer is 0.33–0.37, and it is insensitive to the nature of halogen and interplanar distances. Table IX shows the crystallographic data for the halogen-oxidized polymers. One may readily observe the increase in interplanar spacing as one proceeds from Ni(Pc)I to $\{[Si(Pc)O]I_{1.12}\}_n$ and $\{[Ge(Pc)O]I_{1.07}\}_n$. Finally, in a study (Ciliberto et al., 1984) of monomeric and cofacial dimeric SiPc complexes, it was concluded that the principle charge-transfer pathway in the $[Si(Pc)O]_n$ polymer is via the phthalocyanine π system.

An interesting development has been that $[Si(Pc)O]_n$ polymers can be processed by extruding acid solutions into a precipitation medium. Extrusion into fibers necessitates solubility in strong acids, which $[Si(Pc)O]_n$ possesses. Solutions of $[Si(Pc)O]_n$ in CF_3SO_3H in combination with the plastic Kevlar can be extruded to give fibers that are relatively strong

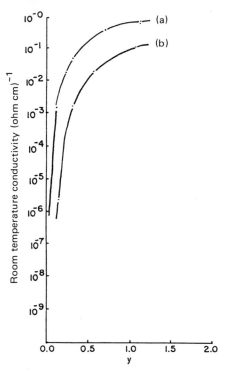

Fig. 24.   Electrical conductivity versus dopant level y for polycrystalline samples of (a) $\{[Si(Pc)O]I_y\}_n$ and (b) $\{[Ge(Pc)O]I_y\}_n$. [From Dirk (1983).]

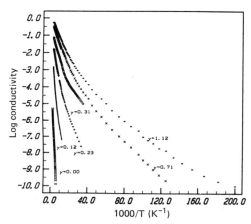

Fig. 25.   Variable-temperature electrical conductivity data for $\{[Si(Pc)O]I_y\}_n$. [From Dirk (1983).]

TABLE IX

Crystallographic Data for Halogen-Oxidized Phthalocyanine Materials[a]

Compound	Space group	Z	Cell parameters (Å)	Density (g cm^{-3})		Interplanar spacing (Å)	Staggering angle between adjacent rings
				Calculated	Found		
Ni(Pc)I[b]	$P4/mcc$	2	$a = 13.936(6)$ $c = 6.488(3)$	1.84	1.78(4)	3.244(2)	39.5°
{[Si(Pc)O]I$_{1.12}$}$_n$	$P4/mcc$	2	$a = 13.97(5)$ $c = 6.60(4)$	1.802(24)	1.744(10)	3.30(2)	39(3)°
{[Si(Pc)O]Br$_{1.12}$}$_n$	$P4/mcc$	2	$a = 13.97(5)$ $c = 6.60(4)$			3.30(2)	39(3)°
{[Ge(Pc)O]I$_{1.07}$}$_n$	$P4/mcc$	2	$a = 13.96(5)$ $c = 6.96(4)$	1.805(23)	1.774(10)	3.48(2)	40(4)°

[a] Taken from Dirk et al. (1983).
[b] Single-crystal data from Schramm et al. (1980).

(weight for weight have six times the modulus of steel) and consist of oriented $[(\!-\!Si\!-\!O]_n$ and Kevlar chains aligned along the extrusion direction. Doping can be carried out before or after extrusion to yield air-stable, electrically conductive fibers (Inabe *et al.*, 1983a,b; Inabe *et al.*, 1984).

Kevlar

Table X shows the electrical conductivities for several $[Si(Pc)O]_n$/Kevlar and M(Pc)/Kevlar formulations. The conductivities are thermally activated (semiconduction) and follow the equation $\sigma = A \exp -(\Delta/kt)$. Conductivities as high as 5 (ohm cm)$^{-1}$ may be reached. Raman spectra

TABLE X

Four-Probe Electrical Conductivity Data for $[Si(Pc)O]_n$/Kevlar and M(Pc)/Kevlar[a]

Composition[b]	Conductivity (300 K)[c] (ohm cm)$^{-1}$	Activation energy[d] (Δ) (eV)
$[Si(Pc)O]_n$ materials		
$\{[Si(Pc)O]I_{1.76}\}_n{}^e$	1.8	2.10×10^{-2}
$\{[Si(Pc)O](K)_{0.59}I_{1.45}\}_n{}^f$	9.1×10^{-2}	4.79×10^{-2}
$\{[Si(Pc)O](K)_{0.98}I_{1.57}\}_n{}^e$	8.5×10^{-2}	5.13×10^{-2}
$\{[Si(Pc)O](K)_{4.38}I_{3.69}\}_n{}^f$	2.9×10^{-2}	5.27×10^{-2}
$\{[Si(Pc)O](K)_{0.73}\}_n$	2.5×10^{-3}	12.0×10^{-2}
$[Si(Pc)O]_n{}^g$	5.5×10^{-6}	29.0×10^{-2}
$\{[Si(Pc)O]I_{1.15}\}_n{}^g$	8.5×10^{-1}	2.80×10^{-2}
M(Pc) materials		
$[Ni(Pc)(K)_{4.36}I_{1.66}]_n{}^e$	1.4	11.4×10^{-3}
$[Ni(Pc)(K)_{1.58}I_{1.27}]_n{}^e$	1.8	6.6×10^{-3}
$[Ni(Pc)(K)_{0.67}I_{1.07}]_n{}^e$	2.5	12.0×10^{-3}
$[Ni(Pc)(K)_{0.43}I_{1.56}]_n{}^h$	4.7	17.2×10^{-3}
$[H_2(Pc)(K)_{0.54}I_{1.24}]_n{}^e$	1.2	13.4×10^{-3}
$Ni(Pc)I^g$	7.9	—[i]

[a] Taken from Inabe *et al.* (1984).
[b] Pc = phthalocyaninato; K = Kevlar monomer unit = $-COC_6H_4CONHC_6H_4NH-$.
[c] Four-probe measurement.
[d] From least-squares fit to equation $\sigma = A \exp -(\Delta/RT)$ where Δ is activation energy.
[e] Doping with C_6H_6/I_2.
[f] Doping with aqueous KI/I_2.
[g] Polycrystalline compaction measurement from Diel *et al.* (1983).
[h] Doping in aqueous KI/I_2 followed by electrochemical oxidation (10 V, 3h) in 0.8 M aqueous HI.
[i] $d\sigma/dt < 0$ above 250 K.

of the iodine-doped Si(Pc)–Kevlar formulations demonstrate the presence of the I_3^- moiety. A typical formulation can be written as $\{[Si(Pc)O](K)_{0.98}I_{1.57}\}_n$.

These results demonstrate that metallomacrocyclics may be blended with Kevlar to produce strong, processible, and orientable polymers with high environmental and temperature stability and reasonable electrical conductivities. Studies of related hybrid systems with other polymer hosts are underway (Inabe et al., unpublished). It should be pointed out that Kevlar can be blended with numerous materials to produce strong, processible, and oriented polymers.

VIII. NONPLANAR, NONOXIDIZED METALLOCYCLES (METALLOTETRABENZO[b,f,j,n][1,5,9,13]-TETRA-AZACYCLOHEXADECINE)

The palladium and platinum complexes of tetrabenzo[b,f,j,n][1,5,9,13]-tetra-azacyclohexadecine (TAAB) have been prepared and iodinated. They represent molecules in which drastic distortions from planarity are encountered. They form highly conducting complexes upon reaction with I_2 (Mertes and Ferraro, 1979; Jircitano et al., 1979), compared to the non-iodinated TAAB. The resulting lustrous, metallic, dark-red needles exhibit the stoichiometry [M(TAAB)][I$_8$] where M is palladium(II) or platinum(II). In contrast to the previously discussed complexes only integral oxidation states are produced. As expected from the considerations discussed herein for previous materials, the conductivities are greatly (vide infra) decreased when compared with other materials discussed in this chapter. X-ray data has confirmed the presence of the octaiodide dianion (Jircitano et al., 1979) indicating that the macrocyclic complex has not undergone oxidation, unlike the previously discussed systems. Within the macrocyclic cation the immediate coordination sphere of the metal is nearly planar; however, the overall geometry of the macrocyclic complex conforms to a hyberbolic parabaloid with cavities of 2.68 Å about the axial-metal sites. The anion consists of discrete Z-shaped nonplanar chains of I_8^{2-} composed of two I_3^- units weakly associated with an elongated I_2 molecule. The conductivity is of the semiconductor type and is thought to occur via macrocycle–I_3^- interactions. These complexes illustrate the importance of a nonintegral oxidation state, as previously discussed, in order to achieve metallic conductivities. The nonplanarity of the complexes also plays a role and these materials become only semiconductors on iodination.

Single-crystal conductivity measurements indicate relatively low conductivities compared to the partially oxidized tetracyanoplatinates and phthalocyanines. An interesting effect is observed upon the application of pressure, however. For the Pd–TAAB complex up to 1000-fold increases in conductivity are observed at pressures of 10 kbar as shown in Table XI.

IX. SUMMARY

As described in this chapter, several prerequisites appear necessary in order to produce electrically conducting metallomacrocycles. Among these are the following:

(1)   The arrangement of the molecules making up the composite molecular array must be in close proximity (preferably stacked), so that an extended path for electron movement is created. Extensive intra- and interstack delocalization of electrons tends to suppress insulating transitions. This is best accomplished by the use of planar conjugated molecules.

(2)   The molecular array must be in a nonintegral or formal fractional oxidation state (incomplete charge transfer). Studies by Moguel and Marks (1985) have indicated the importance of the role of donor–acceptor interactions in determining the chemical and physical properties of charge-transfer conductors discussed in this chapter. For example, using stronger organic acceptors (compared to the inorganic acceptors such as halogens and nitrosyl salts) such as 2,3-dichloro-5,6-dicyano-p-benzoquinone and 2,3,5,6-tetrafluorocyanoquinodimethane to dope silicon phthalocyanine polymers, they have prepared a family of low-dimensional materials with oxidation states that vary continuously from 0 to $+1$. As doping concentration increases the conductivity increases and the polymer begins to

TABLE XI

Conductivity Measurements for M(TAAB)$^{n+}$ Complexes[a]

Complex	Single crystal	Powder (1 bar)	Powder (5 kbar)	Powder (10 kbar)
		Conductivity (ohm cm)$^{-1}$		
[Pd(TAAB)][BF$_4$]$_2$	—	1×10^{-9}	1×10^{-8}	8×10^{-10}
[Pd(TAAB)][I$_3$]$_{2.7}$	10^{-7}	1×10^{-7}	4×10^{-6}	5×10^{-5}
[Pt(TAAB)][BF$_4$]$_2$	—	1×10^{-9}	1×10^{-7}	2×10^{-7}
[Pt(TAAB)][I$_3$]$_{2.7}$	10^{-4}	6×10^{-7}	6×10^{-6}	6×10^{-6}

[a] Taken from Mertes and Ferraro (1979).

show reflectance properties of a metal up to a partial oxidation state of $+0.40$. Thereafter, the conductivity decreases and when a $+1$ oxidation state is attained, the electrical and optical properties become comparable to those of poor semiconductors. In contrast, the inorganic dopants like nitrosyl and the halogens produce a maximum oxidation state of $+0.37$ per Si(Pc) unit. This type of behavior of electrical conductivity, with respect to achieving a maximum at a given partial oxidation state, may also hold for other charge-transfer conductors, but any definite conclusion requires further research (e.g., those discussed in Chapter 2).

The nonintegral or partial oxidation states that occur for the systems described here may be realized by chemical (e.g., iodine) or electrochemical oxidation. Table XII gives for a series of metallomacrocycle complexes the partial oxidation state formulation and the polyhalide species present. To determine the nature of the acceptor species resonance Raman and ^{129}I Mössbauer spectroscopy have been very important diagnostic techniques. The form of the halogen species incorporated in these materials, and therefore, the degree of incomplete charge transfer can be easily discerned by use of the aforementioned techniques.

TABLE XII

Summary of Metallomacrocyclic Complexes with Respect to Partial Oxidation State Formulation and the Polyhalide Moiety Present

Complex moiety	Partial oxidation formulation	Polyhalide present
$Ni(dpg)_2I$	$[Ni(dpg)_2]^{+0.2}(I_5^-)_{0.2}$	I_5^-
$Pd(dpg)_2I$	$[Pd(dpg)_2]^{+0.2}(I_5^-)_{0.2}$	I_5^-
$Ni(dpg)_2Br$	$[Ni(dpg)_2]^{+0.2}(Br_5^-)_{0.2}$	Br_5^-
$Pd(dpg)_2Br_{1.1}$	$[Pd(dpg)_2]^{+0.22}(Br_5^-)_{0.22}$	Br_5^-
$Ni(bqd)_2I_{0.5}S^b$	$Ni(bqd)_2^{+0.17}(I_3^-)_{0.17}$	I_3^-
$Pd(bqd)_2I_{0.5}S^b$	$Pd(bqd)_2^{+0.17}(I_3^-)_{0.17}$	I_3^-
$NiPcI$	$[NiPc]^{+0.33}(I_3^-)_{0.33}$	I_3^-
$Ni[OMTBP]_{1.08}$	$[NiOMTBP]^{+0.36}(I_3^-)_{0.36}$	I_3^-
$[M(4,5-Me_2)_4Pc]I_x{}^{a,c}$	$[M(4,5-Me_2)_4Pc]^{+x/5}(I_5^-)_{x/5}$	I_5^-
$[MOE_2P_2]I_x{}^c$	$[MOEP]^{+x/5}(I_5^-)_{x/5}$	I_5^-
$Ni[TAA]I_{1.80}$	$[NiTAA]^{+0.6}(I_3^-)_{0.6}$	I_3^-
$Ni(TMTAA)I_{2.44}$	$[Ni(TMTAA)]^{+0.8}(I_3^-)_{0.8}$	I_3^-
$\{[Si(Pc)O]I_{0.50}\}_n$	$\{[Si(Pc)O]^{0.17}(I_3^-)_{0.17}\}_n$	I_3^-
$\{[Ge(Pc)O]I_{1.80}\}_n$	$\{[Ge(Pc)O]^{+0.6}(I_3^-)_{0.6}\}_n$	I_3^-
$\{[Sn(Pc)O]I_{1.2}\}_n$	$\{[Sn(Pc)O]^{+0.4}(I_3^-)_{0.4}\}_n$	I_3^-

^{a} Me = methyl.
^{b} S = solvent.
^{c} M = metal.

Electron spin resonance studies have demonstrated that the partial oxidation is predominantly ligand centered yielding π radical cations, unlike that observed in the Krogmann salts, in which it involves the partial oxidation of a metal. Confirmation for these conclusions also comes from the g values, which are near the free electron value, and indicative of unpaired spin density which is ligand, not metal based.

Starting with $[M(Pc)(OH)_2]$ monomers, $[M(Pc)O]_n$ polymers have been synthesized, where M = Si, Ge, Sn are converted to covalently linked, partially oxidized chain compounds. These materials have high environmental and thermal stability and are soluble without decomposition in strong acids. These polymers can be combined with Kevlar to produce robust processable, orientable fibers and when partially oxidized are good electrical conductors [$\sigma \simeq 100$ (ohm cm)$^{-1}$ (298 K) in the stacking direction]. However, the conductivities indicate thermally activated (semiconducting) behavior. The Ni(Pc)–Kevlar hybrid polymers have also been synthesized.

LIST OF SYMBOLS AND ABBREVIATIONS

mac	Macrocyclic ligand
dpg	Diphenylglyoxime
σ	Conductivity in units (ohm cm)$^{-1}$
bqd	Benzoquinonedioxime
Pc	Phthalocyanine
TAA	Dihydrodibenzo[b,i][1,4,8,11]tetra-azocyclotetradecine
TBP	Tetrabenzophorphyrin
OEP	1,2,3,4,5,6,7,8-Octaethylphorphyrin
OMTBP	1,4,5,8,9,12,13,16-Octamethylphorphyrin
TCNQ	7,7,8,8-Tetracyano-p-quinodimethane
DDQ	Dichlorodicyanodiquinone ($C_6Cl_2(CN)_2O_2$)
TAAB	Tetrabenzo[b,f,j,n][1,5,9,13]-tetra-azacyclohexadecine

REFERENCES

Brown, L. D., Kalina, D. W., McClure, M. S., Schultz, S., Ruby, S. L., Ibers, J. A., Kannewurf, C. R., and Marks, T. J. (1979), *J. Am. Chem. Soc.* **79**, 2937.

Ciliberto, E., Doris, K. A., Pietro, W. J., Reisner, G. M., Ellis, D. E., Fragalá, I., Herbstein, F. H., Ratner, M. A., Marks, T. J. (1984), *J. Am. Chem. Soc.* **106**, 7748.

Cowie, M., Gleizes, A., Grynkewich, G. W., Kalina, D. W., McClure, M. S., Scaringe, R. P., Teitelbaum, R. C., Ruby, S. L., Ibers, J. A., Kannewurf, C. R., and Marks, T. J. (1979), *J. Am. Chem. Soc.* **101**, 2921.

Cueller, E. A., and Marks, T. J. (1981), *Inorg. Chem.* **20**, 3766.

Dabrowiak, J. C., Fisher, D. P., McElroy, F. C., and Macero, D. J. (1979), *Inorg. Chem.* **18**, 2304.

Davison, J. B., and Wynne, K. J. (1978), *Macromolecules* **11**, 186.

Diel, B. N., Inabe, T., Lyding, J. W., Schoch, K. F., Kannewurf, C. R., and Marks, T. J. (1983), *J. Am. Chem. Soc.* **105**, 1551.

Dirk, C. W., Inabe, T., Lyding, J. W., Schoch, K. F., Kannewurf, C. R., and Marks, T. J. (1983), *J. Poly. Sci., Poly. Symp.* **70**, 1.

Edelmann, L. E. (1950), *J. Am. Chem. Soc.* **72**, 5765.

Endres, H., Keller, H. J., Mēgnamisi-Bēlombe, M., Moroni, W., and Nöthe, D. (1974), *Inorg. Nucl. Chem. Lett.* **10**, 467.

Endres, J., Keller, H. J., and Lehmann, R. (1975), *Inorg. Nucl. Chem.* **11**, 769.

Ferraro, J. R. (1982), Structural, conductivity and spectroscopic aspects of one-dimensional transition metal complexes, *Coord. Chem.* **43**, 205–232.

Foust, A. S., and Soderberg, R. H. (1967), *J. Am. Chem. Soc.* **89**, 5507.

Gleizes, A., Marks, T. J., and Ibers, J. A. (1975), *J. Am. Chem. Soc.* **97**, 3545.

Goedken, V. L., Pluth, J. J., Peng. S., and Burstein, B. J. (1976), *J. Am. Chem. Soc.* **98**, 8014.

Heeger, A. J. (1979), *in* "Highly Conductive One-Dimensional Solids" (J. T. Devreese, V. E. Evard, and V. E. Van Doren, eds.), chapter 3, Plenum, New York.

Hoffman, B. M., Phillips, T. E., Schramm, C. J., and Wright, S. K. (1979), *in* "Molecular Metals" (W. E. Hatfield, ed.), p. 393, Plenum, New York.

Inabe, T., Lyding, J. W., Moguel, M. K., and Marks, T. J. (1983a), *J. Phys. (Paris, France)* **C3**, 625.

Inabe, T., Lyding, J. W., and Marks, T. J. (1983b), *J. Chem. Soc. Chem. Commun.*, p. 1084.

Inabe, T., Lomax, J. F., Lyding, J. W., Kannewurf, C. R., and Marks, T. J. (1984), *Synth. Met.* **9**, 303.

Inabe, T., Lyding, J. W., Moguel, M. K., Burton, R. L., Kannewurf, C. R., and Marks, T. J., in preparation.

Jircitano, A. J., Timken, M. D., Mertes, K. B., and Ferraro, J. R. (1979), *J. Am. Chem. Soc.* **101**, 7661.

Joyner, R. D., and Kenney, M. E. (1960), *Inorg. Chem.* **82**, 5790.

Kalina, D. W., Lyding, J. W., Ratajack, M. T., Kannewurf, C. R., and Marks, T. J. (1980), *J. Am. Chem. Soc.* **102**, 7854.

Kalina, D. W., Stojakovic, D. R., and Marks, T. J., unpublished data.

Keller, H. J., and Seibold, K. (1971), *J. Am. Chem. Soc.* **93**, 1310.

Lin, L., Marks, T. J., Kannewurf, C. R., Lyding, J. W., McClure, M. S., Ratajack, M. T., and Whang, T. (1980), *J. Chem. Soc. Chem. Commun.*, p. 954.

Marks, T. J. (1976), *J. Coat. Technol.* **48**, 53.

Marks T. J. (1985), *Science* **227**, 881.

Marks, T. J., and Kalina, D. W. (1982), *in* "Extended Linear Chain Compounds" (J. S. Miller, ed.), Vol. 1, pp. 197–331, Plenum, New York.

Martinsen, J., Pace, L. J., Phillips, T. E., Hoffman, B. M., and Marks, T. J. (1982), *J. Am. Chem. Soc.* **104**, 83.

McClure, M. S., Lin, L., Whang, T., Ratajack, M. T., Kannewurf, C. R., and Marks, T. J. (1980), *Bull. Am. Phys. Soc.* **25**, 315.

Mehne, L. F., and Wayland, B. B. (1975), *Inorg. Chem.* **14**, 881.

Mertes, K. B., and Ferraro, J. R. (1979), *J. Chem. Phys.* **70**, 646.

Meyer, G., and Wöhrle, D. (1974), *Makromol. Chem.* **175**, 714.

Meyer, G., Hartmann, M., and Wöhrle, D. (1975), *Makromol. Chem.* **176**, 1919.

Miller, J. S., and Griffiths, J. H. (1977), *J. Am. Chem. Soc.* **99**, 749.

Moguel, M. K., and Marks, T. J. (1985), *Chem. Eng. News*, 16–17 (September 30); Abstracts of 190th American Chemical Society Meeting, Chicago, IL, Sept. 8–13, 1985, p. 138.

Müller, R., and Wöhrle, D. (1975), *Makromol. Chem.* **176**, 2775.

Onodera, A., Kawaii, N., and Kobayashi, T. (1975), *Solid State Commun.* **17**, 775.

Onodera, A., Shirotani, I., Hara, Y., and Anzai, H. (1979), *High Pressure Sci. Technol. AIRAPT 6th Conf., 1977* **1**, 498.

Peterson, J. L., Schramm, C. J., Stojakovic, D. R., Hoffman, B. M., and Marks, T. J. (1977), *J. Am. Chem. Soc.* **99**, 286.

Phillips, T. E., and Hoffman, B. M. (1977), *J. Am. Chem. Soc.* **99**, 7734.

Phillips, T. E., Scaringe, R. P., Hoffman, B. M., and Ibers, J. A. (1980), *J. Am. Chem. Soc.* **102**, 3435.

Scaringe, R. P., Schramm, C. J., Stojakovic, D. R., Hoffman, B. M., Ibers, J. A., and Marks, T. J. (1980), *J. Am. Chem. Soc.* **102**, 6702.

Schoch, K. F., Kundalkar, B. R., and Marks, T. J. (1979), *J. Am. Chem. Soc.* **101**, 7071.

Schramm, C. J., Stojakovic, D. R., Hoffman, B. M., and Marks, T. J. (1978), *Science* **200**, 47.

Schramm, C. J., Scaringe, R. P., Stojakovic, D. R., Hoffman, B. M., Ibers, J. A., and Marks, T. J. (1980), *J. Am. Chem. Soc.* **102**, 6702.

Stojakovic, D. R. (1977), Ph.D. thesis, Northwestern Univ., Evanston, Illinois.

Teitlebaum, R. C., Ruby, S. L., and Marks, T. J. (1980), *J. Am. Chem. Soc.* **102**, 3322.

Teitlebaum, R. C., Ruby, S. L., and Marks, T. J. (1979), *J. Am. Chem. Soc.* **101**, 7568.

Torrance, J. B. (1977), *in* "Synthesis and Properties of Low-Dimensional Materials" (J. S. Miller and A. J. Epstein, eds.), Vol. 313, p. 210, N. Y. Acad. of Sci., New York.

Torrance, J. B. (1979), *in* "Molecular Metals" (W. E. Hatfield, ed.), p. 7, Plenum, New York.

Torrance, J. B. (1985), *Mol. Cryst. Liq. Cryst.* **126**, 55.

Tsutsui, M., Bobsein, R. L., Cash, G., and Patterson, R. (1979), *Inorg. Chem.* **16**, 305.

Underhill, A. E., Watkins, D. M., and Pethig, R. (1973), *Inorg. Nucl. Chem. Lett.* **9**, 1269.

Wright, S. K., Schramm, C. J., Phillips, T. E., Scholler, D. M., and Hoffman, B. M. (1979/1980), *Synth. Met.* **1**, 43.

Review Articles

Alcácer, L., Novais, H., and Pedroso, F. (1979), Perylene-dithiolate complexes—the platinum salt, *in* "Molecular Metals" (W. E. Hatfield, ed.), pp. 415–418, Plenum, New York.

Allcock, H. R. (1985), Inorganic macromolecules, *Chem. Eng. News* **63** (11), 22–36.

Ferraro, J. R. (1982), Structural, conductivity and spectroscopic aspects of one-dimensional transition metal complexes, *Coord. Chem.* **43**, 205–232.

Marks, T. J. (1985), Electrically conductive metallomacrocyclic assemblies, *Science* **227**, 881–889.

Marks, T. J., and Kalina, D. W. (1982), Highly conductive halogenated low-dimensional materials, *in* "Extended Linear Chain Compounds" (J. S. Miller, ed.), Vol. I, pp. 197–331, Plenum, New York.

Marks, T. J. (1978), Rational synthesis of new unidimensional solids: chemical and physical studies of mixed-valence polyiodides, *in* "Synthesis and Properties of Low-Dimensional Materials" (J. S. Miller and A. J. Epstein, eds.), Vol. 313, pp. 594–616, NY Acad. Sci., New York.

Reis, A. H. (1982), A comprehensive review of linear chain iridium complexes, *in* "Extended Linear Chain Compounds" (J. S. Miller, ed.), Vol. 1, pp. 157–196, Plenum, New York.

Reis, A. H., and Peterson, S. W. (1978), Structure and oxidation states of Ir and Pt one-dimensional inorganic complexes, *in* "Synthesis and Properties of Low-Dimensional Materials" (J. S. Miller and A. J. Epstein, eds.), Vol. 313, pp. 560–579, NY Acad. Sci., New York.

Underhill, A. E., and Wood, D. J. (1979), A comparison of the properties of the isostructural ID conductors $Co_{0.83}[Pt(C_2O_4]_2 \cdot 6H_2O$, and $Mg_{0.82}[Pt(C_2O_4)_2] \cdot 6H_2O$, *in* "Molecular Metals" (W. F. Hatfield, ed.), pp. 377–398, Plenum, New York.

Underhill, A. E., Watkins, D. M., Williams, J. M., and Carneiro, K. (1983), Linear chain bis(oxalato) platinate salts, *in* "Extended Linear Chain Compounds" (J. S. Miller, ed.), Vol. 1, pp. 119–156, Plenum, New York.

Books

Hatfield, W. E., ed. (1979), "Molecular Metals," pp. 1–555, Plenum, New York.

Miller, J. S., ed. (1981–1983), "Extended Linear Chain Compounds," Vols. 1–3, Plenum, New York.

Miller, J. S., and Epstein, A. J., eds. (1978) "Synthesis and Properties of Low-Dimensional Materials," Annals of the NY Acad. Sci., Vol. 313, pp. 1–828, NY Acad. Sci., New York.

6 MISCELLANEOUS CONDUCTORS

I. POLYMER–SALT COMPLEXES

A. Introduction

Another class of electrical conductors is the polymer-salt complexes, which are also termed polymer-ion electrolytes or ionomers. These materials are related to the organic polymeric conductors discussed in Chapter 3, which may be considered to be a polymer *ion*. Although the doped organic polymers may be considered to be polymer-ion-type conductors, the conduction mechanism involves primarily solitons, polarons or bipolarons (see Chapter 7), whereas in the polymer-ion electrolytes, to be discussed in this chapter, the conduction mechanism involves ion conduction. Furthermore, the electrical conductivities of the polymer-ion electrolytes are close to the values obtained for semiconductors, whereas the doped organic polymer-ion conductors manifest as much as $\sim 10^7$ orders of magnitude higher conductivities.

Polymers possessing high ionic conductivity have been known for some time. The materials always contained solvents and were used as polyelectrolyte ion-exchange polymers. In 1978, solvent-free systems were developed consisting of polymers such as poly(ethylene oxide), PEO, and poly(propylene oxide), PPO, complexed with alkali-metal salts (Armand *et al.*, 1978, 1979, 1980; Wright, 1976; James *et al.*, 1979; Fenton *et al.*, 1973; Shriver *et al.*, 1981; Dupon *et al.*, 1982; Papke *et al.*, 1982a,b; Shriver and Farrington, 1985).

B. Synthesis

The polymer for example, poly(ethylene oxide) with average molecular weight of 600,000 is treated by ion exchange and filtration to remove inorganic salts and SiO_2 (Papke et al., 1981). The complexes are made in air using reagent-grade acetonitrile as solvent. However, the reagent-grade alkali-metal salt starting materials are dried at 110°C under vacuum and are weighed in an N_2-filled dry box. The complex is prepared by separately dissolving the polymer and salt in acetonitrile and then mixing them. Upon removal of the solvent, a solid complex is formed, which is insoluble in acetonitrile. Although the precise stoichiometry cannot be defined, an approximate limiting stoichiometry around 4 moles of polyether oxygen per mole of alkali metal salt is obtained for many polymer–salt systems (Fenton et al., 1973; Armand et al., 1978, 1980). Some exceptions are known. For example, in the case of the bulky counter ion $B\phi_4^-$, the ratio is 7:1 of ether–oxygen atoms to salt, presumably because the large size of the counter ion prohibits higher salt concentrations.

Polymer salts have been made using poly(ethylene oxide), poly(propylene oxide), poly(epichlorohydran), poly(ethylene succinate), poly(ethylene sulfide), poly(propylene sulfide), and recently polyphosphazene to which short-chain polyether groups have been attached (Shriver and Farrington, 1985). Figure 1 shows the backbone structures of these polymers. Various inorganic salts have been complexed with the known polymers, and the only requisites are a low lattice energy and a high propensity for complex formation. Complex formation is favored by the alkali metal salts with large anions, such as I^-, ClO_4^-, $SO_3CF_3^-$, BF_4^-, and $B\phi_4^-$. Table I gives a comparison of PEO–salt formation with lattice energies of the pure salts.

The ability to use innumerable polymers and various ionic salt starting materials affords one the opportunity to synthesize a large variety of polymer salts. This provides the opportunity to tailor-make a polymer–salt derivative with desired electrical, mechanical, and chemical properties.

C. Structure of Polymer–Salt Complexes

X-ray diffraction studies combined with vibrational spectroscopy have been used to elucidate the structure of polymer–salt complexes. Poly(ethylene oxide) consists of an extended helix, having a fiber repeat distance of ~19.48 Å. A detailed x-ray analysis of $HgCl_2$–PEO has been made (Takahashi and Tadokoro, 1973; Iwamoto et al., 1968). Similar studies of the alkali metal salts with PEO have not been reported. X-ray studies

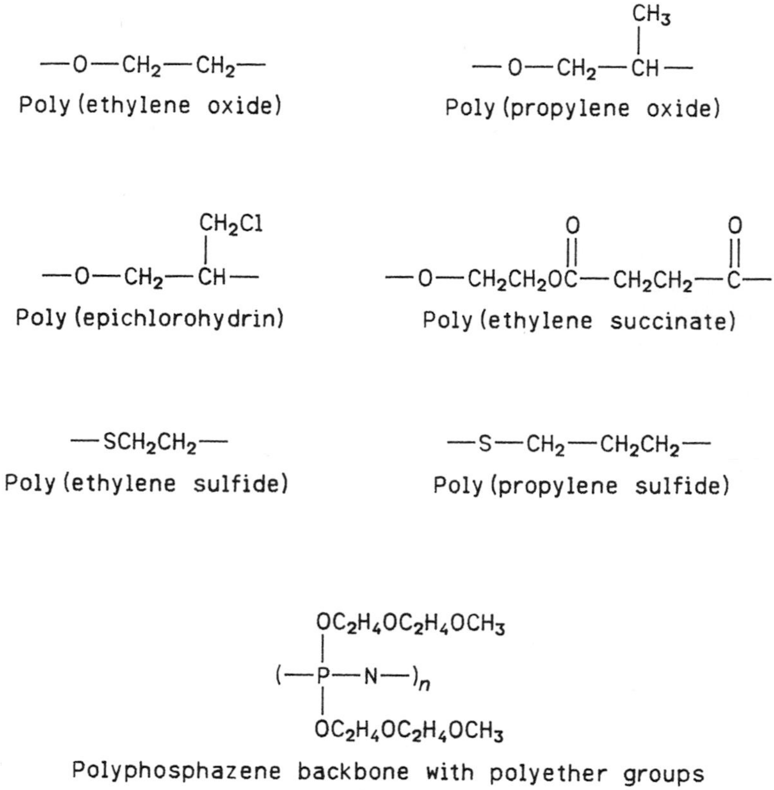

$$-O-CH_2-CH_2-$$

Poly (ethylene oxide)

$$\overset{\displaystyle CH_3}{\underset{\displaystyle |}{-O-CH_2-CH-}}$$

Poly (propylene oxide)

$$\overset{\displaystyle CH_2Cl}{\underset{\displaystyle |}{-O-CH_2-CH-}}$$

Poly (epichlorohydrin)

$$-O-CH_2CH_2O\overset{O}{\overset{\|}{C}}-CH_2CH_2-\overset{O}{\overset{\|}{C}}-$$

Poly (ethylene succinate)

$$-SCH_2CH_2-$$

Poly (ethylene sulfide)

$$-S-CH_2-CH_2CH_2-$$

Poly (propylene sulfide)

$$\overset{\displaystyle OC_2H_4OC_2H_4OCH_3}{\underset{\displaystyle OC_2H_4OC_2H_4OCH_3}{(-\overset{|}{P}-\overset{|}{N}-)_n}}$$

Polyphosphazene backbone with polyether groups

Fig. 1. Backbone structures of various polymers used to form polymer–salt complexes.

with PEO and KSCN show a fiber repeat distance of ~8.1 Å (Fenton *et al.,* 1973). Similarities in spectroscopic data and physical characteristics between PEO·NaX and PEO·LiX complexes suggest that the same conformation exists in both complexes. The configuration is described as trans (CC–OC), trans (CO–CC), gauche (OC–CO), trans (CC–OC), trans (CO–CC), and gauche-minus (OC–CO) and with T (trans) and G (gauche) can be designated as $T_2GT_2G^-$ (Papke *el al.,* 1981). Polar oxygen atoms are directed inward in a $(T_2GT_2G^-)$ conformation. These line the polar cavity while the CH_2 groups all face outward. Space filling molecular models indicate that the conformation has a tunnel radius of ~1.3–1.5 Å, which is large enough to accommodate Na^+ cations (radius 0.99 Å) or Li^+ cations (radius 0.56 Å) (Chabagno, 1980). In the helix the positions of the anions appear to lie outside the helix with no direct cation–anion contact for most cases. As we shall see, ion pairing does occur in some polymer

TABLE I

Comparison of PEO–Salt Complex Formation with Lattice Energies
(kj/mole) of the Pure Salts[a,b]

Anion	Cation				
	Li^+	Na^+	K^+	Rb^+	Cs^+
F^-	No	No	No	No	No
	1036	923	821	785	740
Cl^-	Yes	No	No	No	No
	853	786	715	689	659
CH_3COO^-	—	No	—	—	—
	881	763	682	656	(682)
NO_3^-	—	No	—	—	—
	848	756	687	658	625
NO_2^-	—	No	—	—	—
	—	768	664	765	(598)
Br^-	Yes	Yes	No	No	No
	807	747	682	660	631
N_3^-	—	No	—	—	—
	818	731	658	632	604
BH_4^-	—	Yes	—	—	—
	(771)	(703)	(665)	(649)	(628)
I^-	Yes	Yes	?	No	No
	757	704	644	630	604
SCN^-	Yes	Yes	Yes	Yes	Yes
	805	682	616	619	568
ClO_4^-	Yes	Yes	—	—	—
	723	648	602	582	542
$CF_3SO_3^-$	Yes	Yes	Yes	Yes	Yes
	(<725)	(<650)	(<605)	(<585)	(<550)
BF_4^-	Yes	Yes	—	—	—
	(699)	619	631	605	(556)
$B\phi_4^-$	Yes	Yes	Yes	Yes	Yes
	(<700)	(<630)	(<630)	(<600)	(<550)

[a] Taken from Shriver et al. (1981). These are the pure inorganic salts, which are used
to form the polymer-ion salts.
[b] No means no solvent-free complex formed; yes, solvent-free complex formed;
theoretical or estimated lattice energies are in parentheses.

salts, lowering the electrical conductivity. Figure 2 shows molecular
models of PEO in a $T_2GT_2G^-$ conformation as proposed for the Na^+ and
Li^+ salts.

D. Conductivities of Polymer–Salt Complexes

The more crystalline polymer salts demonstrate the highest conductivities. The sodium salt of PEO is highly crystalline and shows an ionic

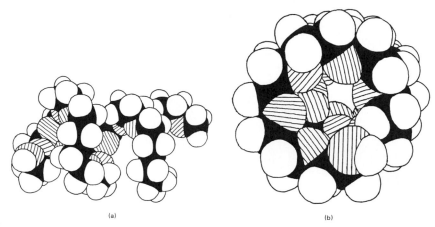

Fig. 2.   Molecular models of poly(ethylene oxide) in a $T_2GT_2G^-$ (see text) conformation as proposed for complexation with sodium and lithium cations: (a) side view and (b) end view. Hydrogen atoms are not shaded, carbon atoms are black, and oxygen atoms are cross-hatched. [From Papke *et al.* (1982a). Reprinted by permission of the publisher, "The Electrochemical Society, Inc."]

conductivity of $\sim 10^{-7}$ (ohm cm)$^{-1}$ (Papke *et al.*, 1981). Table II summarizes typical results for several PEO salts. Although they show typical ionic conductor behavior, i.e., an increase in conductivity by several orders when heated, they are still, at best, semiconductors. The crystalline polymer salts generally give linear plots of log σ versus $1/T$. Recently, a polymer salt based on a flexible polyphosphazene backbone, to which short-chain polyether groups are linked, has shown conductivities 10^3 greater than the PEO analogues shown in Table II (Shriver and Farrington, 1985). Figure 3 demonstrates where the polymer salts fall on a general

TABLE II

Comparison of Conductivities for Several Polymer Salts

PEO·LiBH$_4$	PEO·LiBF$_4$	PEO·LiNO$_3$
σ $\sim 10^{-9}$ (ohm cm)$^{-1}$ (25°C)	1.2×10^{-7} (ohm cm)$^{-1}$ (30°C)	8×10^{-10} (ohm cm)$^{-1}$ (25°C)
σ $\sim 10^{-7}$ (ohm cm)$^{-1}$ (90°C)	7×10^{-5} (ohm cm)$^{-1}$ (90°C)	6×10^{-6} (ohm cm)$^{-1}$ (80°C)
Mechanism of Conduction 1D transport	3D transport	1D transport

Conformation of Polymers
All conductors show 2 strong IR bands at $\sim$880 and $\sim$940 cm^{-1}, indicative of a gauche (G) confirmation about the O–(H$_2$C)$_2$–O linkage.

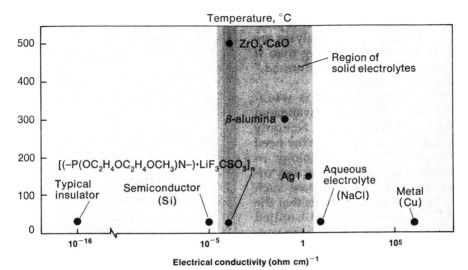

Fig. 3. Electrical conductivities of various materials. The conductivities of solid electrolytes fall between those of semiconductors and aqueous electrolytes: the electrical conductivities of several common substances and representative solid electrolytes are shown here at the temperature where the materials might be used. The oxide conductor $ZrO \cdot CaO$ is used at temperatures at or above 500°C. The data point for β-alumina is that for the sodium form, in which Na^+ is the mobile species. In silver iodide, Ag^+ is responsible for the electrical conductivity. The polymeric electrolyte $[(-P(OC_2H_4OC_2H_4-OCH_3)N-) \cdot LiF_3CSO_3]_n$, based on a polyphosphazene polymer host, contains a mobile cation, Li^+, and an anion, $F_3CSO_3^-$. The ionic conductors have conductivities that fall between those of a typical semiconductor, silicon, and a typical aqueous electrolyte, sodium chloride. [From Shriver and Farrington (1985).]

conductivity scale indicating conductivities close to those of other semiconductors.

Ionic conductivity measurements of $PEO \cdot NaBF_4$ and $PEO \cdot NaBH_4$ with varying salt concentrations have been reported (Papke *et al.*, 1982). A plot of $\ln(\sigma T)$ versus $1000/T$ for these systems is shown in Fig. 4. The 4.5:1 $PEO \cdot NaBF_4$ complex exhibits conductivities like other $PEO \cdot Na$ salt complexes (Papke *et al.*, 1981; Armand *et al.*, 1979). The presence of higher salt concentrations results in lower conductivity. The 4.5:1 complex shows a conductivity of 1.2×10^{-7} (ohm cm)$^{-1}$, which increases to 7.0×10^{-5} (ohm cm)$^{-1}$ at 90°C. A linear Arrhenius-type behavior is observed. In polymer salts where ion pairing (vide infra) occurs between the mobile alkali metal ion and the anion, the conductivity drops (Dupon *et al.*, 1982) as compared to complexes where no ion pairing occurs. Similar ion pairing has been found in PEO-Li salts (Papke *et al.*, 1982a).

The identification of the occurrence of ion-pairing processes has been

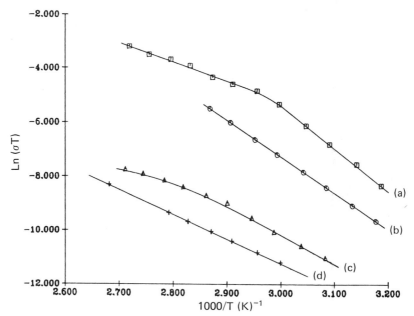

Fig. 4.   Variable temperature conductivity values for $PEO \cdot NaBF_4$ complexes at (a) 4.5:1 and (b) 3:1 stoichiometry, and $PEO \cdot NaBH_4$ complexes at (c) 4.5:1 and (d) 3.4:1 stoichiometry. [From Papke *et al.*, (1981).]

provided by molecular spectroscopy. For example, one assumes that an unperturbed anion BH_4^- (BD_4^-) exists in aqueous solution and for comparison purposes, its Raman and infrared assignments are tabulated in Table III along with results obtained for $PEO \cdot LiBH_4$ and $PEO \cdot LiBF_4$. Only ν_3 and ν_4 are active in a T_d symmetry group. For the polymer salt $PEO \cdot LiBH_4$ a more complicated spectrum is obtained because the degeneracies of ν_3 and ν_4 are lifted and additional bands appear. Also, ν_1 appears in the infrared, where for a T_d symmetry it should be inactive. Thus, one can conclude that the symmetry has been lowered due to cation–anion interactions in the salt portion of the molecule with a symmetry lower than T_d. No substantial ion pairing occurs for $PEO \cdot LiBF_4$ because of the identical infrared spectrum compared to that for ionic $LiBF_4$. Table III also shows the results for $PEO \cdot LiNO_3$ versus the free NO_3^- anion. The ν_3 vibration (D_{3h} symmetry) in the free ionic NO_3^- at 1380 cm^{-1} splits into two vibrations ν_4 and ν_1 at 1415 and 1324 cm^{-1}, respectively, indicative of a lowering of the nitrate ion symmetry due to the cation–anion interaction. In conclusion, the results for both the $LiBH_4$ and $LiNO_3$ polymer salts are indicative of extensive interaction due to ion pairing with a lowering of conductivity for these materials.

Since the cation is the mobile ion which causes the ionic conductivity in polymer salts, any ion-pairing processes would expectedly lower the

TABLE III

Spectroscopic Results Obtained for Several Polymer Salts[a]

PEO·LiBH$_4$[b] (ion pairing) ($<T_d$)		NaBH$_4$ (ionic) Basic aqueous solutions (T_d)	PEO·LiBF$_4$ (ionic) (T_d)		LiBF$_4$ (ionic) (T_d)	PEO·LiNO$_3$ (ion pairing) ($<D_{3h}$)		Free NO$_3^-$ (ionic) (D_{3h})
2347(vs)	$[\nu_2 + \nu_4]$		1065(m)	$[\nu_3]$	1070(m)	1415(s)	$[\nu_4]$	
								1380(s) $[\nu_3]$
2295(m)	$[\nu_1]$		775(m)	$[\nu_1]$	777(m)	1363(m)		
2232(vs)	$[\nu_3]$	2272(s)				1324(vs)	$[\nu_1]$	
2178(s)						827(m)	$[\nu_2]$	828(m) $[\nu_2]$
2169(s)						729(vv)	$[\nu_3]$	
	$[2\nu_4]$	2200(sh)				719(vvw)		716(m) $[\nu_4]$
IR data			Raman data			IR data		
Extensive interaction, ion pairing			Slight interaction, no ion pairing			Extensive interaction, ion pairing		

[a] Frequencies in cm^{-1}.
[b] PEO = Poly(ethylene oxide).

conductivity. Figure 5 shows a schematic illustration of the influence of ion pairing on ionic conductivity (Dupon *et al.*, 1982). Wherever one-dimensional ion transport in the helical tunnels exists, ion pairing blocks the movement of the cation. In the case of three-dimensional transport the ion-pairing does not block the ion transport. Figure 6 depicts the motion

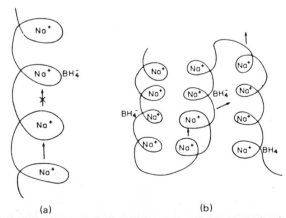

(a)   (b)

Fig. 5. Schematic illustration of the influence of ion pairing on conductivity: (a) one-dimensional ion transport in helical tunnels is blocked by ion pairing with BH$_4^-$ and (b) three-dimensional ion transport is not blocked by ion pairs. [From Dupon *et al.* (1982).]

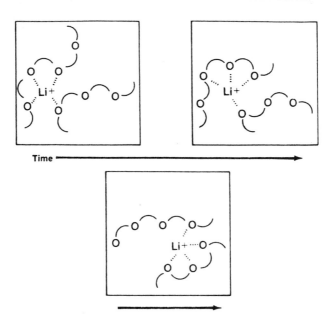

Fig. 6.   The motion of ions in polymers appears to occur by a liquidlike mechanism in which the local motion of the polymer assists the migration of ions through the solid. In this schematic representation of this liquidlike motion, a Li$^+$ ion migrates from the left to the right in a small volume of polymer. The motion of the ion is assisted by the large-amplitude motion of the polymer, which permits the Li$^+$ to remain 3- or 4-coordinated (4-coordination shown) to the polymer's O atoms throughout the migration. [From Shriver and Farrington (1985).]

of the ions in polymers, which appears to be liquidlike with the ion motion aided by the local motion of the polymer. The motion of the polymer allows the cation to maintain its coordination number with, for example, oxygen atoms and to remain four-coordinate during its migration (Shriver and Farrington, 1985). Conduction between helical chains appears to be the dominant process for cation transport (Dupon *et al.,* 1982).

For additional information on ion transport in these materials, see Papke *et al.,* (1982).

E.   Applications of Polymer–Salt Complexes

An advantage of polymer electrolytes over crystalline platinum chain metal and organic charge-transfer salts is that they are pliable and deform under stress. They are, therefore, promising for use as electrolytes in solid state, high-energy–density batteries. Furthermore, these materials can be cast into films. These films are useful since they lower the resistance of the electrolyte and reduce its volume and weight. The materials are thus promising for fuel cells, sensors, and electrolytic cells.

II.   GRAPHITE AND ITS INTERCALATED COMPOUNDS

Graphite is a very excellent room temperature electrical conductor and is considered to be a semimetal at 298 K. When it becomes intercalated with AsF_5 it becomes metallic and exhibits superconductivity when intercalated with potassium, rubidium, and cesium (Hannay *et al.*, 1965; Wolski *et al.*, 1986). For an extensive review of the intercalation compounds of graphite see Dresselhaus and Dresselhaus (1981).

Graphite itself has a layer structure with the separation between layers being 3.35 Å (equal to the sum of van der Waals radii of carbon). Forces between layers are weak, thereby providing the properties of softness and lubricity attributed to graphite. Because of the weak forces between layers, slippage between layers occurs easily. Additionally, the spacing between layers can expand in order to accommodate other atoms or molecules. The structure of graphite is such that each carbon atom is surrounded by three others. After forming a sigma bond with each neighbor, from sp^2 hybridized orbitals, each carbon atom still has one unpaired electron to use in bond formation.

Two types of graphite are known with the most stable form being the hexagonal modification with ABAB $\cdots$ stacking. The ABAB $\cdots$ sequence is illustrated in Fig. 7. A three atom unit is shown with the lattice parameters $a_0 = 2.46$ and $c_0 = 6.70$ Å. The in-plane closest approach between carbon atoms is 1.35 Å with plane separations being 3.35 Å [see

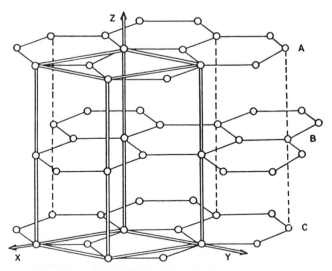

Fig. 7.   Graphite crystal structure illustrating the layered structure. (a_0 axis is labeled X, b_0 axis is Y, and c_0 is Z).

Vogel *et al.* (1977) and Vogel (1979)]. Graphite is highly anisotropic which results from the differences in bond strength within and between the hexagonal planes. The form of graphite which is unstable has a rhombohedral structure where the stacking order is ABC···ABC with every third layer being superposed.

The electrical conductivities of natural and synthetic (HOPG) graphite crystals are tabulated in Table IV (Vogel, 1979). The anisotropies σ_a/σ_c are also illustrated. The high conductivities at room temperature and at liquid helium temperatures are observed to be comparable to the conductivities of silver and copper metal. See Table I in Chapter 7 for a comparison of graphite with other conductors.

The electrical and physical properties of graphite are significantly modified by intercalation. The very loose, layer structure of graphite allows ions and molecules to be inserted between the layers, forming intersitial, lamellar or intercalation compounds (Fischer and Thompson, 1978; Stumpp, 1977; Vogel, 1979; Fischer, 1979; Vogel and Zeller, 1979; Bartlett *et al.,* 1979; Zeller *et al.,* 1977; Vogel *et al.,* 1977). The graphite intercalation compounds consist of stacks of one or more layers of hexagonally arranged carbon atoms, alternating with monolayers of intercalating atoms, ions, or molecules. The intercalation of these substances in graphite causes marked changes in the properties of both host and guest. The anisotropy in these intercalants is caused by the differences in strength of the interlayer interactions as well as the type of intercalated species between layers.

Intercalants may be divided into two groups. The first group involves an alkali metal. In these cases, the guest acts as an electron donor. The second group involves acids, halogens, metal chlorides, and other electrophilic molecules which act as acceptors and behave as electron withdrawing groups. In effect, these intercalants are charge-transfer compounds, which may be described as involving contributions from all the classical types of bonding (ionic, covalent, etc.).

TABLE IV

Conductivities (ohm cm)$^{-1}$ of HOPG and Ticonderoga Graphites
at 298 K and Liquid He Temperatures

Conductivity	Synthetic (HOPG)[a]	Natural (Ticonderoga)
At 298 K		
σ_a	2.3×10^{4}[b]	2.6×10^{4}[c]
σ_a/σ_c	4000	2700
At liquid He temperature		
σ_a	4.1×10^{5}	1.2×10^{6}[c]

[a] HOPG = Highly oriented pyrolytic graphite.
[b] Measured by 4-probe method (d.c.) (Spain, 1973; Spain *et al.,* 1967).
[c] Measured by an rf induction method (Zeller and Pendrys, unpublished data).

A. Donor Intercalation Compounds of Graphite

When the inserted specie is an alkali metal, the graphite lattice attracts electrons from the alkali metal and behaves as a giant anion. The reaction may be considered to be a reduction of the graphite. The composition is described in terms of stages in Fig. 8 for the potassium-graphite intercalant. Each layer becomes filled and a change in the graphite sequence occurs so that the AB sequence between adjacent layers is changed to an AIA sequence on either side of the intercalated layer, where I is the inserted intercalant layer. Table V tabulates the increase in conductivity upon potassium insertion, as well as with other alkali–metal–graphite intercalants.

Ternary compounds are also formed with alkali metals. The general stoichiometry is $M_x M'_{1-x} C_8$ ($O < x < 1$) [see Herold *et al.* (1977)]. New graphite intercalants have also been made such as $C_{4n}MH_g$, where M is K or Rb and $n = 1,2$ [see Iye and Tanama (1982)].

Stage 2 KC_{24}	Stage 3 KC_{36}	Stage 4 KC_{48}	Stage 5 KC_{60}
oo oo oo A	oo oo oo A	oo oo oo A	oo oo oo A
--------	--------	--------	--------
oo oo oo A	oo oo oo A	oo oo oo A	oo oo oo A
oo oo o B	oo oo o B	oo oo o B	oo oo o B
--------	oo oo oo A	oo oo oo A	oo oo oo A
oo oo o B	--------	oo oo o B	oo oo o B
oo oo oo A	oo oo oo A	--------	oo oo oo A
--------	oo oo o B	oo oo o B	--------
oo oo oo A	oo oo oo A	oo oo oo A	oo oo oo A

Fig. 8. The stage structure for potassium graphite; C layer indicated by circles and K layer by dashed line.

TABLE V

Conductivities of Several Graphite–Alkali-Metal Intercalcalation Compounds[a]

Compound	Stage	σ_a (ohm cm)$^{-1}$
K	1	1.1×10^5
	2	1.7×10^5
	3	2.1×10^5
Rb	1	1.0×10^5
	2	1.5×10^5
Cs	1	1.0×10^5
	3	1.2×10^5
Li	1	1×10^5

[a] Regardless of electronegativity of alkali metal all σ's are around 1×10^5 (ohm cm)$^{-1}$; σ_a/σ_c in range of 10 to 100. Stage is indicated in Fig. 8.

B. Acceptor Intercalation Compounds

The acceptor molecules that have been used are HNO_3, $HClO_4$, FSO_3^-, AsF_5, SbF_5, metal halides (e.g., $FeCl_3$, WF_6, ReF_6, OsF_6, IrF_6, PtF_6), and halogens. Table VI summarizes various metal halides intercalated into graphite (Vogel, 1979; Fischer and Thompson, 1978; Vogel *et al.*, 1977; Lagrange *et al.*, 1980; Stumpp, 1977). The reaction can be considered to involve graphite oxidation. Figure 9 illustrates conductivity results for two acceptor molecules, namely, AsF_5, HNO_3 versus the donor molecule K. The AsF_5 intercalant of graphite reaches a maximum conductivity greater than that observed for copper metal.

One of the problems in intercalating acceptor species into graphite is that the exact nature of the intercalation reaction is not entirely known.

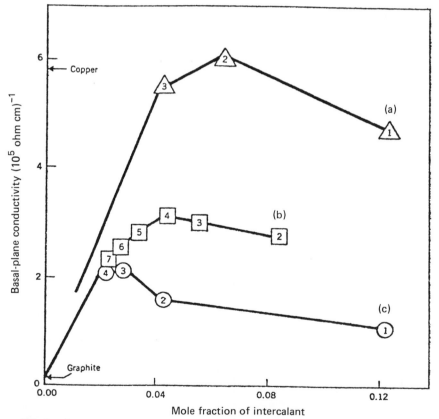

Fig. 9. Room-temperature conductivity along the basal plane as a function of stage for three intercalants: (a) AsF_5, (b) HNO_3, and (c) K. The number inside the data points denotes the stage of the compound, characterized by x-ray diffraction.

TABLE VI
Some Halides Claimed to Form Graphite Intercalant Compounds[a,b]

Chloride	Composition	Stage	I_c (pm)[c]
Group Ib			
$CuCl_2$	$C_{4.9}CuCl_2$	1	940
$AuCl_3$	$C_{12.6}AuCl_3$	1	680
	$C_{25.2}AuCl_3$	2	1015
	$C_{37.8}AuCl_3$	3	1350
	$C_{50.4}AuCl_3$	4	1685
Group IIb			
$MgCl_2$	$C_{12-18}MgCl_2$	1 (+2)	950
	$C_{26-27}MgCl_2$	2 (+3)	1285
	$C_{40}MgCl_2$	3	1626
$ZnCl_2$	$C_{16.6}ZnCl_2$	3	1629
$CdCl_2$	$C_{11.1}CdCl_2$	1	951
	$C_{6.6}CdCl_2$	1	963
$HgCl_2$	$C_{21-26}HgCl_{2.1}$	3	1645
Group IIIa			
$AlCl_3$	$C_9AlCl_{3.3}$	1	954
	$C_{18}AlCl_{3.3}$	2	1283
	$C_{31-50}AlCl_{3.3}$	4	1967
$GaCl_3$	$C_9GaCl_{3.3}$	1	956
	$C_{18}GaCl_{3.4}$	2	1283
	$C_{35}GaCl_{3.5}$	4	1969
$InCl_3$	$C_{16-20}InCl_3$	2	1283
	$C_{25-31}InCl_3$	3	1623
	$C_{32-45}InCl_3$	4	1969
$TlCl_3$	$C_{8.5}TlCl_{3.3}$	1	977
	$C_{18.5}TlCl_{3.2}$	2	1312
Group IVb			
$ZrCl_4$	$C_{23.8}ZrCl_{4.15}$	3	1632
$HfCl_4$	$C_{45.7}HfCl_{4.77}$	3	1587
Group Va + b			
$SbCl_5$	$C_{12}SbCl_5$	1	992
	$C_{24}SbCl_5$	2	1272
	$C_{36}SbCl_5$	3	1607
	$C_{48}SbCl_5$	4	1942
	$C_{12}SbCl_{3.9}$	1	950
	$C_{25}SbCl_{4.3}$	2	1288
$NbCl_5$	$C_{40}NbCl_{5.2}$	3	1621
$TaCl_5$	$C_{27}TaCl_5$	3	1619
Group VIb			
$CrCl_3$	$C_{21}CrCl_3$	2	1280
	$C_{22-29}CrCl_3$	3	1615
$MoCl_5$	$C_{18.6}MoCl_5$	2	1254
	$C_{27}MoCl_5$	3	1602
WCl_6	$C_{70}WCl_6$	5	2302
Group VIIb			
$MnCl_2$	$C_{6.4}MnCl_{2.06}$	1	951
	$C_{12}MnCl_{2.07}$	2	1288
$ReCl_4$	$C_{13}ReCl_{4.3}$	1	1178

continued

TABLE VI (continued)

Chloride	Composition	Stage	I_c (pm)^c
Group VIII			
$FeCl_2$	C_9FeCl_2	1	951
	$C_{15.8}FeCl_2$	2	1285
$FeCl_3$	$C_{5-9}FeCl_3$	1	937
	$C_{20}FeCl_3$	2	1280
	$C_{23-29}FeCl_3$	3	1621
$CoCl_2$	$C_{5.5}CoCl_{2.07}$	1	950
	$C_{15}CoCl_{2.07}$	2	1285
$NiCl_2$	$C_{13}NiCl_{2.04}$	2	1271
$RuCl_3$		2	1260
$OsCl_3$	$C_{12.2}OsCl_3$	2	1270
$PdCl_2$	$C_{14.8}PdCl_2$	3	1670
$PtCl_4$	$C_{42-51}PtCl_{4.5}$	3	1606

Bromides

Bromide	Composition	Stage	I_c (pm)
$AlBr_3$	$C_9AlBr_3 \cdot Br_2$	1	1024
	$C_{18}AlBr_3$	2	1340
	$C_{24}AlBr_{3.3}$	2	1335
	$C_{33}AlBr_3$	4	2010
$GaBr_3$	$C_{13-16.5}GaBr_3 \cdot Br_{2.5}$	2	1338
$AuBr_3$		1	690
$TlBr_3$	$C_{18.6}TlBr_{3.4}$	2	1340
$CdBr_2$	$C_{15}CdBr_{2.06}$	2	1330
	$C_{28.6}CdBr_{2.1}$	3	1662
$HgBr_2$	$C_{23.8}HgBr_2$	3	1684
UBr_5	$C_{38}UBr_{5.1}$	2(+3)	1328
$FeBr_2$	$C_{14.2}FeBr_{2.1}$	1(+2)	1325 (990)
$FeBr_3$	$C_{23}FeBr_3$	2	1290

Rare earth chlorides

	Composition	Stage	I_c (pm)
$SmCl_3$	$C_{56.6}SmCl_{3.2}$		
$EuCl_3$	$C_{37.1}EuCl_3$	3	1630
$GdCl_3$	$C_{23.3}GdCl_{3.1}$	3	1642
$TbCl_3$	$C_{18.7}TbCl_{3.2}$	2	1301
$DyCl_3$	$C_{19}DyCl_3$	2	1290
$HoCl_3$	$C_{20.3}HoCl_{3.1}$	2	1291
$ErCl_3$	$C_{23.3}ErCl_{3.1}$	3	1635
$TmCl_3$	$C_{23.7}TmCl_3$	3(+2)	1647 (1310)
$YbCl_3$	$C_{25.4}YbCl_{3.1}$	3	1629
$LuCl_3$	$C_{34.8}LuCl_3$	4	1948
YCl_3	$C_{28.6}YCl_3$	3(+2)	1654 (1298)

[a] From Stumpp (1977).
[b] I_c is the identity period along the c axis.
[c] pm $= 10^{-2}$ Å.

For example, in the case of AsF_5, the oxidation reaction is

$$3AsF_5 + 2e^- \rightarrow AsF_3 + 2AsF_6^- \qquad (1)$$

If the oxidation occurs in the graphite substrate, the $2e^-$ would appear to come from graphite and the species in the gap between layers would expectedly be AsF_3 and AsF_6^- ions. Conflicting data as to the nature of the moiety has arisen and x-ray data are consistent with the oxidation products being AsF_3 and AsF_6^-. However, Ebert and Selig (1982) by use of NMR data suggest that AsF_5 is the major intercalant, which reacts with AsF_6^- and intercalates in graphite as dimetallic anions ($As_2F_{11}^-$) (see Ebert and Selig).

Intercalation of graphite with other acceptor molecules can result in very high in-plane electrical conductivities. In some cases these values are greater than those for copper and silver metal.

C. Intercalation without Oxidation or Reduction

A number of cases exist where the intercalation reaction involves un-charged guest species. A group of these compounds involve the noble gas fluorides (e.g., KrF_2) and halogen fluorides (e.g., $XeOF_4$, BrF_3, IF_5, IOF_5) [see Selig (1975) and Selig et al. (1981)].

D. Donors versus Acceptors

Table VII compares several donor and acceptor species in graphite with respect to their electrical conductivities. The c axis conductivity of graphite is smaller than σ_a because of the weak interlayer interactions.

TABLE VII

Comparison of Conductivities for Several Donor and Acceptor
Graphite Intercalants at 300 K

Compound	σ_a (ohm cm)$^{-1}$	σ_c (ohm cm)$^{-1}$
C_6Li	2.5×10^4	10
C_6Ba	2.6×10^5	30
C_8K	9.5×10^4	4.8×10^4
C_8Rb	1.0×10^5	—
C_8Cs	1.0×10^5	—
C_6HNO_3	3.0×10^5	2.0
C_8AsF_5	4.0×10^5	0.28
$C_{12}HNO_3$	2.7×10^5	1.96
$C_{16}AsF_6$	2.2×10^5	0.23

TABLE VIII

Superconducting Parameters for Graphite–Alkali-Metal
Intercalation Compounds

Compound	T_c (K)	Reference
C_8K [a]	0.55	Hannay *et al.* (1965)
C_8K	0.39	Hannay *et al.* (1965)
Cs_8Rb	0.023–0.151	Hannay *et al.* (1965)
Cs_8Cs	0.020–0.135	Hannay *et al.* (1965)
C_8K	0.0128–0.198	Koike *et al.* (1978, 1980)
C_8K [b]	0.125	Kobayashi and Tsujikawa (1981)
C_8K	<0.3	Poitrenaud (1970)
C_4K	1.3	Wolski *et al.* (1986)

[a] Excess K intercalant from stoichiometry.
[b] Graphite host material was grafoil (graphite foil).

E. Superconductivity

Hannay *et al.* (1965) reported the first graphite–alkali metal system
which became superconductive. The intercalant KC_8 demonstrated
superconductivity with T_c = 0.12–0.18 K (Koike *et al.*, 1978, 1980; Kobayashi and Tsujikawa, 1981). A new report (Wolski *et al.*, 1986) describes
superconductivity occurring in KC_4 with T_c = 1.3 K. Table VIII tabulates
superconductor parameters for graphite–alkali-metal intercalation compounds [see Dresselhaus and Dresselhaus (1981)].

LIST OF SYMBOLS AND ABBREVIATIONS

PEO	Poly(ethylene oxide)
PPO	Poly(propylene oxide)
T	Trans conformation
G	Gauche conformation
$T_2GT_2G^-$	Trans, trans, gauche, trans, trans, gauche minus configuration
HOPG	Highly ordered pyrolytic graphite
grafoil	graphite foil

REFERENCES

Armand, M. B., Chabagno, J. M., and Duclot, M. J. (1978), *Ext. Abstr. Int. Conf. Solid
Electrolytes, 2nd St. Andrews, Scotland,* No. 6.5
Armand, M. B., Chabagno, J. M., and Duclot, M. J. (1979), *in* "Fast Ion Transport in
Solids" (P. Vashista, J. N. Mundy, and G. K. Shenoy, eds.), pp. 131–136, North-
Holland Publ., Amsterdam.
Armand, M. B., Duclot, M. J., and Rigand. Ph. (1980), *Int. Meet. Solid Electrolytes Solid
State Ionics Galvanic Cells, 3rd, Tokyo, Japan,* No. C116.

Bartlett, N., Biagioni, R. N., McCarron, G., McQuillan, B., and Tanzella, F. (1979), *in* "Molecular Metals" (W. E. Hatfield, ed.), p. 293, Plenum, New York.

Chabagno, J. M. (1980), Thesis, Grenoble, France.

Dresselhaus, M. S., and Dresselhaus, G. (1981), *Adv. Phys.* **30**, 139, and references therein.

Dupon, R., Papke, B. L., Ratner, M. A., Whitmore, D. W., and Shriver, D. F. (1982), *J. Am. Chem. Soc.* **104**, 6247.

Ebert, L. B., and Selig, H. (1982), *Chem. Eng. News* **60**(40), 27.

Fenton, D. E., Parker, J. M., and Wright, P. V. (1973), *Polymer* **14**, 589.

Fischer, J. E. (1979), *in* "Molecular Metals" (W. E. Hatfield, ed.), p. 281, Plenum, New York.

Fischer, J. E., and Thompson, T. E. (1978), *Phys. Today,* pp. 36–45, (July).

Hannay, N. B., Gebralle, T. M., Mattias, B. T., Andres, K., Schmidt, P., and NacNair, D. (1965), *Phys. Rev. Lett.* **14**, 225.

Herold, A., Billaud, D., Guerard, D., and Lagrande, P. (1977), *Mat. Sci. Eng.* **31**, 25.

Iwamoto, R., Saito, Y., Ishihana, H., and Tadokoro, H. (1968), *J. Poly. Sci. Part A-2* **6**, 1507.

Iye, Y., and Tanuma, S. (1982), *Solid State Commun.* **44**, 1.

James, D. B., Stein, R. S., and Macknight, W. J. (1979), *Bull. Am. Phys. Soc.* **24**, 479.

Kobayashi, M., and Tsujikawa, I. (1981), *J. Phys. Soc. Jpn.* **50**, 3245.

Koike, Y., Suematsu, H., Higuchi, K., and Tanuma, S. (1978), *Solid State Commun.* **27**, 623; (1980), *Physica* **99B**, 503.

LaGrange, P., Makrini, M. El., Guerard, D., and Herold, A. (1980), *Physica B* **19B**, 473; *Synth. Met.* **2**, 191.

Papke, B. L., Dupon, R., Ratner, M. A., and Shriver, D. F. (1981), *Solid State Ionics* **5**, 685.

Papke, B. L., Ratner, M. A., and Shriver, D. F. (1982a), *J. Electrochem. Soc.* **129**, 1434.

Papke, B. L., Ratner, M. A., and Shriver, D. F. (1982b), *J. Electrochem. Soc.* **129**, 1694.

Poitrenaud, J. (1970), *Rev. Phys. Appl.* **5**, 275.

Selig, H. (1975), *Inorg. Nucl. Chem. Lett.* **11**, 75.

Selig, H., Pron, A., Druy, M. A., MacDiarmid, A. G., and Heeger, A. J. (1981), *J. Chem. Soc. Chem. Comm.,* p. 1288.

Shriver, D. F., and Farrington, G. C. (1985), *Chem. Eng. News* **63**(20), 42–57.

Shriver, D. F., Papke, B. L., Ratner, M. A., Dupon, R., Wong, T., and Brodwin, M. (1981), *Solid State Ionics* **5**, 83.

Spain, I. L. (1973), *Chem. Phys. Carbon* **8**, 1.

Spain, I. L., Ubbelohde, A. R., and Young, D. A. (1967), *Proc. R. Soc. London Ser. A* **262**, 345.

Stumpp, E. (1977), *Mat. Sci. Eng.* **31**, 53.

Takahashi, Y., and Tadokoro, H. (1973), *Macromolecules* **6**, 672.

Vogel, F. L. (1979), *in* "Molecular Metals" (W. E. Hatfield, ed.), p. 261, Plenum, New York.

Vogel, F. L., and Zeller, C. (1979), *in* "Molecular Metals" (W. E. Hatfield, ed.), p. 289, Plenum, New York.

Vogel, F. L., Foley, G. M. T., Zeller, C., Falardeau, E. R., and Gan, J. (1977), *Mat. Sci. Eng.* **31**, 2611.

Wolski, E. P., Zharikov, O. V., Palnichenko, A. U., Audeev, V. V., Mordkovich, V. Z., and Semenko, K. N. (1986), *Solid State Commun.* **57**, 421.

Wright, P. V. (1976), *J. Br. Polymer* **7**, 319.

Zeller, C., Foley, G. M. T., Falardeau, E. R., and Vogel, F. L. (1977), *Mat. Sci. Eng.* **31**, 2551.

Zeller, C., and Pendrys, L. A., unpublished data.

Review Articles

Dresselhaus, M. S., and Dresselhaus, G. (1981), *Adv. Phys.* **30**, 139.

Shriver, D. F., and Farrington, G. C. (1985), Solid ionic conductors, *Chem. Eng. News,* (May 20), pp. 42–57.

Books

Hagenmuller, P., Van Gogh, W., eds. (1978), "Solid Electrolytes—General Principles, Characterization, Materials, Applications," pp. 1–549, Academic Press, New York.

Mahan, G. D., and Roth, W. L., eds. (1976), "Superionic Conductors," pp. 1–438, Plenum, New York.

Nakao, K., and Solin, S. A., eds. (1985), "Graphite Intercalation Compounds," Proc. Int. Symp. on Graphite Intercalation Compounds, *Synth. Met.* **12**, 1–541.

Vashista, P., Mundy, J. N., and Shenoy, G. K., eds. (1979), "Fast Ion Transport in Solids," pp. 131–136, North-Holland, Amsterdam.

"Nomenclature and Technology of Graphite Intercalation Compounds," report by a subgroup of the International Committee for Characterization and Terminology of Carbon and Graphite on Suggestions for Rules for the Nomenclature and Terminology of Graphite Intercalation Compounds, H. P. Boehm, R. Setton, and E. Stumpp (1985), *Synth. Met.* **11**, 363.

7

INTRODUCTION TO THE PHYSICS OF LOW-DIMENSIONAL SYSTEMS

I. INTRODUCTION

In previous chapters we have attempted to discuss the general properties of low- or one-dimensional (1D) conductors. The initial thrust in synthetic metal research since the mid 1960's was directed toward one-dimensional conductors, especially since Little (1964) proposed the possible existence of high-temperature superconductivity in a one-dimensional system. A knowledge of the crystal structures of certain of these materials has been especially helpful in providing an understanding of the often highly anisotropic character of their electrical conductivities. In many cases it has become possible to relate subtle structural differences to electronic properties. Since the metal–insulator (M→I) transition is so important in many one-dimensional conductors, considerable attention has been devoted to this area. X-ray diffuse and neutron scattering experiments, in addition to the conventional crystal structure determinations based on Bragg diffraction data, have proven invaluable in characterizing these transitions.

In this chapter we will attempt to address the nature of the conduction behavior of the systems discussed in previous chapters. A simple discussion of one-dimensional band theory will be presented first because of its importance to the understanding of one-dimensional systems.

A. One-Dimensional Band Theory

A simple theoretical basis for one-dimensional conduction is given by the tight-binding band approximation in which one considers a one-dimensional lattice of N equally spaced molecules with a repeat distance equal to c (see Fig. 1a). Each molecule has an orbital with energy E_0 (in the isolated molecule) that overlaps with the equivalent orbitals on each of its two neighbors. The energies for the orbitals constituting the one-dimensional array are derived from the tight-banding band theory such that

$$E_i(\mathbf{k}) = E_0 - W \cos(k \cdot c) \equiv E_i(\mathbf{k}) \tag{1}$$

where the allowed values of wave vector $\mathbf{k}$, and therefore, the energies E_i are quantized. The energies are so closely spaced that an apparent continuum of allowed energies above and below the isolated molecule E_0 exists, as shown in Fig. 1b. This continuum is bounded within a bandwidth of $2W$ and is composed of N orbitals capable of holding $2N$ electrons. The bandwidth is related to the nearest-neighbor transfer integral t by $2W = 4t$. The wave vector $\mathbf{k}$ has N values evenly spaced between $-\pi/c$ and $+\pi/c$, such that the density of states per unit energy per sum $D(E)$ gives the curve illustrated in Fig. 1b (Stucky et al., 1977).

It has been pointed out by Peierls (1955) and Frohlich (1954) that one-dimensional conductors are inherently unstable towards lattice distortions. For a Peierls transition to occur in a one-dimensional crystal with exactly a half-filled band, as shown in Fig. 1, a lattice distortion in which the molecules dimerize results in a band gap at the Fermi wave vector $\mathbf{k}_F = \pi/2c$. The metal–insulator transition occurs at $T > 0$ and results from coupling between the conduction band electrons and the phonon mode at wave vector $2\mathbf{k}_F$, which produces the Kohn (Kohn, 1959) anomaly as shown in Fig. 1c. A static lattice distortion is produced when the energy $h\omega$ of the $2\mathbf{k}_F$ phonons becomes zero at the transition temperature.

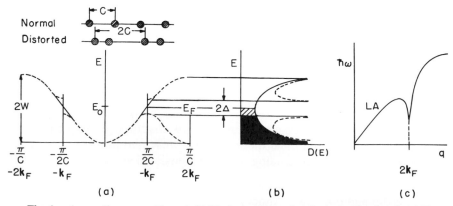

Fig. 1.   Anomalies caused by a half-filled one-dimensional conduction band. (a) Formation of a band gap at $\mathbf{k}_F$ due to a Peierls distortion of one-dimensional lattice. (b) Plot of density of states per unit energy per spin. (c) Giant Kohn anomaly at $q = 2\mathbf{k}_F$ in the LA branch of the phonon dispersion curve. [From Stucky et al. (1977).]

Frohlich (1954) proposed the existence of charge-density waves due to the periodic potential produced by the $2\mathbf{k}_F$ lattice vibration. If the waves are not "pinned" at some point in the lattice, propagating waves produce a traveling potential that could result in enhanced conductivity. The metal–insulator transition occurs at $T_{M \rightarrow I}$ when the waves become "pinned" due to a static lattice distortion.

It should be pointed out that the electron–phonon coupling is stronger for cases where the first reciprocal lattice vector G ($= 1/c$) and the wave vector $\mathbf{k}_F$ are commensurate than for the cases where they are incommensurate (Sham and Patton, 1975). A one-dimensional conductor will remain metallic only if the elastic restoring forces of the lattice overcome the distortions caused by electron–phonon coupling. Stronger coupling results in a higher $T_{M \rightarrow I}$.

B. Electronic Properties in Energy Band Systems

The electrical conductivity of any solid is determined by its electronic structure. The electronic distribution in any electrically conductive system is described by its orbitals and the orbital defines the spatial distribution of electron charge. Orbital overlap is a requirement to provide facile transport of electrons from site to site in the solid lattice. For a solid, each electron occupies a particular orbital with a certain energy associated with it. Each orbital is capable of accommodating two electrons and for an atom in its ground state, the first two electrons fill the lowest lying energy orbital. Any additional electrons must be accommodated in a higher energy orbital. Because of the confinement of the electrons in orbitals the energy distribution involves discrete energy levels, with all intermediate levels being forbidden.

Whenever two identical atoms are brought to close proximity, whereby orbital overlap can occur, each energy level is split into two new levels with one above and one below the original level. The extent of orbital overlap determines the magnitude of splitting. If a third atom is added, the atomic levels are split into three levels. Figure 2 illustrates how energy bands are formed ranging from an isolated atom to a polyatomic solid. For a solid with $\sim 10^{23}$ atoms each energy level splits, but intervals between energy levels become so small that an effectively continuous band results, as illustrated in Fig. 1b.

Figure 3 depicts various band structures of solids. Conduction is also impossible if a band is full because no net motion of electrons is possible. In a metal, electrons in the highest filled energy level can be readily promoted to an unoccupied level since energy differences (gaps) are small. Similar conditions are present in a semimetal. An insulator has only completely filled bands and completely empty ones, and the energy gap or

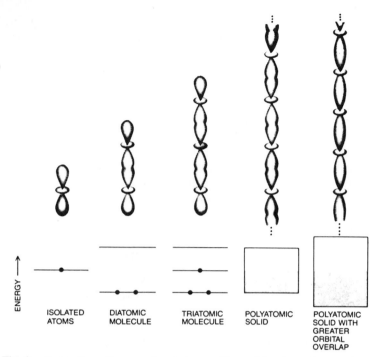

Fig. 2.   Formation of energy bands in a solid resulting from the overlapping of the orbitals of adjacent atoms. In an isolated atom each orbital, which can hold no more than two electrons, has a well-defined energy. When two atoms are brought together close enough for their orbitals to overlap, each energy level is split in two; adding a third atom splits the atomic levels into three components. In a solid made up of, perhaps, 10^{23} atoms the levels are so numerous and so closely spaced that they effectively form a continuous band. The width of the band is determined by the extent to which the orbitals overlap. In a solid made up of n atoms each band can accommodate 2n electrons. If each of the atoms in a solid has only one electron at some energy level (rather than the maximum of two) the corresponding band will be only half full. [From Epstein and Miller (1979).]

band gap is extremely large, thus preventing conduction. A semiconductor resembles the electron distribution found in an insulator, except for a smaller gap. In a semimetal the conduction band is empty and the valence band is filled, but they overlap with one another, resulting in electron redistribution to create two partially filled bands.

The extent of occupation of the energy levels and the magnitude of the gap determines the electrical properties of the material. When the proper conditions are met the electrons are said to be delocalized and may propagate throughout the solid. However, if any electrical current is to be sustained there must be a net movement of electrons in one direction. No net movement results in insulating materials because the energy levels are continuously filled from the lowest available to the highest levels.

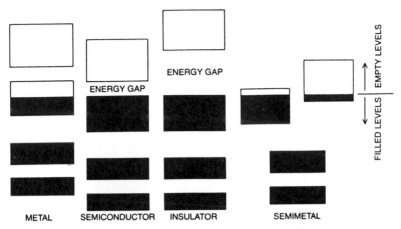

Fig. 3. Band structure of a solid which determines most of its electrical properties. No conduction can take place in an empty band because it contains no electrons. Conduction is also impossible in a full band because the total population of electrons in such a band can have no net motion. A metal is a material with a partially filled band, where the electrons can be given a net velocity by shifting some of them to infinitesimally higher energies within the band. An insulator has only completely filled bands and completely empty ones, with a large energy gap between them. A semiconductor has a band structure much like that of an insulator but has some mobile charge carriers. The carriers can be introduced by impurities, by defects, by departures from stoichiometry or by the excitation of electrons from the highest filled band (the valence band) to the lowest empty one (the conduction band). What distinguishes a semiconductor from an insulator is that the gap between bands is comparatively small in a semiconductor. In a semimetal the valence band is filled and the conduction band is empty, but these bands overlap in energy. As a result electrons redistribute themselves to create two partially filled bands. [From Epstein and Miller (1979).]

C. The Bardeen–Cooper–Schrieffer (BCS) Theory

A short discussion of superconductivity as it relates to one-dimensional materials might prove helpful for later discussions in this chapter. The BCS theory (Bardeen *et al.*, 1957) is based on an electron-pairing mechanism driven by phonons which occurs at sufficiently low temperatures. Electrical conduction in a superconductor is caused by the net motion of the centers of mass of an electron pair (called Cooper pairs). Energy requirements that hinder Cooper pair formation are greater than those for ordinary electrons, and at very low temperature this energy may not be available. In a metal, electrical conduction is caused by the net motion of individual valence electrons. The superconducting transition temperature can be predicted, according to the BCS theory by

$$T_c \simeq (\omega_0/k_b) \exp[-1/D(\epsilon_F)V] \qquad (2)$$

where ω_0 is the Debye frequency of the lattice, $D(\epsilon_F)$ is the density of states at the Fermi surface, and V is the effective electron–electron attraction. At the present time, the highest observed values of T_c are <24 K. If an attempt is made to raise T_c by increasing V, at some point a lattice instability results. Little (1964, 1970) proposed changing the frequency term by replacing it with a much higher frequency of an electronically polarizable system. In such a model, electrons traveling along a two-dimensional backbone, surrounded by polarizable side groups or ligands would pair up via an electron–exciton–electron mechanism, where an exciton is defined as an electron-hole pair. Little's theory predicts very high critical temperatures (~2000 K), which as yet have not been realized (Miller and Epstein, 1976).

II. LATTICE INSTABILITIES IN ONE-DIMENSIONAL CONDUCTORS

Having considered some elementary concepts concerning electrical conduction in terms of orbital theory and band structure, it is worth considering why electron delocalization ceases in solids in certain cases as the temperature is lowered. This is one of the often encountered characteristics of one-dimensional conductors, and must be circumvented in order to create conductive and superconductive materials.

Lattice instabilities in one-dimensional solids are most frequently caused in three ways:

(1) Peierls instability [electron interactions with phonons, called a charge density wave (CDW)];

(2) instabilities in one- or two-dimensional conductors caused by spin-density wave formation (SDW) arising from magnetic interactions; and

(3) anion-ordering transitions at low temperature.

A. Peierls Instabilities

The so-called Fermi surface is a mathematical construction related to the dynamical properties of the conduction electrons in a metal. The surface provides all of the electron states that can play any part in the ordinary transport properties of the metal, or in such thermal properties as the specific heat (Ziman, 1963). In one-dimensional solids the Fermi surface will shrink to two points at $\pm\mathbf{k}_F$ (wave vector of lattice). This would normally produce metallic behavior if the system were stable. However, the ground state of a system of interacting electrons in one-dimension does not possess a Fermi surface. There is always a gap involved, and the gap is dependent on details of interactions between the electrons and the phonons in the lattice. Peierls (1955) has shown that the phonon–electron interaction produces periodic distortions of the lattice whereby the wavelength will be given by $2\pi/\lambda = 2k_F$ and $\lambda = \pi/k_F$. Figure 4 shows the electronic energy for a one-dimensional conductor with states filled for $|\mathbf{k}| \leq k_F$. Lattice distortions cause the wave vector to equal $2k_F$ and this causes a gap 2Δ to open in the electronic spectron at $\pm\mathbf{k}_F$, which lowers the energy of all the occupied states (see Fig. 5). The size of Δ is proportional to μ, the amplitude of the $2k_F$ distortion. Because of this gap that opens as one lowers the temperature, the conduction ceases, and a transition to an insulating state (or semiconducting) ensues; depending on the size of the gap. If, for example, sufficient thermal energy can be provided to overcome this gap, conduction (a semiconductor) can be maintained. In Chapter 8 it will be shown that increasing the external pressure is one way to decrease this gap and also maintain conduction. In normal situations

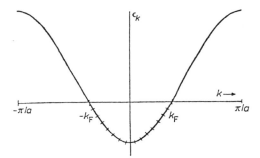

Fig. 4.   Unstable Fermi surface in one-dimensional conductors at low temperatures. [From Berlinsky (1979).]

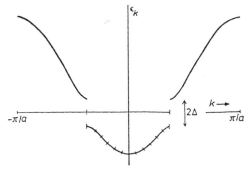

Fig. 5.   Peierl's distortion at low temperatures (material can become a semiconductor or insulator). Gap $= 2\Delta$ depends on amplitude of $2k_F$ distortion. [From Berlinsky (1979).]

the interaction between the electrons and the phonons of the lattice, is unpinned and periodic. In other cases the interaction causes strain to the lattice, thereby increasing the energy. Figure 6 illustrates the difference between a pinned and unpinned CDW, where the periodicity is disturbed in the former case. The Peierls transition occurs only if the energy of the electrons is reduced to compensate for the strain of the lattice and in one-dimensional solids this condition occurs at low temperatures. At high temperatures the uniform lattice configuration is preferred. To present a further elaboration, consider that in the distortion that occurs the electrons tend to become concentrated in the region of greater positive charge, where the chain is contracted. This aperiodic electron distribution is called a charge-density wave (CDW).

B. Spin-Density Wave (SDW) Instabilities

Instabilities in one- or two-dimensional solids may also be caused, for example, by a competition between superconducting and magnetic (anti-ferromagnetic–SDW) ground states. The antiferromagnetic phase is insulating due to a $2k_F$ periodicity of the magnetic order, which destroys the Fermi surface. Separation of positive charges and unpaired electrons (spins) are stabilized. This type of distortion is exhibited by many of the charge-transfer conductors (Bechgaard salts) discussed in Chapter 2, and the distortion has been studied by NMR as well as ESR and magnetic susceptibility measurements (Walsh et al., 1980; Scott et al., 1984). Confirmation of a spin-density wave driving a distortion in $(TMTSF)_2PF_6$ came from Walsh et al. (1982), who found that the electron spins order antifer-romagnetically (alternating spin up and spin down). This magnetic order causes an interruption of the electron flow by introducing an energy gap in the electronic band. Application of external pressure on these materials can sometimes suppress the SDW distortion and can, as has been shown for $(TMTSF)_2PF_6$, result in a superconducting ground state.

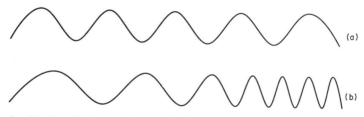

Fig. 6. Unpinned (a) and pinned (b) charge-density wave (CDW) formation. An unpinned CDW shows the periodicity of the lattice. A pinned CDW shows a disturbance of the periodicity.

C. Anion Ordering

Another cause of metal–insulator transitions in conductive solids may involve anion ordering that occurs at some low temperature. Often the anions are disordered at higher temperatures. For example, in $(TMTSF)_2ReO_4$ the metal–insulator transition is caused by anion ordering and occurs at ~ 180 K. In this system, pressure suppresses the transition, and it becomes superconducting. See Chapter 2 for a further discussion of this topic.

III. TRANSITION BEHAVIOR OF VARIOUS SYNTHETIC METAL CONDUCTORS

Table I summarizes data relative to the observed transition temperatures of various types of conductors. The various conductors can be generally classified into four general groups based on their conduction behavior (see Ferraro, 1982). Class I_A involves one-dimensional organic charge-transfer conductors, which all exhibit a metal–insulator transition caused by charge-density wave formation. Class I_B involves the organic charge-transfer conductors which are structurally two-dimensional and at low temperature become three-dimensional and exhibit generally lower $T_{M\rightarrow I}$'s than those conductors in Class I_A. Spin-density wave formation and anion ordering cause many of the metal–insulator transitions in such systems. In the case of $(TMTSF)_2ReO_4$ and $(TMTSF)_2BF_4$, anion ordering is responsible for the metal–insulator transitions as previously indicated. Class II_A involves the one-dimensional tetracyanoplatinates and metallic macrocyclic conductors. Here a CDW mechanism is the origin of the metal–insulator transition. Class II_B comprises the one- and two-dimensional conductors like $Hg_{2.86}AsF_6$ and $Hg_{2.91}SbF_6$, which become three-dimensional at lower temperatures and also become metallic and superconductive at <4.2 K. However, the superconductivity in these systems is believed to arise from the presence of elemental mercury which is released during cooling. Class III_A is represented by the one-dimensional organic polymers such as $[CH^{+y}(X)_y^-]_n$, which may involve CDW mechanisms for the metal–insulator transitions. Class III_B is represented by polysulfur nitride and its doped polymers, which are semimetals at room temperature, and which become superconductors at $T_c \sim 0.3$ K. Class IV contains graphite and doped graphite which are considered to be two-dimensional semimetals at room temperature, and which become metallic at lower temperatures. Undoubtedly, this classification scheme will

TABLE I

Metal–Insulator Transition Temperatures for Various Conductors

Type of conductors	$T_{M \to I}$ (K)	Origin	Temperature (K) (SC)[a]	Pressure (kbar)
Class I_A: 1D				
TTF–I	~230	CDW		
TTF–TCNQ	58	CDW		
TSF–TCNQ	28	CDW		
TMTTF–TCNQ	60	CDW		
TMTSF–TCNQ	~57	CDW		
HMTTF–TCNQ	48	CDW		
HMTSF–TCNQ	16	Resembles a metal CDW		
TMTSF–DMTCNQ	41 (ambient p)	Metallic at 10 kbar and 1 K		
Class I_B: 2D → 3D (TMTSF)₂X				
X = PF₆	12–17	SDW	0.9	6.5
AsF₆	12–16	SDW	1.1	12
SbF₆	17	SDW	0.4	11
TaF₆	—	SDW	1.4	12
NbF₆	12	—	Absent	12
TeF₆	Conductor at 5 K	—		
ClO₄	Absent	Anion ordering	1.4	Ambient
ReO₄	180	Anion ordering	1.3	9.5
BF₄	40	—		
BrO₄	Metallic	—		
IO₄	Semiconductor	—		
NO₃	12	SDW, anion ordering	Absent	12
F₂PO₂	137	Metallic at 14.5 kbar and 20 K		
SO₃F	86–90	—	2.5	6.5

Compound	Property	CDW	T_c / value	Ref.
CF$_3$SO$_3$	Narrow metallic region at H.T.	—		
(ET)$_2$X		—		
(ET)$_2$ClO$_4$	Metallic to 1.4	—		2
(ET)$_2$ReO$_4$	81	—		4
β-(ET)$_2$I$_3$	Superconductive at 1.4–1.5, ambient pressure			
β-(ET)$_2$IBr$_2$	Superconductive at 2.8, ambient pressure			
β-(ET)$_2$AuI$_2$	Superconductive at 5, ambient pressure			
(TMTTF)$_2$X				
X = PF$_6$	230 (10 at 30 kbar)			
ClO$_4$	230 (30 at 30 kbar)			
Br	(10 at 22 kbar)	—		
Class II$_A$: 1D				
KCP	200	CDW		
NiPcI	~50	CDW		
Class II$_B$: 1D → 2D				
Hg$_{2.86}$AsF$_6$	3D ordering at L.T. metallic, SC at <4.2 K	—		
Hg$_{2.91}$SbF$_6$	Hg deposits out at L.T.			
Class III$_A$: 1D				
[CH^{+y}(X)^-_y]$_x$ (Polymeric)		CDW		
Class III$_B$ (Semimetal at 298 K)				
(SN)$_x$ (Polymeric)	Conductor 298 K to superconductive	—	~0.3	—
(SNBr$_{0.4}$)$_x$		—	~0.36	—
Class IV: 2D → 3D				
Graphite	Semimetal 298	—		
Graphite–AsF$_5$	Metallic	—		
Graphite–K				
(KC$_8$)		—	Superconductive, $T_c = 0.12$–0.18 K	—
(KC$_4$)		—	Superconductive, $T_c = 1.3$ K	—

^{a}SC = superconductivity.

undergo modifications as new conductors are synthesized. Figure 7 depicts the temperature dependence of the conductivity of several of these conductors.

Epstein and Miller (1979) have examined the conductivity behavior in various solid conductors and proposed three classes of behavior in addition to the metallic class (e.g., Cu, etc.). They have proposed a model to explain this behavior applicable to temperatures down to 70 K. The model is based on the concentration of charge carriers (n) and the mobility of the carriers (μ).

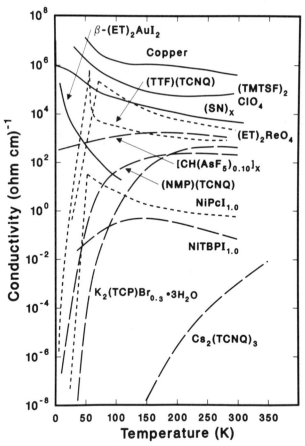

Fig. 7.   Temperature dependence of conductivity for various electrical conductors. [Modified from Epstein and Miller (1979).]

IV. NATURE OF ELECTRONIC DELOCALIZATION IN VARIOUS CONDUCTORS

A. Introduction

In this section we attempt a limited presentation of the nature of electronic delocalization occurring in some of the conductors previously discussed. It should be noted that considerable controversy exists concerning projected mechanisms of conduction especially in such cases as the superconducting charge-transfer salts. Therefore, our presentation must be viewed as one which will undoubtedly change with the accumulation of new information on many of these systems.

Many problems exist with some of these conductors which contribute toward the difficulty in establishing the method of conduction. In the case of polymeric conductors, the problem of nonuniform distribution of dopant is one example. The dependence of the conductivity on the dopant concentration is also a problem. This necessitates proposing different mechanisms of conduction at different doping levels. Another serious problem occurring for the new superconductors is the necessity of postulating a mechanism for the conduction from room temperature to the onset of a superconductive state at low temperature and then a second mechanism in the superconductive state. In the latter case, one is presented with the lack of definitive knowledge of whether or not some of these materials that become superconductive follow the BCS theory of superconductivity in metals (Bardeen *et al.*, 1957). Some evidence has accumulated which indicates that $(TMTSF)_2X$ salts form Cooper pairs along the donor chains even at temperatures considerably above the onset of bulk superconductivity (Bryce and Murphy, 1984). With these problems in mind, the reader must realize that much of our discussion is speculative.

B. One-Dimensional Metal Conductors—Krogmann Salts, M(mac), and $[Si(PcO)]_n$

The Krogmann salts are metal-centered systems in which electron delocalizaton occurs along overlapping d_z^2 orbitals of the closely spaced platinum atoms ($d_{Pt-Pt} \simeq 2.9$ Å) in the $[Pt(CN)_4]^{n-}$ chains. Partial oxidation of the metal is necessary (forming a nonintegral oxidation state), which creates partially filled bands. Holes in the Pt–Pt chains form and this occurs with a smaller activation energy than if an integral oxidation state were involved with higher activation energy. Charge transport occurs with movement of holes or electrons between degenerate configurations. The movement between other columns or chains, typically separated by 10–12 Å distances, is much less efficient causing these solids to have highly anisotropic electrical conductivities.

For the organometallic complexes involving transition metals and macrocyclic ligands the ligand π-orbital interactions are undoubtedly the major contributors to the conductivity. The π-systems in the ligand columns are the pathway for the conduction. Table II summarizes some physical properties and the predominant conductive mechanisms in these conductors.

In the case of metallomacrocyclic polymeric conductors, the conduction band of the partially oxidized $[Si(PcO)]_n$ conductor is composed principally of carbon $p\pi$ orbitals located near the macrocyclic core. Mixing between C-π orbitals with Si or O orbitals occurring in the highest occupied molecular orbital is minimal. Thus, it is unlikely that the $(SiO)_n$ chains play any direct role in the conduction process, and the conduction involves π-orbital overlaps in the macrocylic ligand. In these cases variable conductivity data have been best fitted to a fluctuation-induced carrier tunneling model. This mechanism is based on tunneling between small metal particles in an insulating medium. This is consistent with the inhomogeneous nature of doping is these conductors (Marks, 1985).

C. Organic Charge-Transfer Metals

The TCNQ molecule is planar and stacks easily in the solid state. The highest molecular orbital of $TCNQ^-$ is a π-orbital and one obtains a linear combination of atomic p_z orbitals where z is in the direction normal to the molecular plane. Interaction of these π-molecular orbitals leads to the formation of a conduction band in the solid. It is along these overlapping π-orbitals that electron motion occurs.

In (TTF)TCNQ both molecules are planar and two separated stacks, in a zigzag arrangement, are found. Carbon p-π orbitals of each molecule overlap in the stacks. Little contact between orbitals of different stacks exist and interchain electronic communication is quite limited resulting in insulating transitions in both stacks at low temperatures. Therefore, one observes primarily a one-dimensional conduction where the conductivity parallel to each stack is 500 times greater than in the perpendicular direction. The amount of charge transfer is important in these compounds. In (TTF)TCNQ, 0.59 electrons are transferred from each TTF molecule to a TCNQ acceptor. This results in emptying a filled band and partially filling an empty band.

For the Bechgaard salts, $(TMTSF)_2X$, and presumably the new $(ET)_2X$

TABLE II

Some Physical Properties of One-Dimensional Conductors[a]

Conductor	Color	$d(M\text{-}M)$ Å	Δ^{b} (Å)	Oxidized σ (ohm cm)$^{-1}$	Nonoxidized σ (ohm cm)$^{-1}$	R^{c}	Predominant conductive mechanism[d]
One-dimensional tetracyanoplatinates	Bronze, copper, gold	<2.96	0.7	10^3	10^{-7}	10^{10}	Metal–metal centered
Pt–Pt in metal		2.775	—	9.4×10^4	—	—	
Planar organometallic [e.g., $[NiPc](I_3)_{0.33}$]	Silver–gold	3.244	e	1	10^{-11}	10^{11}	Ligand–ligand centered
Ni–Ni in metal		2.49	—	1.46×10^3			
Distorted organometallic [e.g., $Ni(dpg)_2(I_5)_{0.2}$]	Black–gold	3.22	0.32	$2\text{–}11 \times 10^{-3}$	$<8 \times 10^{-9}$	$\sim 10^7$	Ligand–ligand centered
Pd–Pd in metal		2.751	—	9.26×10^4	—		

[a] Taken from Ferraro (1982).

[b] $\Delta = d(M\text{-}M)(\text{nonoxidized}) - d(M\text{-}M)(\text{oxidized})$.

[c] $R = \sigma$ oxidized$/\sigma$ nonoxidized.

[d] Martinsen et al. (1985) have found that in (CoPc)I the conductivity is in the direction of the stacking. Thus, this compound behaves more like a Krogmann salt insofar as the conductivity path is concerned (metal–metal centered). Hanack (1984) described polymers of Fe, Co, Ru phthalocyanines that are bridged with pyrazine, 4,4'-bipyridine and 1,4-diisocyanobenzene, whereby both the metal and the macrocycle (phthalocyanine) are participating in the conductivity. Collman et al. (1986) prepared conductive polymers of Fe, Ru, and Os of octaethylporphyrin (OEP) with pyrazine bridges and a stoichiometry of M:pyrazine:OEP of 1:1:1. Electron transport based on electrochemical and ESR measurements appears to be along the metal–pyrazine backbone with OEP not participating. [Martinsen, J., Stanton, S., Greene, J., Tanaka, J., Hoffman, B., and Ibers, J. (1985). J. Am. Chem. Soc. 107, 6915; Hanack, M. (1984). Mol. Cryst. Liq. Cryst. 105, 133; Collman, J. P., McDevitt, J. T., Yee, G. T., Leidner, C. R., McCullough, L. G., Little, W. A., and Torrance, J. B. (1986). Proc. Natl. Acad. Sci. USA, 83, 4581.]

[e] Comparison meaningless since Ni(Pc) does not stack.

conductors, the electrical conduction is based on the overlap of the selenium or sulfur orbitals, respectively. The intermolecular distances are frequently larger than the van der Waals radius sum of carbon, but shorter than that for selenium or sulfur, respectively, and this provides considerably added dimensionality to these systems especially at low temperature. Additionally, the interstack chalcogenide distances are often much shorter than the van der Waals radius sum and may, in some cases, result in extensive orbital overlap. In these systems, as has already been alluded to in Chapter 2, the anions play an important role through Coulomb interactions or anion-ordering transitions. There is also some evidence that in the $(TMTSF)_2X$ salts, Cooper pairing may take place along the donor chains at temperatures as high as 30 K. Below T_c the superconducting behavior may follow the BCS mechanism (Bardeen *et al.*, 1957; Bryce and Murphy, 1984). However, this has not been fully confirmed nor accepted. Needless to say, the question of the exact mechanisms of superconduction in these systems is one of the major controversies in the field of organic superconductors [for discussions see Jérome and Schultz (1982) and Greene and Chaikin (1984)].

D.　Organic Polymeric Conductors

Since the conduction mechanisms for conjugated polymers has centered on discussions of solitons, polarons, and bipolarons, it is appropriate to discuss these concepts from a qualitative point of view.

The soliton theory has been discussed quite extensively from a mathematical perspective (Su *et al.*, 1980; Rice, 1979; Takayama *et al.*, 1980; Brazovskii, 1978, 1980; Brazovskii and Kirova, 1981; Brédas and Street, 1985; Davydov, 1985). These studies have been conducted with the hope of better understanding the physics of polyacetylene, and in order to explain the unusual electrical and magnetic properties observed in the polyacetylene polymer.

Polyacetylene exists in two isomeric forms, or as a mixture of the two, depending on the temperature at which it is synthesized. The structures of the two forms (cis and the trans) are

cis　　　　　　　　　　　　*trans*

The cis form is converted to the trans isomer on heating to 150–200°C or by chemical or electrochemical doping at room temperature.

One can interchange the double and single bonds in trans-$(CH)_x$ without changing the energy. Thus, there are two degenerate lowest energy states, A and B, having two distinct bonding structures (Su *et al.*, 1979).

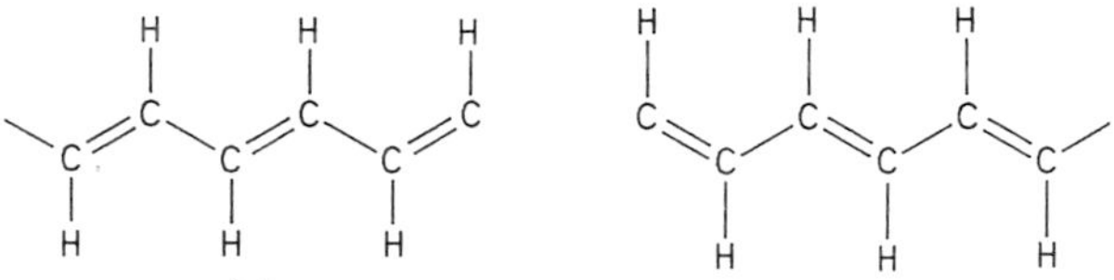

(a)   (b)

The two-fold degeneracy leads to the existence of nonlinear topological excitations, bond-alternation domains, or solitons, which appear to be responsible for the remarkable properties of $(CH)_x$. This occurs by joining together the two lowest energy states a and b. If the localized state contains one electron, the soliton is neutral, with spin $\frac{1}{2}$, and is therefore paramagnetic.

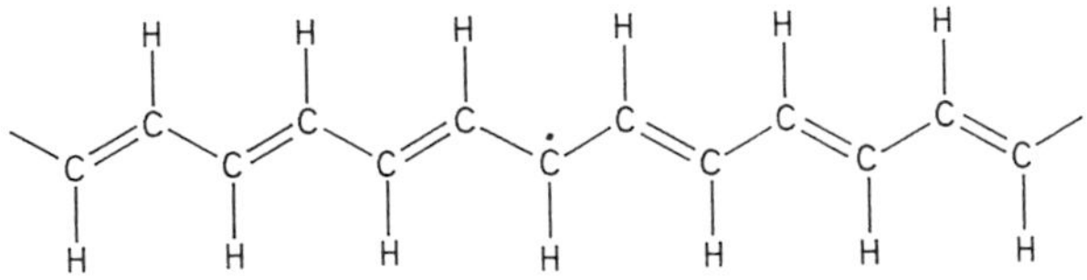

nuetral soliton

When the electron in the localized state is removed, for example by *p*-doping, the soliton is positively charged with spin $= 0$, and is nonmagnetic.

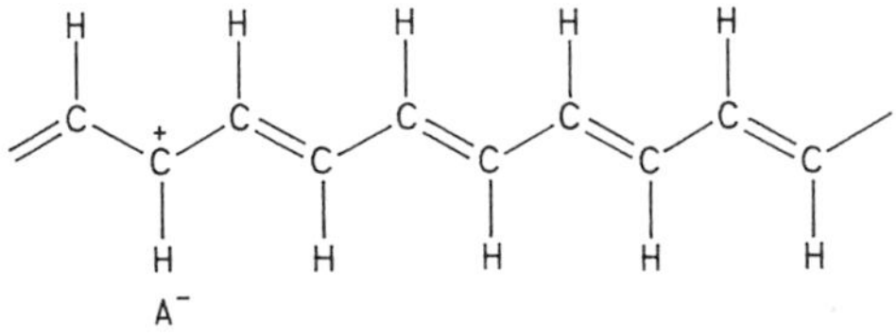

positive soliton

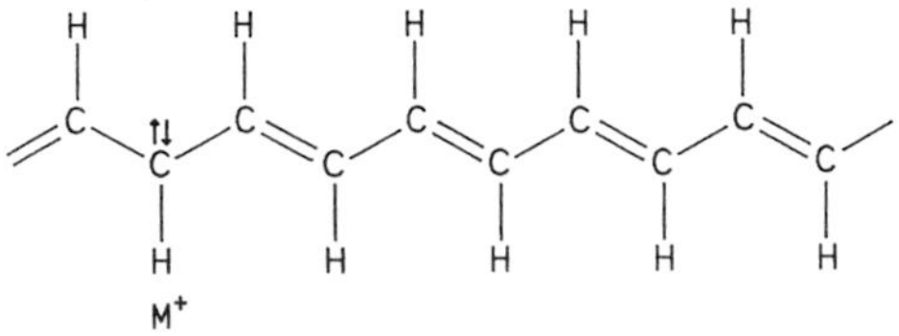

negative soliton

Similarly, if n-doping occurs, a negative soliton is obtained with spin = 0 and the system is nonmagnetic. Effectively, a negative soliton is a stabilized carbanion. The ESR studies demonstrate that neutral solitons are highly mobile, whereas the positive and negative solitons are believed to have their unit charge localized and distributed over about 15 CH units at the center of which is located the dopant cation or anion (Su et al., 1980). The charge on each CH unit decreases as its distance from the dopant ion increases. The C–C bond lengths at the center of the soliton are equal, but the bond lengths revert to their characteristic alternating lengths (as found in parent $(CH)_x$ as the distances from the dopant ion increases. Figures 8–10 demonstrate the three types of solitons relative to their location between the conduction and valence bands.

It is generally believed that conduction in doped polyacetylene proceeds by a hopping mechanism. The dominant process may involve the capture of a mobile electron from a neutron solition in an adjacent chain by means of an interhopping mechanism (Suzuki et al., 1980). Electrical conductivity

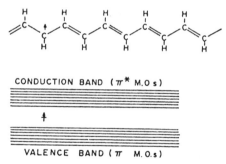

Fig. 8.   Diagrammatic representation of a neutral soliton (free radical, denoted as ↑ located on a nonbonding π molecular orbital) in trans $(CH)_x$. [From Heeger and MacDiarmid (1981).]

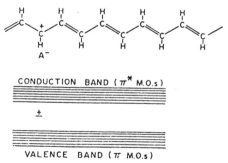

Fig. 9.   Diagrammatic representation of a positive soliton (carbonium ion located on a nonbonding π molecular orbital) in trans $(CH)_x$. [From Heeger and MacDiarmid (1981).]

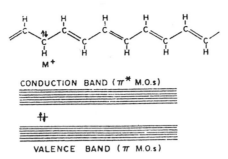

CONDUCTION BAND (π^* M.O.s)

VALENCE   BAND (π M.O.s)

Fig. 10. Diagrammatic representation of a negative soliton (carbanion located on a nonbonding π molecular orbital) in trans $(CH)_x$. [From Heeger and MacDiarmid (1981).]

and thermopower measurements are in excellent agreement with Kivelson's theory of intersoliton electron hopping (ISH) (Kivelson, 1981a,b, 1982; Kivelson *et al.*, 1982; Heeger and MacDiarmid, 1981).

Other mechanisms are, of course, possible and these have been discussed by Chien (1984). Chien has also discussed the probable transport mechanisms as a function of dopant concentration for polyacetylene. Whereas, in polyacetylene one is dealing with a conjugated polymer with degenerate ground states, most conjugated polymers have nondegenerate ground states. The polaron concept was conceived in order to treat an electron moving through an ionic crystal for polymers with nondegenerate states. The electron attracts the positive ions and repels the negative ions, as expected. The ions which are in displaced positions change the periodic potential of the crystal lattice and provide a potential well of lower energy for the electron. The electron can then be considered to be in a bound state of the well. This combination of the electron with its induced lattice deformation is called a polaron. It may also be viewed as the bound state of a charged soliton and a neutral soliton, after the removal of a single charge from a chain segment of the polymer, or in chemical terminology, a radical cation (Fig. 11a) as illustrated for polypyrrole. The bipolaron, or dication, is formed by the removal of a second charge from the polaron (Fig. 11b). Polarons carry a spin $S = \frac{1}{2}$, and bipolarons are spinless (Scott *et al.*, 1984).

Conduction by polarons and bipolarons is believed to be more dominant in conjugated polymers with nondegenerate ground states, such as polypyrrole. In this system polarons and bipolarons appear to extend over about four pyrrole rings (Brédas *et al.*, 1984). The presence of both of these species has been detected in polypyrrole by use of ESR and optical experiments (Scott *et al.*, 1984). Polarons are observed at low and intermediate doping concentrations of polypyrrole (Greene and Street, 1984). The ESR measurements showed that at low dopant concentrations, the

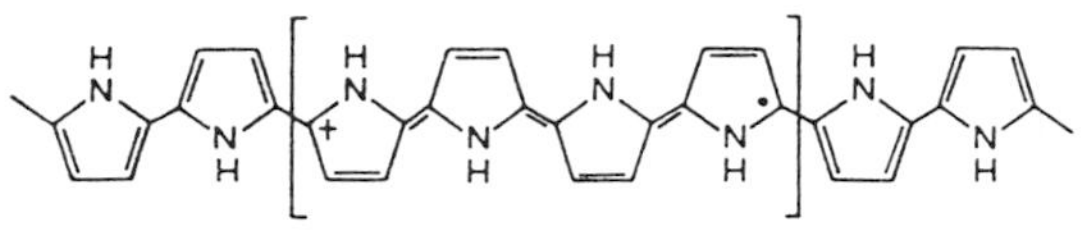

(a)

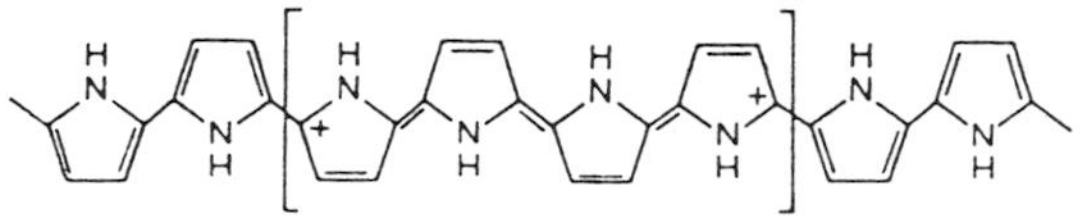

(b)

Fig. 11. Valence bond representation of (a) polaron and (b) bipolaron on a polypyrrole chain. More accurately, the charge (and spin) is delocalized over the whole defect.

number of spins increase with the conductivity consistent with polaron carriers (Scott *et al.,* 1984). At higher dopant concentrations, the ESR signal saturates, and then decreases, suggesting that the polarons combine to form spinless bipolarons. In the highly conducting state no ESR signal is observed, indicating that the bipolarons are the chief charge carriers.

In order for the polyacetylene polymer to accommodate the dopant a different geometry from the undoped polymers is adopted. This geometric modification causes electronic energy levels to be removed from both the top of the valence band and the bottom of the conduction band resulting in an increase in the gap. Thus, conduction by a hole mechanism is ruled out, and it is believed that formation of radical cations (polarons and bipolarons) become the chief carriers (Brazomskii and Kirova, 1981; Bishop *et al.,* 1981; Brédas *et al.,* 1981). In the case of polyacetylene, which has degenerate ground states, the bipolaron can dissociate into two isolated spinless charged solitons, referred to as the charged soliton. For polymers with nondegenerate ground states the geometric rearrangement of the ground state produced on separating the bipolaron into two solitons requires energy (see Fig. 12). Due to this energy requirement, the solitons are confined in pairs as bipolarons. Thus the soliton must be viewed as a spinless carrier, of importance to polymers with degenerate ground states, and the polaron–bipolaron model a general model for conduction in all conjugated polymers; those with degenerate and nondegenerate ground states (Greene and Street, 1984).

The degree of charge transfer is also of importance for the conductive polymers as it is for the organic charge-transfer conductors. The efficiency of charge transfer is believed to be high and each dopant could provide ~0.8 charged carriers into $(CH)_x$ (Chien, 1984). If one considers the doped polymer possessing equal amounts of I_3^- and I_5^-, such as

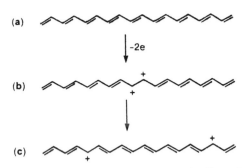

Fig. 12. Instability of bipolarons relative to charged solitons in *trans*-polyacetylene. (a) Removal of two electrons from polyacetylene to form (b) a bipolaron and dissociation of the bipolaron to form (c) two charged solitons.

$[CH(I_3^-)_{0.025}(I_5^-)_{0.015}]_x$, one can conclude that there exists ~80% charge transfer per dopant.

For most of the polymeric materials discussed in this chapter, it should be borne in mind that great uncertainties exist about the structures of the doped and undoped phases, degree of polymerization (in case of polymers), inhomogenity in doping, and changes in structures occurring in the doping process. Other conductors, which require doping, have similar problems (e.g., M(mac) complexes). We, therefore, again emphasize the speculative nature of the discussion in this chapter. As more research accumulates, and as some of the aforementioned problems are solved, improved understanding of conduction mechanisms will result.

LIST OF SYMBOLS AND ABBREVIATIONS

CDW	Charge-density wave
SDW	Spin-density wave
1D	One-dimensional with respect to physical properties (e.g., electrical conductivity)
$\mathbf{k}_F$	Fermi wave vector of lattice
$2\mathbf{k}_F$	$2\pi/\lambda$
Δ	Semiconducting gap parameter
KCP	Krogmann's salt = $K_2[Pt(CN)_4]Br_{0.3} \cdot 3H_2O$
(NMP)(TCNQ)	*m*-Methyl phenazinium salt of 7,7,8,8-tetracyano-*p*- quinodimethane
$Ni(TBP)I_{1.0}$	Iodine-doped nickel tetrabenzophorphyrin
$\{CH(AsF_5)_{0.10}\}_x$	AsF$_5$-doped polyacetylene
$Ni(Pc)I_{1.0}$	Iodine-doped nickel phthalocyanine
σ	Conductivity in (ohm cm)$^{-1}$
n	Concentration of electrons
e	Electron charge
μ	Mobility of carriers
E_g	Energy gap

ISH   Intersoliton electron hopping
ω_0 Debye frequency
$D(\epsilon_F)$ Density of states at the Fermi level
V Effective electron–electron attraction
kB   0.086 meV/k

REFERENCES

Bardeen, J., Cooper, L. N., and Schrieffer, J. R. (1957), *Phys. Rev.* **108**, 1175.
Berlinsky, A. J. (1979), *Rep. Prog. Phys.* **42**, 1243.
Bishop, A. R., Campbell, D. K., and Fesser, K. (1981), *Mol. Cryst. Liq. Cryst.* **77**, 253.
Brazovskii, S. (1978), *JETP Lett. (Engl. Transl.)* **28**, 656; *JETP Lett. (Engl. Transl.)* **78**, 677 (1980).
Brazovskii, S., and Kirova, N. (1981), *JETP Lett. (Engl. Transl.)* **33**, 4.
Brédas, J. L., and Street, G. B. (1985), *Acc. Chem. Res.* **18**, 309.
Brédas, J. L., Chance, R. R., and Silbey, R. (1981), *Mol. Cryst. Liq. Cryst.* **77**, 319.
Brédas, J. L., Scott, J. C., Yakushi, K., and Street, G. B. (1984), *Phys. Rev. B* **30**, 1023.
Bryce, M. R., and Murphy, L. C. (1984), *Nature* **309**, 119.
Chien, J. W. (1984), "Polyacetylene Chemistry, Physics and Material Science," pp. 1–364, Academic Press, Orlando, Florida.
Davydov, A. S. (1985), "Solitons in Molecular Systems," pp. 1–319, Reidel Publ., Dordrecht, Holland.
Epstein, A. J., and Miller, J. S. (1979), *Sci. Am.* **241**, 52.
Ferraro, J. R. (1982), *Coord. Chem. Rev.* **43**, 205–232.
Frohlich, H. (1954), *Proc. R. Soc. London Ser. A.* **223**, 296.
Greene, R. L., and Chaikin, P. M. (1984), *Physica* **126B**, 431.
Greene, R. L., and Street, G. B. (1984), *Science* **226**, 651.
Heeger, A. J., and MacDiarmid, A. G. (1981), *Mol. Cryst. Liq. Cryst.* **77**, 1.
Jérome, D., and Schultz, H. J. (1982), *Adv. Phys.* **31**, 249.
Kivelson, S. (1981a), *Mol. Cryst. Liq. Cryst.* **77**, 65.
Kivelson, S. (1981b), *Phys. Rev. Lett.* **46**, 1344.
Kivelson, S. (1982), *Phys. Rev. B* **25**, 3798.
Kivelson, S., Lee, T. K., Lin-Liu, Y. R., Perschel. I., and Yu, L. (1982), *Phys. Rev. B* **25**, 4173.
Kohn, W. (1959), *Phys. Rev. Lett.* **2**, 393.
Little, W. A. (1964), *Phys. Rev. A.* **134**, 1416.
Little, W. A. (1970), *J. Polym. Sci., Part C* **29**, 17.
Marks, T. J. (1985), *Science* **227**, 881.
Miller, J. S., and Epstein, A. J. (1976) *in* "Progress in Inorganic Chemistry" (S. J. Lippard, ed.), Vol. 20, pp. 1–151, Wiley, New York.
Peierls, R. E. (1955), "Quantum Theory of Solids," p. 107, Oxford Univ. Press, London and New York.
Rice, M. J. (1979), *Phys. Lett. A* **71**, 152.
Scott, J. C., Brédas, J. L., Pfluger, P., Yakushi, K., and Street, G. B. (1984), *Synth. Met.* **9**, 165.
Sham, L., and Patton, B. R. (1975), *in* "One-Dimensional Conductors" (H. G. Schuster, ed.), p. 272, Springer-Verlag, Berlin.
Stucky, G. D., Schultz, A. J., and Williams, J. M. (1977), *Ann. Rev. Mat. Sci.* **7**, 301.

Su, W. P., Shrieffer, J. R., and Heeger, A. J. (1979), *Phys. Rev. Lett.* **42**, 1698; *Phys. Rev. B* **22**, 2099 (1980).

Suzuki, N., Ozaki, M., Etemad, S., Heeger, A. J., and MacDiarmid, A. G. (1980), *Phys. Rev. Lett.* **45**, 1209.

Takayama, H., Lin-Liu, Y. R., and Maki, K. (1980), *Phys. Rev. B* **21**, 2388.

Walsh, W. M., Wudl, F., Thomas, G. A., Nalewajek, D., Hauser, J. J., Lee, P. A., and Poehler, T. O. (1980), *Phys. Rev. Lett.* **45**, 829.

Walsh, W. M., Wudl, F., Aharon-Shalom, E., Rupp, L. W., Vandenberg, J. M., Andres, K., and Torrance, J. S. (1982), *Phys. Rev. Lett.* **49**, 885.

Ziman, J. M. (1963), "Electrons in Metals," pp. 1–80, Taylor and Francis, Ltd. London.

Review Articles
Review articles listed for Chapters 2–6 are topical for this chapter.

Books
Books listed for Chapters 2–6 are of interest for this chapter.

8 EFFECTS OF PRESSURE AND TEMPERATURE ON ELECTRICAL CONDUCTORS

I. INTRODUCTION

A number of problems arise when attempting to understand and explain the changes which a solid conductor undergoes when it is subjected to pressure and/or low temperatures. In some cases these involve the inhomogeneity of the solid, an unknown degree of polymerization, and uncertainties in the structures at high pressure and/or low temperatures. All of these factors make it difficult to arrive at quantitative conclusions when studying these materials under these conditions. It is important to note that we have not found any systematic study of any one class of conductors reported in this volume that has been made both at nonambient temperatures and pressures. Frequently, only one of these two variables is used in physical properties studies of conducting materials. Additionally, often little is known of the solid phases existing at nonambient conditions. Diffraction studies incorporating low temperature simultaneously with applied pressure are few indeed and would be useful.

Numerous physical properties studies of conducting solids have been reported as a function of either temperature or pressure and in the next section we summarize and discuss the most important findings.

II. LOW-TEMPERATURE EFFECTS

We have discussed the effects of decreased temperature on conductivity and structure in Chapters 2–6 and will not again dwell on these topics. In

Table I, we summarize the effects of temperature and pressure on conductivities for the various conductors previously discussed. Of importance here is the fact that as one lowers the temperature, the lattice instabilities in one-dimensional conductors come into play and metal–insulator transitions often occur. In some cases this $T_{M \to I}$ can be lowered or supressed by pressure when the solid demonstrates a change from one- to two- or three-dimensional behavior. For many organic charge-transfer solids one can produce a superconducting state even without pressure [e.g., $(TMTSF)_2ClO_4$, $\beta\text{-}(ET)_2X$ $(X = I_3^-, IBr_2^-, AuI_2^-)]$, see Chapter 2.

III. PRESSURE EFFECTS

A. Introduction

A number of physical parameters in addition to electrical conductivity are affected by the application of high external pressures on solids (Drickamer and Frank, 1973; Ferraro, 1984). Prominent among pressure effects

TABLE I

Summary of P and T Effects on the Electrical Conductivity of Various Materials

	Conductivity (σ)	
Conductivity	Effects of T decrease	Effects of P increase
Cu metal	Increase (but never SC)[a]	Moderate increase
$(SN)_x$	Increase (SC, 0.3 K)	Increases to 9 kbar, then decreases
$(TMTSF)_2ClO_4$	Increase (SC, 1.4 K)	No pressure needed
$(TMTSF)_2AsF_6$	Max. at 12, then decreases	11 kbar and 1 K (SC)
$(ET)_2ReO_4$	Max. at 81, then decreases	>4 kbar and 2 K (SC)
$\beta\text{-}(ET)_2X$, $(X = I_3^-, IBr_2^-, AuI_2^-)$	Increase (SC) at 1.4–1.5 K, 2.8 K, 5 K, respectively	No pressure needed
$NiPcI_{1.0}$	Max. at 90 then rapid decrease	Little effect
KCP	Broad max., decrease	Max. 25 kbar, then decrease
$[CH(AsF_5)_{0.10}]_x$	Slight decrease	Little effect
$[\phi CH(I_5)]_x$	Decrease	2–4 orders increase in σ
$Cs_2(TCNQ)_3$	Large decrease	Not investigated
TTF–TCNQ	Large increases to M → I transitions at 54 K, 38 K	Increase initially then decrease (min. at 43 kbar)
TMTSF–DMTCNQ	Increase to M → I transition at 42 K	Metallic at 10 kbar and 1 K

[a] SC = superconducting.

are those involving electronic transitions in solids, which are very important to the electrical conductivities of various materials. Other pressure effects involve phase transitions and changes in bond distances (e.g., in stacked molecules changes in intra- and interstack distances). These effects, too, cause conduction changes in electrical conductors.

Figure 1 shows the change in structure and the effect on the band gap on solid I_2 with pressure to 200 kbar.* The gap between the conduction and valence bands is seen to narrow. At ~210 kbar, there is a transition to I_2 metal (Jayaraman, 1984). Ultimately, in theory, all materials could show a metal transition with pressure, if sufficient pressure is reached. Xenon has been reported to become metallic at 330 kbar (Ruoff, 1979; Ruoff and Chan, 1979; Ruoff and Nelson, 1978). Ross and McMahan (1980) have reported that pressures to 1.3 Mbar are necessary for metallic xenon to form (Chan *et al.*, 1982). Much present research is devoted to producing metallic hydrogen. To date no metallic phase of hydrogen has been prepared, although pressures to 5.5 Mbar are now possible (Bell *et al.*, 1984).

B.   Effects on Electrical Conductors

Decreased interatomic distances accompanying increased pressure should broaden energy bands in solids. Ultimately, this broadening could result in greater overlap between a filled valence band and an empty conduction band (decreasing the band gap), and this should result in a transition from an insulator to metal. The electrical resistivity is affected by pressure. Electrical resistivity can be given by

$$\rho = 1/n\mu e \qquad (1)$$

where n is the number of carriers, e the charge on the electron, and μ the mobility of carriers. Electrical conduction in semiconductors is carrier limited and the number of carriers is given by

$$n = \exp - (E_g/2RT) \qquad (2)$$

where E_g is the gap between the conduction and valence bands, T the temperature, and R the gas constant. As the temperature is decreased, n decreases. For a metal, electron transport is mobility limited and at ordinary temperatures the mobility is limited by lattice scattering. In a metal the conductivity will increase linearly with a decrease in temperature, up to the metal–insulator transition if one exists. Compression tends to reduce the lattice amplitude and one obtains a decrease in resistance with pressure. The following experimental results are thus obtained:

(a)   lower temperature: band gap decreases, conductivity increases, and
(b)   increased pressure: band gap decreases, conductivity increases.

*1 kbar $= 10^8 Nm^{-2} = 10^8 Pa$.

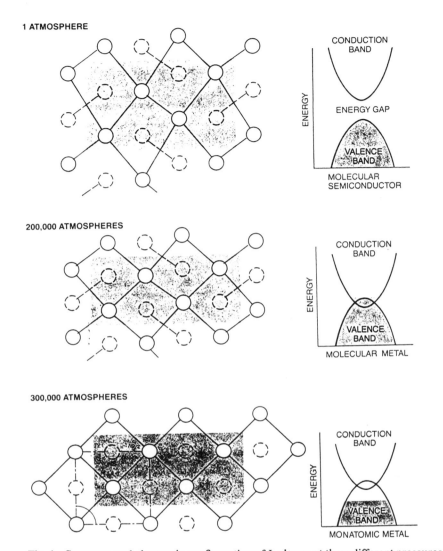

Fig. 1. Structure and electronic configuration of I_2 shown at three different pressures. The configuration of the I atoms in the crystalline solid is shown at the left and the corresponding electronic energy diagram is shown at the right. At atmospheric pressure the basic unit of the crystal is the diatomic molecule (I_2); the solid is a semiconductor because there is a gap between the valence band, which is filled with electrons (black), and the conduction band, which is empty (white). As the pressure is increased to 200,000 atmospheres the molecules assume a more ordered, tightly packed configuration and the gap between the bands narrows until the bands overlap. The molecular crystal becomes an electrically conducting metal. At about 210,000 atmospheres there is an abrupt transition to an atomic, metallic phase; at 300,000 atmospheres the atoms in this phase form a highly regular crystal and the valence band becomes partially filled, like the most energetic energy band in an ordinary metal. Atoms and bonds in the plane of the paper are shown as solid circles and lines; atoms and bonds not in the plane of the paper are shown as broken circles and lines. The high-pressure structures of iodine were determined by K. S. Takemura, S. Minomura, O. Shimomura and Y. Fujii of the Institute for Solid State Physics in Tokyo. [From Jayaraman (1984).]

1.   Krogmann Salts

Some pressure studies made on Krogmann salts were reported in Chapter 4, Section IV.H. These materials exhibit electrical conductivity along the metal chain as we have previously noted. The structural basis for the interesting electrical conductivity in these conductors is based on the stacking of the square–planar moieties. Differences are observed in Pt–Pt distances depending on the degree of partial oxidation (DPO). Decreases in Pt–Pt distances can increase the electrical conductivity along the chain. Figure 2 illustrates the chains in $K_2[Pt(CN)_4]$, a poor conductor and in $K_2[Pt(CN)_4]Br_{0.30} \cdot 3H_2O$ (KCP), a good conductor near 298 K. One can observe the shorter Pt–Pt distances in the latter compound.

The nature of these stacked conductors makes them of interest from a standpoint of pressure sensitivity. Based on the limited amount of experimental data available, it appears that pressure should increase the d_z^2 overlap, causing increases in conductivity, at least initially, but further pressure causes repulsive forces to appear and conductivity could decrease.

Interrante and Bundy (1972) studied the Krogmann salt $K_2Pt(CN)_4$-$Br_{0.3} \cdot 3H_2O$ at pressures up to 100 kbar. The a and c unit cell dimensions decreased with pressure as determined by x-ray diffraction measurements made under pressure. The a parameter changed by 0.47 Å at 70 kbar. The c axis (Pt–Pt chain axis) was less pressure sensitive and much less sensitive than the chain axis in $Pt(NH_3)_4PtCl_4$ [Magnus green salt (MGS)] (Interrante

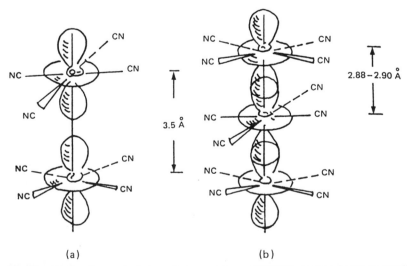

(a)   (b)

Fig. 2.   Comparison of chains of square planar $Pt(CN)_4^{2-}$ groups in (a) $K_2[Pt(CN)_4]$ with $\sigma = 5.0 \times 10^{-7}$ $(ohm \ cm)^{-1}$ and (b) $K_2[Pt(CN)_4] \ Br_{0.3} \cdot 3H_2O$ with $\sigma = 4.83$ $(ohm \ cm)^{-1}$ (298 K).

and Bundy, 1971). This may be due to the shorter metal–metal distance in the Krogmann salt (2.89 Å) as compared to 3.25 Å in MGS, as well as the differences in charge on the platinum atoms. This trend is consistent with results obtained for other solids with varying bond distances. Ionic compounds manifest greater bond distance shortening with pressure than do covalent molecules. This is presumably due to the longer bond distances in ionic salts as compared to the shorter distances found in covalent bonds in covalent molecules (Ferraro, 1984).

A metal–semiconductor phase transition in $K_2[Pt(CN)_4]Br_{0.3} \cdot 3H_2O$ at ~20 kbar was observed by Müller and Jérome (1974) and Thielemans *et al.* (1976) and its pressure dependency followed (Thielemans *et al.*, 1976). These results were interpreted as being due to the increased interchain coupling occurring with pressure.

A difference in pressure response was observed between a cation- and an anion-deficient tetracyanoplatinate (Hara *et al.*, 1975). The cation-deficient salt with longer $d_{Pt–Pt}$ showed a decrease in resistance followed by a slight drift of resistance upward at 180 kbar. The anion-deficient compound with shorter $d_{Pt–Pt}$ showed an increase in resistance by a factor of 100 between 40 and 100 kbar after displaying a minimum at 225 kbar (Interrante and Bundy, 1972) (see Chapter 4, Section IV.H for a tabulation of $d_{Pt–Pt}$ in several cation- and anion-deficient Krogmann salts).

Although few studies have been made, it appears that greater pressure effects are noted for the compounds where $d_{Pt–Pt}$ is longer; as determined by conductivity measurements. Additionally, intra- and interstack distances can be affected with pressure. Extremely high pressure studies have not been made with the Krogmann salts, and it would appear that such studies would be worthwhile.

2. Metal Macrocyclic Materials

Pressure effects on M(mac) compounds were discussed in Chapter 5 Sections II.D and V.E. Metal–metal distances are generally greater in the M(mac) conductors than in the Krogmann salts. Therefore, we might expect pressure to produce significant effects, and the M(mac) materials do demonstrate large increases in conductivity in related compounds such as the metal phthalocyanines (see Table VI, Chapter 5).

3. Charge-Transfer Compounds

Pressure–conductivity results obtained for $(TMTSF)_2X$ and $(ET)_2X$ materials were discussed in Chapter 2, Section VI.C (Tables X and XII). Some of the difficulties observed with pressure studies are exemplified by the organic charge-transfer complexes. These involve chemical reactions occurring with a pressure increase. These reactions are demonstrated by the following results. In the case of the charge transfer compounds involving TCNQ, Aust *et al.* (1964) studied $[(C_2H_5)_3NH^+]TCNQ_2^-$, (quinolinium$^+$) $TCNQ_2^-$, and KTCNQ salts up to pressures of 140–150 kbar. A decrease in resistance was observed, followed by a large rise with drift. The changes were irreversible and infrared studies showed that a reaction took place involving the $C\equiv N^-$ groups. Above these pressures increases in resistivity occur. Onodera *et al.* (1979) studied TTF–TCNQ under applied pressure and found that the resistivity showed a minimum at ~43 kbar. Thereafter, the resistance drifts upward when the pressure is kept constant. On decreasing the pressure, the resistance rises and the changes are irreversible. This evidence suggests chemical reactions are occurring with applied pressure. The resistance of the salts TTF–Br$_{0.8}$ and TTF–I$_5$ exhibit an irreversible upward drift from the minimum in resistivity, again suggesting chemical reaction occurring under pressure.

DMTTF–TCNQ and TMTTF–TCNQ salts have also been studied under applied pressure and results parallel to those of TTF–TCNQ were obtained, indicating the added methyl groups had little effect on conductivity with pressure. In Table II we have compiled the minimum resistivity for several of these materials.

Pressure effects on TTF–TCNQ have been made by Bernstein *et al.* (1975), Jérome *et al.* (1974), and Bouffard and Zuppiroli (1978). The plasma edge shifts toward higher energy and sharpens with pressure. A 15% increase in plasma frequency is found at 20 kbar for this compound. Since the plasma edge is related to bandwidth, the conductivities are found to increase with pressure (Fig. 3) (see Jérome and Schulz (1982) for a more comprehensive discussion). Debray *et al.,* (1977) and Chu *et al.* (1973) studied TTF–TCNQ under pressure at 20 kbar and 20–273 K. The metal–

TABLE II

Minimum Resistance (ρ_{min}) for Several Charge-Transfer Compounds[a]

Materials	ρ_{min} (ohm cm)	Pressure (kbar)
TTF–TCNQ	8.2×10^{-3}	43.0 ± 3.5
DMTTF–TCNQ	7.2×10^{-3}	43.0 ± 3.5
TMTTF–TCNQ	7.0×10^{-3}	43.0 ± 3.5
TTF–Br$_{0.8}$	1.0×10^{-1}	18.0 ± 1.0
TTF$_7$–I$_5$	4.4×10^{-2}	41.0 ± 3.0
TTF	6.0×10	250
DMTTF	9.0×10^5	250
TMTTF	5.0×10^2	250

[a] Taken from Onodera and Anzai (1979).

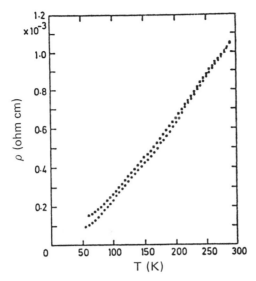

Fig. 3. Resistivity versus temperature for two TTF–TCNQ crystals.

insulator temperature increased 1 K/kbar contrary to general expectations. The conductivity increased by a factor of 4.5 at 20 kbar over the ambient pressure values. Shirotani *et al.* (1975) examined alkali metal salts of TCNQ and found that resistances dropped significantly up to 150 kbar. Conwell (1980), in following the pressure dependence of resistivity of TTF–TCNQ suggested that the charge transfer changed from 0.59 at ambient pressure to $\frac{2}{3}$ this value at ~17 kbar pressure.

Torrance *et al.* (1981) have identified a neutral–ionic phase transition in various charge-transfer organic compounds (TTF donors and some TCNQ type acceptors) upon application of pressure. The experiment was conducted in a diamond anvil cell. Powdered samples dispersed in KBr were used and the change in color observed as the pressure was increased. Neutral molecules (having closed shells) tend to be yellow and absorb only in the ultraviolet. Ionic molecules tend to be highly colored in the visible, associated with unpaired electrons. With pressure, the materials show a distinct change in color from yellow to red (usually), indicative of a reversible phase transition occurring from a neutral to an ionic ground state (e.g., TTF–chloranil at 8 kbar).

Pressure application for the compounds discussed here should effect both intrastack and interstack distances as well. Where increased orbital overlap occurs, the band gaps would diminish and these factors could all contribute toward increasing conductivity. In addition, pressure tends to suppress the Peierls, spin-density wave, and anion ordering lattice instabilities in one-dimensional conductors, and thus $T_{M \to I}$ should be lowered or suppressed. For some of these charge-transfer compounds the application of pressure at low temperature has induced superconductivity, probably due to decreased interchain spacings and greater orbital overlap, thus affectively creating a 2D $\to$ 3D conductor.

4.   Organic Polymeric Conductors

Again, only a minimal number of pressure studies are available for the organic polymeric conductors. Moses *et al.* (1982) measured the absorption edge of *trans*- and *cis*-polyacetylene and found that it was strongly dependent on pressure.

Trans-polyacetylene was subjected to hydrostatic pressures of 44 kbar in a diamond anvil cell (see Yacoby and Roth, 1982). The results showed that phonon frequency changes with pressure were very small. The gap was found to decrease rapidly with pressure but saturates at high pressure.

Few pressure studies are available on $(SN)_x$. Gill *et al.* 1975) found that the conductivity of this polymer increased with applied pressure to 9 kbar and then decreased. The critical temperature (T_c) was found to be 0.6 K, which represents an increase in T_c from 0.3 K. $(SNBr_{0.4})_n$ shows a similar increase in conductivity with pressure, but T_c decreases monotonically under 10 kbar pressure (Dee *et al.*, 1977). Friend *et al.* (1977) measured the conductivity of $(SN)_x$ up to 40 kbar.

Table I summarizes pressure effects on various conductors and compares them to temperature effects.

IV.   SUMMARY

The application of high external pressure to induce solid state phase transitions from nonconductive and semiconductive phases to the metallic regime has been known for some time. For materials that are semimetals such as $(SN)_x$ and $Hg_{2.86}AsF_6$ and $Hg_{2.91}SbF_6$ which are quasi-two-dimensional, respectively, decreasing the temperature is sufficient to produce superconductive states (at 0.3 and 4.2 K, respectively).

The application of pressure at low temperature to produce conductive materials has been a recent development. Andrieux *et al.* (1979) found that for the one-dimensional TMTSF–DMTCNQ the Peierls instability could be suppressed with pressure. At 1 K it transforms at 10 kbar from a semiconductor to a metal. This result suggested that application of pressure to the Bechgaard salts could prove fruitful. Indeed, $(TMTSF)_2PF_6$

demonstrated that the superconductive state was attainable at ~6.5 kbar and 0.9 K (Jérome *et al.*, 1980; Galois *et al.*, 1985). Even in Bechgaard salts with high $T_{M \to I}$ (180 K for (TMTSF)$_2$ReO$_4$) it was possible to achieve the superconductive state at ~1.5 K and 12 kbar. It was demonstrated that (TMTSF)$_2$ClO$_4$ became superconductive at 1.4 K at ambient pressure. As discussed in Chapter 2, (TMTSF)$_2$ClO$_4$ has a smaller room temperature unit cell volume than other (TMTSF)$_2$X conductors, V_c (694.3 Å^3), which correlates well with the average interstack Se$\cdots$Se distance [d_{avg} = $\frac{1}{3}(2d_7 + d_9)$]. As V_c decreased, the d_{avg} also decreased (see Fig. 18 and Tables VII and IX in Chapter 2). (TMTSF)$_2$PF$_6$ has a V_c value of 714.3 Å^3 (298 K) and (TMTSF)$_2$ReO$_4$ a V_c value of 710.5 Å^3. The volume compressibility for the PF$_6^-$ salt is estimated to be 0.5% kbar^{-1} (Morosin *et al.*, 1982). Thus it may be calculated that it requires ~6 kbar to compress (TMTSF)$_2$PF$_6$ into the same volume of (TMTSF)$_2$ClO$_4$ (Williams and Carneiro, 1985), comparing well with the experimental results. For these materials, it may be concluded that pressure decreases the intra- and interstack distances providing added dimensionality leading, in certain special cases, to a superconductive phase.

Although, the (TMTTF)$_2$X materials are isostructural with the (TMTSF)$_2$X salts, conductivities are considerably lower in the former salts. This difference may be explained by significant dimerization often found to occur in the TMTTF stacks. Increased Coulomb repulsions between electrons also occur in the TMTTF salts. These factors all contribute to an increased gap in the electron spectrum causing a decrease in electrical conductivity (Williams and Carneiro, 1985). Figure 35 in Chapter 2 illustrates the temperature–pressure phase diagram for (TMTSF)$_2$X salts.

Although the Peierls instability is often suppressed with the application of pressure, no superconductive phases have been confirmed to date in the TMTTF family (see Chapter I and Table I of Chapter 7). The S$\cdots$S interactions in TMTTF salts are more important in the intrastack directions than in the interstack directions (Bechgaard, 1982; Coulon *et al.*, 1982). Comparison of interstack and intrastack distances in the (TMTSF)$_2$X and (TMTTF)$_2$X salts are of interest. Table III shows a tabulation of structural data. It may be observed that for the TMTTF salts the interstack S$\cdots$S distances are always longer than the van der Waals radius sum (~3.60 Å), whereas for the TMTSF salts the Se$\cdots$Se distances are often shorter than the van der Waals sum of ~4.0 Å. These differences contribute to the observed variations in the pressure measurements. The TMTSF salts appear to respond to pressure by producing superconductive phases, whereas for TMTTF salts the pressures used to date only suppresses the Peierls instability.

Comparison of the aforementioned pressure results for the TMTSF and

TABLE III

Comparison of Interstack and Intrastack Distances for the
TMTSF and TMTTF Salts[a]

	TMTSF salts			TMTTF salts						
	PF_6^-	AsF_6^-	ClO_4^-	SCN^-	ClO_4^-	PF_6^-	BF_4^-	I_3^-	NO_3^-	Br^-
Closest Se $\cdots$ Se or S $\cdots$ S interstack distance (Å)	3.88	3.90	3.77	4.01	3.99	3.99	3.88	3.80	3.79	3.78
Intrastack distances										
d_1 (Å)	3.66	3.65	3.63	3.56	3.59	3.62	3.56	3.54	3.56	3.58
d_2 (Å)	3.63	3.62	3.62	3.49	3.52	3.52	3.54	3.50	3.50	3.50

[a] Taken from Flandrois et al. (1982).
[b] See Fig. 1 in Flandrois article.

TMTTF salts, with those obtained for the ET family, is of considerable interest. In the ET:X system, the short $S \cdots S$ contacts always occur in the interstack distances (Leung et al., 1985). These $S \cdots S$ contacts are often, but not always, significantly shorter than the sum of the van der Waals radii (3.6 Å), indicative of considerable intermolecular interactions occurring at room temperature assuming considerable orbital overlap, and which demonstrates that the salts are structurally two-dimensional in nature. In the ET systems the unit cell volume (V) plays an important role when comparing electrical properties of isostructural salts, as in the TMTSF salts. For the $(ET)_2X$ salts where X is I_3^-, IBr_2^-, and AuI_2^- the unit cell volumes are all similar (see Table XIV, Chapter 2) and as the temperature is lowered, pressure is *not* required in order for superconductivity to occur. For $(ET)_2ReO_4$, a pressure of 4–6 kbar at 1.3 K is required to induce the superconducting state.

LIST OF SYMBOLS AND ABBREVIATIONS

$T_{M \to I}$	Metal–insulator transition temperature
P_{SC}	Pressure at which superconducting phase is formed
KCP	Krogmann's salt—$K_2[Pt(CN)_4]Br_{0.3} \cdot 3H_2O$
(TTF)TCNQ	Tetrathiafulvalene complex of 7,7,8,8-tetracyano-p-quinodimethane
$(TMTTF)_2X$	Tetramethyltetrathiafulvalene conductors, where X = inorganic anion (e.g., ClO_4^-)
kbar	$10^8 Nm^{-2} = 10^8 Pa$
σ	Conductivity in (ohm cm)$^{-1}$
n	Concentration of charge carriers

μ	Mobility of carriers
e	Charge of electron
E_g	Energy gap
1D	One-dimensional with respect to physical properties (e.g., electrical conductivity)
2D, 3D	Two- or three-dimensional with respect to physical properties
(DMTTF)TCNQ	Dimethyltetrathiafulvalene complex with 7,7,8,8-tetracyano-p-quionodimethane
(TMTTF)TCNQ	Tetramethyltetrathiafulvalene complex with 7,7,8,8-tetracyano-p-quionodimethane
V_c	Unit cell volume in $Å^3$
D_{avg}	Equal to $\frac{1}{3}(2d_7 + d_9)$ average interstack Se$\cdots$Se distance in $(TMTSF)_2X$ metals where d_7 and d_9 are individual interstack Se$\cdots$Se distances
SDW	Spin-density wave
M(mac)	Metal macrocyclic
DPO	Degree of partial oxidation

REFERENCES

Andrieux, A., Duroure, C., Jérome, D., and Bechgaard, K. (1979), *J. Phys. (Paris) Lett.* **40**, 381.

Aust, R. B., Samara, G. A., and Drickamer, H. G. (1964), *J. Chem. Phys.* **41**, 2003.

Bechgaard, K. (1982), *Mol. Cryst. Liq. Cryst.* **79**, 1.

Bell, P. M., Mao, H. K., and Goettel, K. (1984), *Science* **226**, 542.

Bernstein, U., Chaikin, P. M., and Pincus, P. (1975), *Phys. Rev. Lett.* **34**, 271.

Bouffard, S., and Zuppiroli, L. (1978), *Solid State Commun.* **28**, 113.

Chan, K. S., Huang, T. L., Grzybowski, T. A., Whetten, T. J., and Ruoff, A. L. (1982), *Phys. Rev. B* **26**, 7116.

Chu, C. W., Harper, J. M. E., Geballe, T. H., and Greene, R. L. (1973), *Phys. Rev. Lett.* **31**, 1491.

Conwell, E. M. (1980), *Solid State Commun.* **33**, 17.

Coulon, C., Delhaes, P., Flandrois, S., Lagnier, R., Bonjour, E., and Fabre, J. M. (1982), *J. Phys. (Paris)* **43**, 1059.

Debray, D., Millet, R., Jérome, D., Barisic, S., Giral, L., and Fabre, J.M. (1977), *J. Phys. Lett. (Orsay, Fr.)* **38**, L227.

Dee, R. H., Dollard, D. H., Turrell, B. G., and Carolan, J. F. (1977), *Solid State Commun.* **24**, 1977.

Drickamer, H. G., and Frank, C. W. (1973), "Electronic Transitions and the High Pressure Chemistry and Physics of Solids," Chapman and Hall, London.

Ferraro, J. R. (1984), "Vibrational Spectroscopy at High External Pressures—The Diamond Anvil Cell," pp. 1–264, Academic Press, New York.

Flandrois, S., Coulon, C., Delhaes, P., Chasseau, D., Hauw, C., Gaultier, J., Fabre, J. M., and Giral, L. (1982), *Mol. Cryst. Liq. Cryst.* **79**, 307.

Friend, R. H., Jérome, D., Rehmatullah, S., and Yoffe, A. D. (1977), *J. Phys. C* **10**, 1001.

Gallois, B., Gaultier, J., Hauw, C., Chasseau, D., Meresse, A., Filhol, A., and Bechgaard, K. (1985), *Mol. Cryst. Liq. Cryst.* **119**, 225.

Gill, W. D., Greene, R. L., Street, G. B., and Little. W. A. (1975), *Phys. Rev. Lett.* **35**, 1732.

Hara, Y., Shirotani, I., and Onodera, A. (1975), *Solid State Commun.* **17**, 27.

Interrante, L. V., and Bundy, F. P. (1971), *Inorg. Chem.* **13**, 1162.

Interrante, L. V., and Bundy, F. P. (1972), *Solid State Commun.* **11**, 1641.

Jayaraman, A. (1984), *Sci. Amer.* **150**, 54–62.

Jérome, D. (1981), *Int. Conf. Low-Dimension Conduct., Boulder, Colo. 1981.*

Jérome, D., Mazaud, A., Ribault, M., and Bechgaard, K. (1980), *J. Phys. (Paris) Lett.* **41**, 95.

Jérome, D., and Schulz, H. J. (1982), *Adv. Phys.* **31**, 299.

Jérome, D., Müller, W., and Weger, M. (1974), *J. Phys. Lett.* **35**, L77.

Leung, P. C. W., Beno, M. A., Emge, T. J., Wang, H. H., Bowman, M. K., Firestone, M. A., Sowa, L. M., and Williams, J. M. (1985), *Mol. Cryst. Liq. Cryst.* **125**, 113.

Morosin, B., Schriber, J. E., Greene, R. L., and Engler, E. M. (1982), *Phys. Rev. B* **26**, 2660.

Moses, D., Feldblum, A., Ehrenfreund, E., Heeger, A. J., Chung, T. C., and MacDiaramid, A. G. (1982), *Phys. Rev. B* **26**, 3361.

Müller, W. H.-G., and Jérome, D. (1974), *J. Phys. Lett. (Orsay, Fr.)* **35**, L103.

Onodera, A., Shirotani, I., Hara, Y., and Anzai, H. (1979), *High Pressure Sci. Technol. AIRAPT Conf. 1977* **1**, 498.

Ross, M., and McMahan, A. K. (1980), *Phys. Rev. B* **21**, 1658.

Ruoff, A. L. (1979), *Cornell Q. Eng.* **14**, 2–10 (Cornell University internal publication).

Ruoff, A. L., and Chan, K. S. (1979), *High Pressure Sci. Technol. AIRAPT Conf., 6th, 1977* **1**, 779–784.

Ruoff, A. L., and Nelson, D. A. (1978), *Chem. Eng. News* **46**(47), 22.

Shirotani, I., Onodera, A., and Sakai, N. (1975), *Bull. Chem. Soc. Jpn.* **48**, 167.

Thielemans, M., Deltour, R., Jérome, D., and Cooper, J. R. (1976), *Solid State Commun.* **19**, 21.

Torrance, J. B., Vasquez, J. E., Mayerle, J. J., and Lee, V. Y. (1981), *Phys. Rev. Lett.* **46**, 253.

Williams, J. M., and Carneiro, K. (1985), *Adv. Inorg. Chem. Radiochem.* **29**, 249–296.

Xu, J. A., Mao, H. K., and Bell, P. M. (1986), *Science* **232**, 1404.

Yacoby, Y., and Roth, S. (1985), *Solid State Commun.* **56**, 319.

Review Articles

Ferraro, J. R. (1979), *Coord. Chem. Rev.* **29**, 1.

Ferraro, J. R. (1982), *Coord. Chem. Rev.* **43**, 205.

Ferraro, J. R., and Basile, L. J. (1974), *Appl. Spectroc.* **28**, 505.

Jayaraman, A. (1983), *Rev. Mod. Phys.* **55**, 65.

Jayaraman, A. (1984), *Science* **250**, 54.

Sherman, W. F. (1982), *Bull. Soc. Chim. Fr.* **9**(10), 347.

Books

Drickamer, H. G., and Frank, C. W. (1973), "Electronic Transitions and the High Pressure Chemistry and Physics of Solids," pp. 1–220, Chapman and Hall, London.

Ferraro, J. R. (1984), "Vibrational Spectroscopy at High External Pressures—The Diamond Anvil Cell," pp. 1–264, Academic Press, New York.

9 SUMMARY

I. INTRODUCTION

In Chapters 2–6, we have attempted to discuss the synthesis, structure, and physical properties of the various types of synthetic electrical conductors presently of interest to numerous international research groups. In this chapter we will briefly summarize these results with special emphasis on the similarities and differences between these systems. Additionally, we will attempt to cite promising directions for future research.

The studies of electrical conductors has been conducted by use of a multidisciplinary approach, in which scientists from quite different specialties have carried out research on these novel materials. Extensive cross fertilization of ideas has also occurred, which has further stimulated progress. In several instances groups from large laboratories have joined forces in an attempt to solve problems incurred in the study of electrical conductors that have widely ranging chemical and physical properties. This has worked well inasmuch as no single group is equipped with the wide range of expertise and instrumental tools required to make such studies.

Considerable effort has been devoted to the development of new organic superconductors, and this has occurred in the area of organic charge-transfer compounds (see Chapter 2). In the course of these investigations many new compounds have been synthesized many of which are essentially semiconductors ($d\sigma/dT > 0$). It is our contention that new semiconductors, and insulators, merit further investigation even though they may not appear as glamorous as the superconductors. From a device application potential, the new semiconductors may be found to be useful, and, therefore, they most certainly deserve future attention. The study of insulators, semiconductors, and superconductors also adds greatly to our overall knowledge of the physics and chemistry of these materials. As a result, we can better fine tune the design of new conducting systems.

II. SIMILARITIES AND DIFFERENCES BETWEEN THE VARIOUS CLASSES OF CONDUCTORS

Certain commonalities exist between families or classes of conductors. For example, the Krogmann salts (Chapter 4) and the metal macrocyclic compounds (Chapter 5) form nonintegral oxidation states (partial charge transfer) in order to become conducting materials. The organic charge transfer compounds like TMTSF–TCNQ or TTF–TCNQ demonstrate partial charge transfer (see Chapter 2). For these classes of conductors, the better electrical conductors are found in which the charge transfer is incomplete. On the other hand, the organic superconductors demonstrate complete charge transfer.

Additionally, these materials are either one- or two-dimensional materials and often exhibit lattice instabilities which result in metal–insulator transitions at $T_{M \to I}$. The instabilities may be of the Peierls type (CDW) as for the Krogmann salts, metal macrocycles, and organic polymers. The charge-transfer organic conductors often demonstrate spin-density wave (SDW) or anion ordering lattice instabilities, also frequently resulting in a MI transition.

Comparisons of the inter- and intrastack distances, and resulting chalcogenide atom overlaps, in the TMTSF, TMTTF, and ET salts, and their relationships to the conducting properties of these conductors, were discussed in Chapter 8 (see Table III of Chapter 8). The interstack distances are the shortest in the ET salts, and the trend follows TMTTF < TMTSF > ET. Only where the two-dimensional interstack interactions become predominant, as often occurs in these systems at very low temperature, does superconductivity appear as in the TMTSF and ET salts. In the case of $(TMTSF)_2ClO_4$, superconductivity occurs at ambient pressure. For all other superconducting TMTSF salts, applied pressure is required to achieve this state. For numerous ET salts, the superconductive phases occur at ambient pressure. For both families, the *inter*stack interactions cause added dimensionality ($1D \to 2D \to 3D$) and a superconducting ground state is achieved at lower temperatures and/or high pressure.

III. FUTURE DIRECTIONS

A. Charge-Transfer Organic Conductors

As was discussed in Chapter 2, the charge-transfer organic systems have provided several ambient pressure superconductors. This area of research has proved to be very promising, and it appears that it will remain so for the immediate future. In these systems, the anions have been found to play an important role in providing the structures which yield superconductive phases. Only inorganic anions have been found in superconductors of both the TMTSF and ET families. A small number of inorganic anions of varying symmetry have been used for the TMTSF salts. However, for the ET salts, it has been found that only linear, symmetrical, polyatomic, inorganic ions were suitable for achieving the superconducting state (e.g., I_3^-, IBr_2^-, and AuI_2^-) at ambient pressure; $(ET)_2 ReO_4$ being an exception. For a list of anions used for both the TMTSF and ET salts, see Table XIV, Chapter 2. In the TMTSF and ET systems, the anion size and symmetry control the average $Se \cdots Se$ or $S \cdots S$ distances, respectively, and the latter lead to added dimensionality. This tends to suppress MI transitions. For the ET conductors, the fact that only a few linear, symmetrical anions of appropriate size to yield conducting salts exists, limits the preparation of new systems. It appears that attempting to prepare variants of the ET donor will prove more fruitful in the future. Figure 1 shows several such promising ET variants that have been synthesized.

In the future, x-ray crystallographic studies on these materials should be made at liquid helium temperatures as well as high pressure in order to match the conditions under which the novel properties often occur. The results would be useful in determining the structural aspects of the superconductive phases in order to facilitate comparisons with the structures of the metal-alloy superconductors.

Attempts should also be made to prepare both fibers and thin films of some of these organic charge-transfer salts. Although this may be difficult to achieve, a variety of novel properties would likely result. Approaches such as those previously discussed for the M(mac) conductors could be attempted. Although mixing semiconductors and conductors with commercial polymers, and subsequently extruding them will lower conductivities, the materials have increased strength and are not brittle. Nevertheless, these systems may have application potential, and obtaining these materials as fibers could enhance this potential.

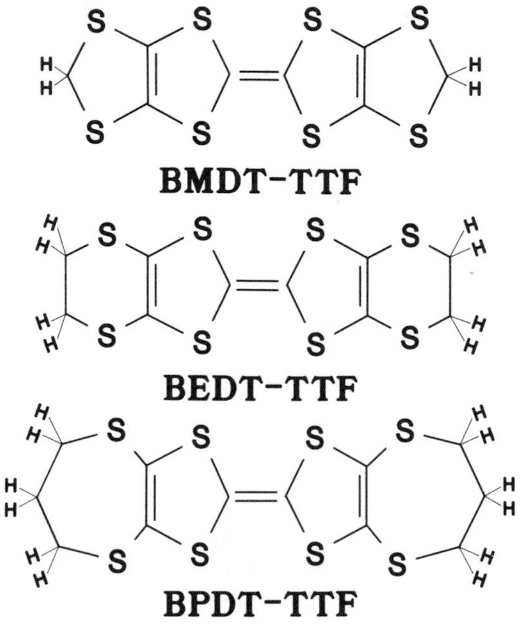

Fig. 1.   Structures of various ET variants.

B. Organic Polymers

Further activity in developing synthetic methods to provide new soluble polymers would appear to be useful. This could also be helpful in improving doping procedures and compound homogeneity. Homogeneous doping might be possible if doping were accomplished in solution, and the present problems which occur with inhomogeneous and incomplete doping could be improved or eliminated.

Only a minimal number of x-ray investigations of polymeric organic conductors have been conducted because of the limited data that can be obtained. However, such studies are useful in characterizing these materials. Continued research in the direction of improving environmental stability, and mechanical and electrical properties, is necessary.

C. Krogmann Salts

Research on Krogmann salts appears to have reached a plateau, since only platinum seems to be capable of forming the majority of these salts. Likewise, the number of anions that can give Pt–Pt chains in these salts is limited (e.g., CN^-, $C_2O_4^-$). This limits formation of new Krogmann-type salts. In addition, all of these materials are semiconducting or insulating in nature. It would be useful to investigate the potential applications of the semiconducting salts.

D. M(mac) Compounds

Research in this area should continue with the preparation of new macrocyclic ligands. This would, in turn, afford the preparation of new M(mac) compounds. Investigations on improving the environmental stability, mechanical, and electrical properties presently underway should be continued. Further studies on the potential applications of these systems are also necessary.

IV. CONCLUDING REMARKS

To date, the only superconductors that have been prepared have been derived from the organic charge-transfer class [$(SN)_x$ being the exception]. This trend will likely continue in the near future.

The metal macrocycle [$Si(Pc)O]_n$ can be combined with Kevlar and extruded into robust fibers (tensile strength—greater than steel). Doping of the material can also be accomplished before or after extrusion. The fibers are also found to have high thermal stability (up to 513 K for the bromine-doped polymer), are soluble in strong acids, and are environmentally stable. Although the electrical conductivity is reduced when diluted with Kevlar, the materials demonstrate conductivities of ~5 (ohm cm)$^{-1}$ with metal-like properties (e.g., $d\sigma/dT < 0$).

Although many of the materials previously discussed are semiconductors, these properties make certain of them particularly suited for device applications. For example, the organic polymers (e.g., polyacetylene) can be obtained in sheets or films, which when doped in an appropriate electrolyte makes them suitable for batteries. Recent results with polyaniline polymer by MacDiarmid (unpublished results, 1986) indicate that this polymer is a better battery candidate than polyacetylene. Polyaniline polymer has also been suggested as a corrosion-resistant material which can be applied to stainless steel. The stainless steel coated with polyaniline films remains passive for long periods of time in acid solutions [see DeBerry (1985) and MacDiarmid (1986)]. Poly(p-phenylene) films have also been used as the cathode material in a rechargeable battery utilizing the electrolyte $LiAsF_6$/propylenecarbonate with Li anode [see Satoh *et al.* (1986)]. The polymer–salt electrolytes appear promising in high-energy-density battery applications. These materials can also be cast as pliable films.

In the case of the organic charge-transfer compounds, many technological applications appear possible because they are lightweight, compared to copper metal, and they might, for example, be used in electronics (especially of space vehicles). They have been proposed as futuristic components in compact high-efficiency electric motors, for zero-resistance electrical transmission lines, and in computers where because of the very low heat production of superconducting synmetals, they could provide circuits of extremely high density. Some synmetals have also been used as photoresists (e.g., TTF-Cl, TTF-Br) (Graff, 1985).

REFERENCES

DeBerry, D. W. (1985), *J. Electrochem. Soc.* **132**, 1022.
Graff, G. (1985), *Ind. Technol.* (November), p. 64.
MacDiarmid, A. G. (1986), unpublished data.
Satoh, M., Tabata, M., Kaneto, K., and Yoshino, K. (1986), *Jpn. J. Appl. Phys.* **25**, L73.

APPENDIX

The purpose of this section is to provide an additional bibliography that is pertinent to the previous chapters and which has appeared in the recent literature. Some of these entries have been published after the book was submitted to the publishers.

CHAPTER 1. INTRODUCTION

Cook, W. B., and Perkowitz, S. (1986). Far-infrared properties and characterization of superconducting Nb$_3$Ge, *Phys. Rev. B: Condens. Matter* **33**, 4557.
Pourrahimi, S., Thieme, C. L. H., Foner, S., and Suenaga, M. (1986). Nb$_3$Sn(Ti) powder metallurgy processed high field superconductors, *Appl. Phys. Lett.* **48**, 1808.

CHAPTER 2. ORGANIC CHARGE-TRANSFER METALS INCLUDING NEW SUPERCONDUCTORS

Aharon-Shalom, E., Becker, J. Y., Bernstein, J., Bittner, S., and Shaik, S. (1985). The synthesis, structure, and properties of the complex between a new donor, 2,3,6,7-tetra(ethyltellura)-tetrathiafulvalene, (TETeTTF), and Tetracyanoquinodimethane (TCNQ), *Synth. Met.* **11**, 213.
Anzai, H., Tokumoto, M., and Saito, G. (1985). Effect of the solvent used for electro-crystallization of organic metals, *Mol. Cryst. Liq. Cryst.* **125**, 385.
Baillargeon, P., Roubi, L., Carlone, C., and Truong, K. D. (1985). The influence of temperature and pressure of the optical spectrum of monoclinic tetrathiafulvalene crystals, *Can. J. Phys.* **63**, 1088.
Bando, H., Kajimura, K., Anzai, H., Ishiguro, T., and Saito, G. (1985). Superconducting tunneling in (TMTSF)$_2$ClO$_4$/a-Si/Pb junction, *Mol. Cryst. Liq. Cryst.* **119**, 41.

Bando, H., Tokumoto, M., Murata, K., Anzai, H., Saito, G., Kajimura, K., and Ishiguro, T. (1985). Ordinary and anomalous magneto-resistances in β-(BEDT-TTF)$_2$X (X = IBr$_2$, I$_2$Br, I$_3$), *J. Phys. Soc. Jpn.* **54**, 4625.

Bourbonnais, C., Stein, P., Jérome, D., and Moradpour, A. (1986). Nuclear relaxation and antiferromagnetic critical effects in organic conductors, *Phys. Rev. B: Condens. Matter* **33**, 7608.

Bozio, R., and Pecile, C. (1985). Infrared and Raman study of anion-donor interactions in tetrahedral anion (TMTSF)$_2$X and (TMTTF)$_2$X salts, *Mol. Cryst. Liq. Cryst.* **119**, 211.

Bryce, M. R., Hammond, G. E., and Weiler, L. (1985). 2,3,Dimethyl-TCNQ: A new electron acceptor for organic metals, *Synth. Met.* **11**, 305.

Buravov, L. J., Kartsovnik, M. V., Kaminskii, V. F., Kononovich, P. A., Kostuchenko, E. E., Laukhin, V. N., Makova, M. K., Pesotskii, S. I., Shchegolev, I. F., Topinkov, V. N., and Yagubskii, E. B. (1985). Superconducting transitions in β-(BEDT-TTF)$_2$I$_3$, *Synth. Met.* **11**, 207.

Carlson, K. D., Crabtree, G. W., Nuñez, L., Wang, H. H., Beno, M. A., Geiser, U., Firestone, M. A., Webb, K. S., and Williams, J. M. (1986). Ambient pressure superconductivity at 4–5 K in β-(BEDT-TTF)$_2$AuI$_2$, *Solid State Commun.* **57**, 89.

Chaikin, P. M., and Greene, R. L. (1986). Superconductivity and magnetism in organic metals, *Phys. Today* **39**, 24–32.

Chaikin, P. M., Mele, E. J., Chamberlin, R. V., Chiang, L. Y., Naughton, M. J., and Brooks, J. J. (1986). On the Kwak transition: Field-induced states in two-dimensional organic conductors, *Synth. Met.* **13**, 45.

Challener, W. A., and Richards, P. L. (1984). Far infrared measurements (TMTSF)$_2$ClO$_4$, *Solid State Commun.* **51**, 765.

Chappell, J. S., Bloch, A. N., Bryden, W. A., Maxfield, M., Polhler, T. O., and Cowan, D. O. (1981). Degree of charge transfer in organic conductors by infrared absorption spectroscopy, *J. Am. Chem. Soc.* **103**, 2442.

Choi, M.-Y., Burns, M. J., Chaikin, P. M., Engler, E. M., and Greene, R. L. (1985). Magnetothermopower of tetramethyltetrathiafulvalenium phosphorus hexafluordide [(TMTSF)$_2$PF$_6$], *Phys. Rev. B: Condens. Matter* **31**, 3576.

Conwell, E. M., and Howard, A. (1986). Bipolarons in Qn(TCNQ)$_2$ and (NMP)$_x$(Phen)$_{1-x}$ (TCNQ) for $x \simeq 0.5$, *Synth. Met.* **13**, 71.

Cooper, J. R., Forró, L., Korin-Hamzic, B., Bechgaard, K., and Moradpour, A. (1986). Magnetoresistance of the organic conducting tetramethyltetraselenafulvalene salts, (TMTSF)$_2$ClO$_4$, and (TMTSF)$_2$SF$_6$: Search for the coherent-diffusive transition or localization effects with increasing temperature, *Phys. Rev. B: Condens. Matter* **33**, 6810.

Coppens, P., Li, L., Petricek, V., and White, J. A. (1986). Crystallographic studies of one-dimensional polyiodide conductors 3. Structure of nonstoichiometric coumarin–iodine complexes by x-ray analysis and EXAFS, *Synth. Met.* **14**, 215.

Coulon, C., Parkin, S. S. P., and Laversanne, R. (1985). Structureless transition and strong localization effects in bis-tetramethyltetrathiafulvalene salts [(TMTTF)$_2$X], *Phys. Rev. B: Condens. Matter* **31**, 3583.

Coulon, C., Scott, J. C., and Laversanne, R. (1986). Antiferromagnetic resonance in tetramethyltetrathiafulvalene (TMTTF) salts, *Phys. Rev. B: Condens. Matter* **33**, 6235.

Creuzet, F., Creuzet, G., Jérome, D., Schweitzer, D., and Keller, H. J. (1985). Homogeneous superconducting state at 8.1 K under ambient pressure in the organic conductor β-(BEDT–TTF)$_2$I$_3$, *J. Phys. Lett.* **46**, L-1079.

Creuzet, F., Bourbonnais, C., Jérome, D., Schweitzer, D., and Keller, H. J. (1986).

Proton NMR relaxation in the high T_c organic superconductor β-(BEDT–TTF)$_2$I$_3$, *Europhys. Lett.* **1**, 467.

Daoben, Z., Ping, W., Meixiang, W., Zhaolou, Y., and Naijue, Z. (1986). Synthesis, structure, and electrical properties of a two-dimensional organic conductor α-(BEDT-TTF)$_2$BrI$_2$, *Solid State Commun.* **57**, 843.

Eldridge, J. E., Homes, C. C., Bates, F. E., and Bates, G. J. (1985). Far-infrared powder absorption measurements of some tetramethyltetraselenafulvalene salts, [(TMTSF)$_2$X], *Phys. Rev. B: Condens. Matter* **32**, 5156.

Emge, T. J., Williams, J. M., Leung, P. C. W., Schultz, A. J., Beno, M. A., and Wang, H. H. (1985). Neutron diffraction study of (TMTSF)$_2$BF$_4$ after slow and rapid cooling to 20 K, *Mol. Cryst. Liq. Cryst.* **119**, 237.

Emge, T. J., Leung, P. C. W., Beno, M. A., Wang, H. H., Firestone, M. A., Webb, K. S., Carlson, K. D., and Williams, J. M. (1986). Correlations of anion size and symmetry with the structure and electronic properties of β-(BEDT-TTF)$_2$X conducting salts with trihalide anions X = I$_3^-$, I$_2$Br$^-$, IBr$_2^-$, *Mol. Cryst. Liq. Cryst.* **132**, 363.

Emge, T. J., Leung. P. C. W., Beno, M. A., Wang, H. H., Williams, J. M., Whangbo, M.-H., and Evain, M. (1986). Structural characterizations and band electronic structure of α-(BEDT-TTF)$_2$I$_3$ below its 135 K phase transitions, *Mol. Cryst. Liq. Cryst.* **138**, 393.

Emge, T. J., Hammond, C., Reed, P. E., Williams, J. M., Turner, D. J., Underhill, A. E., and Carneiro, K. (1986). A new organic metal (TMTSF)(PO$_2$F$_2$), *Chemtronics* **1**, 32.

Emge, T. J., Wang, H. H., Geiser, U., Beno, M. A., Webb, K. S., and Williams, J. M. (1986). Neutron diffraction evidence for unusual cohesive, H-bonding interactions in β-(BEDT–TTF)$_2$X organic superconductors, *J. Am. Chem. Soc.* **108**, 3849.

Endres, H., Hiller, M., Keller, H. J., Bender, K., Gogu, E., Heinen, I., and Schweitzer, D. (1985). Preparation, structure and investigations of BEDT–TTF trihalides, *Z. Naturforsch. B: Anorg. Chem., Org. Chem.* **40**, 1664.

Enoki, T., Imaeda, K., Kobayashi, M., Inokuchi, H., and Saito, G. (1986). ESR studies of organic conductors with bis(ethylenedithio)tetrathiafulvalene (BEDT–TTF), (BEDT–TTF)$_2$ClO$_4$ (C$_2$H$_3$Cl$_3$)$_{0.5}$ and (BEDT–TTF)$_3$(ClO$_4$)$_2$ and their two-dimensionality, *Phys. Rev. B: Condens. Matter* **33**, 1553.

Fabre, J. M., Emad-Mahnal, and Giral, L. (1986). Synthesis and electrical properties of two new conducting materials containing unsymmetrical tetraselenafulvalenes, *Synth. Met.* **13**, 339.

Ferraro, J. R., Beno, M. A., Thorn, R. J., Wang, H. H., Webb, K. S., and Williams, J. M. (1986). Spectroscopic and structural characterization of tetrabutylammonium trihalides: *n*- Bu$_4$I$_3$, *n*-Bu$_4$NI$_2$Br, *n*-Bu$_4$NIBr$_2$, and *n*-Bu$_4$NAuI$_2$; precursors to organic conducting salts, *J. Phys. Chem. Solids* **47**, 301.

Fitzky, H. G., and Hocker, J. (1986). Simultaneous ESR detection of the radical cation and radical anion in a solution of the CT complex of tetrathiafulvalene and tetracyanoquinodimethane (TTF–TCNQ), *Synth. Met.* **13**, 335.

Geiser, U., Wang, H. H., Beno, M. A., Firestone, M. A., Webb, K. S., Williams, J. M., and Whangbo, M.-H. (1986). Crystal and band electronic structures of orthorhombic γ- (BEDT-TTF)$_2$AuI$_2$, *Solid State Commun.* **57**, 741.

Heid, R., Endres, H., Keller, H. J., Gogu, E., Heinen, I., Bender, K., and Schweitzer, D. (1985). Radical cation salts of an unsymmetrical BEDT–TTF derivative: Molecular structure and physical properties of (DiMET)$_2$ClO$_4$ · *x*THF, *Z. Naturforsch. B: Anorg. Chem., Org. Chem.* **40**, 1703.

Heidmann, C.-P., Veith, H., Andres, K., Fuchs, H., Polborn, K., and Amberger, E.

(1986). Diamagnetic evidence of superconductivity in the organic conductors β-(BEDT-TTF)$_2$IBr$_2$, and β-(BEDT-TTF)$_2$AuI$_2$, *Solid State Commun.* **57**, 161.

Hirsch, J. E., and Scalapino, D. J. (1986). Enhanced superconductivity in quasi-two-dimensional systems, *Phys. Rev. Lett.* **56**, 2732.

Hodina, A. J., ed. (1984–1986). Recent publications in the area of organic conductors and related phenomena, *Synth. Met.;* titles and source given. Subjects are listed under the following headings: (a) organic conductors and related topics, general; (b) (CH)$_x$ (polacetylene and derivatives); (c) polypyrrole, polythiophene, polyethylene, polyphenylene, and their derivatives; (d) TTF, TCNQ, TMTSF (tetrathiafulvalene, tetracyanoquinodimethane, and tetramethyltetraselenafulvalene compounds); (e) other conducting polymers; (f) one-dimensional systems and charge and spin density wave dynamics, general; (g) NbSe$_3$, TaS$_3$ (niobium triselenide, tantalum trisulphide, and similar compounds); (h) transition metal dichalcogenide compounds, intercalated and pristine; (i) graphite intercalation compounds; (j) other similar intercalated and layered compounds and related topics; and (k) patents: conducting polymers, intercalated graphite, and transition metal chalcogenides.

Inoue, K., Tasaka, Y., Yamazaki, O., Nogami, T., and Mikawa, H. (1986). Synthesis of bis(dimethylvinylene-dithio)tetrathiafulvalene, BDMVDT–TTF, *Chem. Lett. Chem. Soc. Jpn.,* p. 781.

Iwahana, K., Kezmany, H., Wudl, F., and Aharon-Shalom, E. (1982). Raman scattering from TMTSF-salts, *Mol. Cryst. Liq. Cryst.* **79**, 39.

Jurgenson, C. W., and Drickamer, H. G. (1986). High-pressure studies of the neutral-to-ionic transition in some organic charge-transfer solids, *Chem. Phys. Lett.* **125**, 554.

Kanbara, H., Tajima, H., Aratani, S., Yakushi, K., Kuroda, H., Saito, G., Kawamoto, A., and Tanaka, J. (1986). Crystal structure of α-(BEDT-TTF)$_3$(ReO$_4$)$_2$, *Tech. Rep. Inst. Solid State Phys. Univ. of Tokyo, Ser. A.,* No. 1636, pp. 1–12.

Katayama, C., Honda, M., Kumagai, H., Tanaka, J., Saito, G., and Inokuchi, H. (1985). Crystal structures of complexes between hexacyanobutadiene and tetramethyltetrathiafulvalene and tetramethylthiotetrathiafulvalene, *Bull. Chem. Soc. Jpn.* **58**, 2272.

Kato, R., Kobayashi, H., Kobayashi, A., and Sasaki, Y. (1985). Crystal structure and electrochemical properties of organic donors, BMDT–TTF and BEDSe–TSeF. Two modifications of BEDT–TTF, *Chem. Lett. Chem. Soc. Jpn.,* p. 1231.

Kato, R., Kobayashi, H., Kobayashi, A., and Sasaki, Y. (1985). Three-dimensional intermolecular Se – – – Se network in (BEDSe–TSeF)PF$_6$, *Chem. Lett. Chem. Soc. Jpn.,* p. 1943.

Kinoshita, N., Tokumoto, M., Anzai, H., and Saito, G. (1985). Anisotropy in ESR *g* factors and linewidths for α- and β-(BEDT-TTF)$_2$I$_3$, *J. Phys. Soc. Jpn.* **54**, 4498.

Kistenmacher, T. J. (1985). Structural systematics in the family of (BEDT-TTF)$_2$X salts, *Solid State Commun.* **53**, 831.

Kleman, R. A., and Schrieffer, J. R. (1985). Single domain dynamics of weakly pinned charge density waves, *Synth. Met.* **11**, 307.

Kobayashi, H., Kobayashi, A., Sasaki, Y., Saito, G., and Inokuchi, H. (1984). The crystal structure of (TMTTF)$_2$ReO$_4$, *Bull. Chem. Soc. Jpn.* **57**, 2025.

Kobayashi, H., Kato, R., Mori, T., Kobayashi, A., Sasaki, Y., Saito, G., and Inokuchi, H. (1985). Organic conductors based on multi-sulfur π-donor and/or π-acceptor molecules—BEDT–TTF, BMDT–TTF, BPDT–TTF, and M(dmit)$_2$, *Mol. Cryst. Liq. Cryst.* **125**, 125.

Kobayashi, H., Kato, R., Kobayashi, A., Saito, G., Tokumoto, M., Anzai, H., and Ishiguro, T. (1985). The crystal and electronic structures of (BEDT-TTF)$_2$I$_2$Br, *Chem. Lett. Chem. Soc. Jpn.,* p. 1293.

Kobayashi, H., Kato, R., Kobayashi, A., Nishio, Y., Kajita, K., and Sasaki, W. (1986).

A new molecular superconductor, $(BEDT-TTF)_2(I_3)_{1-x}(AuI_2)_x$ $(x < 0.02)$, *Chem. Lett., Chem. Soc. Jpn.*, p. 789.

Kobayashi, H., Kato, R., Kobayashi, A., Nishio, Y., Kajita, K., and Sasaki, W. (1986). Crystal and electronic structures of layered molecular superconductor, θ-(BEDT-TTF)$_2$(I$_3$)$_{1-x}$(AuI$_2$)$_x$, *Chem. Lett. Chem. Soc. Jpn.*, p. 833.

Kobayashi, H., Kato, R., Kobayashi, A., Saito, G., Tokumoto, M., Anzai, H., and Ishiguro, T. (1986). Crystal stucture of α'-(BEDT-TTF)$_2$BrICl, *Chem. Lett. Chem. Soc. Jpn.*, p. 93.

Kobayashi, H., Kato, R., Kobayashi, A., Saito, G., Tokumoto, M., Anzai, H., and Ishiguro, T. (1986). The crystal structure of β'-(BEDT-TTF)$_2$ICl$_2$. A modification of the organic superconductor, β-(BEDT-TTF)$_2$I$_3$, *Chem. Lett. Chem. Soc. Jpn.*, p. 89.

Kobayashi, H., Kobayashi, A., Sasaki, Y., Saito, G., and Inokuchi, H. (1986). The crystal and molecular structures of bis(ethylenedithio)tetrathiafulvalene, *Bull. Chem. Soc. Jpn.* **59**, 301.

Krauzman, M., Poulet, H., and Pick, R. M. (1986). Resonant Raman scattering in bis-tetramethyltetraselenafulvalene-hexafluorophosphate [(TMTSF)$_2$PF$_6$] single crystal, *Phys. Rev. B: Condens. Matter* **33**, 99.

Kuroda, H., Yakushi, K., Tajima, H., and Saito, G. (1985). Reflectance spectra of BEDT–TTF salts, *Mol. Cryst. Liq. Cryst.* **125**, 135.

Kwak, J. F., Azevedo, L. J., Schirber, J. E., Williams, J. M., and Beno, M. A. (1985). Transport studies of several novel organic conductors, *Mol. Cryst. Liq. Cryst.* **125**, 365.

Kwak, J. F., Schirber, J. E., Chaikin, P. M., Williams, J. M., and Wang, H. H. (1985). Further magnetotransport studies of (TMTSF)$_2$PF$_6$, *Mol. Cryst. Liq. Cryst.* **125**, 375.

Kwak, J. F., Schirber, J. E., Chaikin, P. M., Williams, J. M., Wang, H. H., and Chiang, L. Y. (1986). Spin-density-wave transitions in a magnetic field, *Phys. Rev. Lett.* **56**, 972.

Laversanne, R., Coulon, C., Amiell, J., Dupart, E., Delhaes, P., Morand, J. P., and Manigand, C. (1986). Physical properties of radical cation salts of the dimethylethyl-enedithiotetrathiafulvalene, *Solid State Commun.* **58**, 765.

LeMehaute, A., Perichaud, A., Van Huong, T., and Guyof, A. (1985). Charge transfer complexes between sulfur and polyacetylene-like materials, *Synth. Met.* **11**, 373.

Leung, P. C. W., Emge, T. J., Schultz, A. J., Beno, M. A., Carlson, K. D., Wang, H. H., Firestone, M. A., and Williams, J. M. (1986). The role of anions on the crystal structure and electrical properties of the organic metals and superconductors, (BEDT–TTF)$_2$X (X = Trihalide Anions), *Solid State Commun.* **57**, 93.

Li, Z. S., Matsuzaki, S., Onomichi, M., Sano, M., and Saito, G. (1986). Partial charge transfer in HMTTeF–TCNQF$_4$, *Tech. Rep. Inst. Solid State Phys., Univ. of Tokyo, Ser. A.*, No. 1648, pp. 1–15.

Maniwa, Y., Takahaski, T., and Saito, G. (1986). ^{1}H NMR in organic superconductor β-(BEDT-TTF)$_2$I$_3$, *J. Phys. Soc. Jpn.* **55**, 47.

Matsuzaki, S., Kuwata, R., and Toyoda, K. (1980). Raman spectra of conducting TCNQ salts; estimation of the degree of charge transfer from vibrational frequencies, *Solid State Commun.* **33**, 403.

McCall, R. P., Tanner, D. B., Miller, J. S., Epstein, A. J., Howard, I. A., and Conwell, E. M. (1985). Peierls gap in the large-U quarter-filled band compound quinolinium tetracyanoquinodimethanide [Qn(TCNQ)$_2$], *Synth. Met.* **11**, 231.

Megtert, S., Bjelis, A., Przystawa, J., and Barišić, S. (1985). Symmetry approach to the (P, T) diagram of TTF–TCNQ (tetrathiafulvalene–tetracyanoquinodimethane), *Phys. Rev. B: Condens. Matter* **32**, 6692.

Mengeghetti, M., Girlando, A., and Pecile, C. (1985). Ionicity and electron molecular vibration interaction in mixed stack CT systems: M_2P–TCNQ and M_2P–TCNQF$_4$, *J. Chem. Phys.* **83**, 3134.

Mengeghetti, M., and Pecile, C. (1986). Charge-transfer organic crystals: Molecular vibrations and spectroscopic effects of electron-molecular vibration coupling of the strong electron acceptor TCNQF$_4$, *J. Chem. Phys.* **84**, 4149.

Mori, T., Kobayashi, A., Sasaki, Y., Kato, R., and Kobayashi, H. (1985). Band structure of β-(BEDT-TTF)$_2$PF$_6$ one-dimensional metal along the side-by-side molecular array, *Solid State Commun.* **53**, 627.

Mori, T., Sakai, F., Saito, G., and Inokuchi, H. (1986). Crystal and band structures of an organic conductor β-(BEDT-TTF)$_2$AuBr$_2$, *Chem. Lett. Chem. Soc. Jpn.*, p. 1037.

Mortensen, K., Williams, J. M., and Wang, H. H. (1985). Anisotropic thermopower of the organic metal, β-(BEDT-TTF)$_2$I$_3$, *Solid State Commun.* **56**, 105.

Murata, K., Tokumoto, M., Anzai, H., Bando, H., Kajimura, K., and Ishiguro, T. (1985). Superconductivity with an onset at 8 K in β-(BEDT-TTF)$_2$I$_3$ under pressure, *Synth. Met.* **13**, 3.

Murata, K., Tokumoto, M., Bando, H., Tanino, H., Anzai, H., Kinoshita, N., Kajimura, K., Saito, G., and Ishiguro, T. (1985). High T_c superconducting state in (BEDT–TTF)$_2$ trihalides, *Physica B (Amsterdam)* **135**, 515.

Murata, K., Tokumoto, M., Anzai, H., Bando, H., Saito, G., Kajimura, K., and Ishiguro, T. (1985). Pressure phase diagram of the organic superconductor β-(BEDT-TTF)$_2$I$_3$, *J. Phys. Soc. Jpn.* **54**, 2084.

Nakamura, T., Matsumoto, M., Takei, F., Tanaka, M., Sekiguchi, T., Komizu, H., Manda, E., Kawabata, Y., and Saito, G. (1986). Langmuir–Blodgett film of TMTTF-octadecyl TCNQ. An intrinsic conductor, *Tech. Rep. Inst. Solid State Phys., Univ. of Tokyo, Ser. A.*, No. 1650, pp. 1–11.

Nakamura, T., Takei, F., Tanaka, M., Matsumoto, M., Sekiguchi, T., Manda, E., Kawabata, Y., and Saito, G. (1986). Conducting monolayer of simple charge transfer complex on a glycerin subphase, *Tech. Rep. Inst. Solid State Phys., Univ. of Tokyo, Ser. A.*, No. 1630, pp. 1–6.

Nakasuji, K., Nakatsuku, M., Yamochi, H., Murata, I., Harada, S., Kasai, N., Yamamura, K., Tanaka, J., Saito, G., Enoki, T., and Inokuchi, H. (1986). Three-dimensionality-modified tetracyanoquinodimethanes and their charge-transfer complexes with tetrathiafulvalene derivatives having a wide range of ionicity, *Bull. Chem. Soc. Jpn.* **59**, 207.

Nakasuji, K., Kubota, H., Kotani, T., Murata, I., Saito, G., Enoki, T., Imaeda, K., Inokuchi, H., Honda, M., Katayama, C., and Tanaka, J. (1986). Novel peri-condensed Weitz type donors: synthesis, physical properties and crystal structure of 3,10-Dithiaperylene (DTPR), 1,6-Dithiapyrene (DTPY), and some of their CT complexes, *Tech. Rpt, Inst. Solid State Phys., Univ. of Tokyo, Ser. A.*, No. 1632, pp. 1–30.

Naughton, M. J., Brooke, J. S., Chiang, L. Y., Chamberlin, R. V., and Chaikin, P. M. (1985). Magnetization study of the field-induced transitions in tetramethyltetraselenafulvalene perchlorate, (TMTSF)$_2$ClO$_4$, *Phys. Rev. Lett.* **55**, 969.

Ng, H. K., Timusk, T., Jérome, D., and Bechgaard, K. (1985). Far-infrared spectrum of di-tetramethyltetraselenafulvalene hexafluoroarsenate [(TMTSF)$_2$AsF$_6$], *Phys. Rev. B: Condens. Matter* **32**, 8041.

Okada, N., Saito, G., and Mori, T. (1986). Tetratelluradicyclopenta [b, g] naphthalene (TTeDCN), *Tech. Rep. Inst. Solid State Phys., Univ. of Tokyo, Ser. A.*, No. 1627, pp. 1–17.

Onoda, M., Nagasawa, H., and Kobayashi, K. (1985). EPR study on metal-to-nonmetal transition in the quasi-one-dimensional conductor (TMTTF)$_2$ClO$_4$, *J. Phys. Soc.* **54**, 1240.

Papavassiliou, G. C., Yiannopoulos, S. Y., and Zambounis, J. S. (1986). Pyrazenoethyl-enedithiotetrathiafulvalene: A new unsymmetrical π-donor, *J. Chem. Soc. Chem. Commun.*, p. 820.

Parkin, S. S. P. (1984). A narrow window for superconductivity in organic conductors, *in* "Physics and Chemistry of Electrons and Ions in Condensed Matter" (J. V. Acrivos, ed.), p. 655. Reidel Publ., Boston, Massachusetts.

Parkin, S. S. P., and Acrivos, J. V. (1983). XANES in (TMTSF)$_2$ReO$_4$: Polarization dependence of the Se-K edge, *J. Phys. Colloq.* **44**, C3-1011.

Pesty, F., Garoche, P., and Bechgaard, K. (1985). Cascade of field-induced phase transitions in the organic metal tetramethyltetraselenafulvalenium perchlorate [(TMTSF)$_2$ClO$_4$], *Phys. Rev. Lett.* **55**, 2495.

Rangel-Zamudio, L. I., Rushforth, D. S., and Van Duyne, R. P, (1986). Long-range resonance energy transfer from aromatic hydrocarbons to the anion radical of tetra-cyanoquinodimethane, *J. Phys. Chem.* **90**, 807.

Ravy, S., Moret, R., Pouget, J. P., and Comés, R. (1986). Structural phase transitions in (TMTSF)$_2$X and related compounds, *Synth. Met.* **13**, 63.

Ravy, S., Moret, R., Pouget, J. P., Comés, R., and Parkin, S. S. P. (1986). Competition between organic superconductivity and a displacive structural modulation—bis(ethylenedithio)tetrathiafulvalene perrhennate, (BEDT-TTF)$_2$ReO$_4$, *Phys. Rev. B: Condens. Matter* **33**, 2049.

Rommelmann, H., Epstein, A. J., Miller, J. S., Restle, P. J., Black, R. D., and Weissman, M. B. (1985). Noise-power studies in newly commensurate quasi-one-dimensional conductor (*N*-methylphenazinium)$_x$(phenazine)$_{1-x}$, 7,7,8,8-tetracyano-*p*-quinodime-thane, *Phys. Rev. B: Condens. Matter* **32**, 1257.

Sachs, D., Dormann, E., and Schwoerer, M. (1985). One-dimensionality of the organic conductor (FA)$_2^+$PF$_6^-$, *Solid State Commun.* **53**, 73.

Saito, G., Enoki, T., Inokuchi, H., Kumagai, H., Katayama, C., and Tanaka, J. (1985). Crystal structures and electrical properties of hexacyano-butadiene (HCBD) charge transfer complexes, *Mol. Cryst. Liq. Cryst.* **120**, 1345.

Saito, G., Hayashi, H., Enoki, T., and Inokuchi, H. (1985). The study of charge transfer complexes of BEDT–TTF derivatives, *Mol. Cryst. Liq. Cryst.* **120**, 341.

Saito, G., Sugano, T., Yamochi, H., Kimoshita, M., Oshima, K., Suzuki, M., Katayama, C., and Tanaka, J. (1985). Organic superconductor β-(ET)$_2$IBr$_2$ obtained by diffusion method, *Chem. Lett. Chem. Soc. Jpn.*, p. 1037.

Saito, G., Kumagai, H., Tanaka, J., Enoki, T., and Inokuchi, H. (1985). Organic metals based on hexamethylenetetratellurafulvalene (HMTTeF), *Mol. Cryst. Liq. Cryst.* **120**, 337.

Saito, G., Enoki, T., Kobayashi, M., Imaeda, K., Sato, N., and Inokuchi, H. (1985). Chemical and physical properties of cation radical salts of BEDT–TTF, *Mol. Cryst. Liq. Cryst.* **119**, 393.

Schriber, J. E., Azevedo, L. J., and Williams, J. M. (1985). Phase diagrams of organic superconductors, *Mol. Cryst. Liq. Cryst.* **119**, 27.

Schultz, A. J., Beno, M. A., Wang, H. H., and Williams, J. M. (1986). Neutron-diffraction evidence for ordering in the high-T_c phase of β-di[bis-(ethylenedithio)tetrathiafulvalene]triiodide [β*-(ET)$_2$I$_3$], *Phys. Rev. B: Rapid Commun.* **33**, 7823.

Schumaker, R. R., Lee, V. Y., and Engler, E. M. (1984). Noncoupling synthesis of tetrathiafulvalenes, *J. Org. Chem.* **49**, 564.

Schwenk, H., Heidmann, C. P., Gross, F., Hess, E., Andres, K., Schweitzer, D., and Keller, H. J. (1985). New organic volume superconductor at ambient pressure, *Phys. Rev. B: Condens. Matter* **31**, 3138.

Schwenk, H., Parkin, S. S. P., Schumaker, R., Greene, R. L., and Schweitzer, D. (1986). Magnetic-field-induced transition and quantum oscillations in tetramethyltetraselenafulvalenium perrhenate, *Phys. Rev. Lett.* **56**, 667.

Seki, K., Tang, T. B., Mori, T., Ji, W. P., Saito, G., and Inokuchi, H. (1986). Valence electronic structures of tetrakis(alkylthio)tetrathiafulvalene, *Tech. Rep. Inst. Solid State Phys., Univ. of Tokyo, Ser. A.,* No. 1633, pp. 1–24.

Stewart, G. R., O'Rourke, J., Crabtree, G. W., Carlson, K. D., Wang, H. H., Williams, J. M., Gross, F., and Andres, K. (1986). Specific heat of the ambient-pressure organic superconductor β-di[bis(ethylenedithio)tetrathiafulvalene] triiodide [β-(BEDT-TTF)$_2$I$_3$], *Phys. Rev. B: Condens. Matter* **33**, 2046.

Sugai, S., and Sato, G. (1986). Resonant Raman scattering in organic conductors α- and β-(BEDT-TTF)$_2$X (X = I$_3$ and IBr$_2$), *Tech. Rep. Inst. Solid State Phys., Univ. of Tokyo, Ser. A.,* No. 1649, pp. 1–17.

Sugano, T., Yamada, K., Saito, G., and Kinoshsita, M. (1985). Polarized spectra of the organic conductors: α- and β-modifications of di[bis-(ethylenedithio)tetrathiafulvalene] triiodide, (BEDT-TTF)$_2$I$_3$, *Solid State Commun.* **55**, 137.

Sugano, T., Saito, G., and Kinoshita, M. (1986). Conduction-electron-spin resonance in inorganic conductors: α and β phases of di[bis(ethylenedithiolo)tetrathiafulvalene] [(BEDT-TTF)$_2$I$_3$], *Phys. Rev. B: Condens. Matter* **34**, 117.

Tajima, H., Yakuski, K., Kurada, H., and Saito, G., (1985). Polarized reflectance spectrum of β-(BEDT-TTF)$_2$PF$_6$, *Solid State Commun.* **56**, 251.

Tajima, H., Yakushi, K., Kuroda, H., and Saito, G. (1985). Polarized reflectance spectrum of β'-(BEDT-TTF)$_2$I$_3$ single crystal, *Solid State Commun.* **56**, 159.

Tajima, H., Kanbara, H., Yakushi, K., Kuroda, H., and Saito, G. (1986). Temperature dependence of the reflectance spectrum of β-(BEDT-TTF)$_2$I$_3$, *Tech. Rep. Inst. Solid State Phys., Univ. of Tokyo, Ser. A.,* No. 1634, pp. 1–12.

Takahashi, T., Maniwa, Y., Kawamura, H., and Saito, G. (1986). Determination of SDW characteristics in (TMTSF)$_2$PF$_6$ in ^{2}H-NMR analysis, *J. Phys. Soc. Jpn.* **55** (to be published).

Takigawa, M., and Saito, G. (1986). Evidence for the slow motion of organic molecules in (TMTSF)$_2$ClO$_4$ extracted from the spin echo decay of ^{77}Se NMR, *J. Phys. Soc. Jpn.* **55**, 1233.

Takigawa, M., and Saito, G. (1985). Evidence for the slow motion of organic molecules in (TMTSF)$_2$ClO$_4$ extracted from the spin echo decay of ^{77}Se NMR, *Tech. Rep. Inst. Solid State Phys., Univ. of Tokyo, Ser. A.,* No. 1601.

Tanaka, M. (1986). Optical and magnetic properties of (DBTTF)$_2$(Cu$_2$Cl$_6$) and (DBTTF)$_2$(Cu$_2$Br$_6$) complexes, *Bull. Chem. Soc. Jpn.* **59**, 775.

Tanaka, M., Shimizu, M., Saito, Y., and Tanaka, J. (1986). Raman spectra of radical ion DBTTF complexes; relation between Raman frequency and formal charge, *Chem. Phys. Lett.* **125**, 594.

Tanino, H., Kato, K., Tokumoto, M., Anzai, H., and Saito, G. (1985). X-ray study of compressibilities of β-(BEDT-TTF)$_2$I$_3$ under hydrostatic pressure, *J. Phys. Soc. Jpn.* **54**, 2390.

Terauchi, H., Kozaki, T., Nishihata, Y., Sakashita, H., Yoshikawa, M., and Nakashima, S. (1985). On the phase transitions of purple NH$_4$-TCNQ(I), *J. Chem. Phys.* **83**, 1305.

Thorn, R. J., Carlson, K. D., Wang, H. H., and Williams, J. M. (1985). Asymmetric broadening of Se(3$d_{5/2}$)XPS spectra of (TMTSF)$_2$ClO$_4$ and (TMTSF)$_2$ReO$_4$, *Mol. Cryst. Liq. Cryst.* **119**, 233.

Thorn, R. J., Carlson, K. D., Crabtree, G. W., and Wang, H. H. (1985). Photoelectron spectroscopy in the perchlorate and perrhenate of TMTSF, *J. Phys. C.* **18**, 5502.

Tokumoto, M., Bando, H., Anzai, H., Saito, G., Murata, K., Kajimura, K., and Ishiguro, T. (1985). Critical field anisotropy in an organic superconductor β-(BEDT-TTF)$_2$I$_3$, *J. Phys. Soc. Jpn.* **54**, 869.

Tokumoto, M., Murata, K., Bando, H., Anzai, H., Saito, G., Kajimura, K., and Ishiguro, T. (1985). Ambient-pressure superconductivity at 8 K in the organic conductor β-(BEDT-TTF)$_2$I$_3$, *Solid State Commun.* **54**, 1031.

Tokumoto, M., Anzai, H., Bando, H., Saito, G., Kinoshita, N., Kajimura, K., and Ishiguro, T. (1985). Critical field anisotropy in an organic superconductor β-(BEDT-TTF)$_2$IBr$_2$, *J. Phys. Soc. Jpn.* **54**, 1669.

Tokumoto, M., Bando, H., Anzai, H., Saito, G., Murata, K., Kajimura, K., and Ishiguro, T. (1985). Critical field anisotropy in an organic superconductor β-(BEDT-TTF)$_2$I$_3$, *J. Phys. Soc. Jpn.* **54**, 869.

Tokumoto, M., Bando, H., Murata, K., Anzai, H., Kinoshita, N., Kajimura, K., Ishiguro, T., and Saito, G. (1986). Ambient-pressure superconductivity in organic metals, BEDT–TTF trihalides, *Synth. Met.* **13**, 9.

Tokura, Y., Okamoto, H., Mitani, T., Saito, G., and Koda, T. (1986). Pressure induced neutral-to-ionic phase transition in TTF–*p*-chloranil studied by infrared vibrational spectroscopy, *Tech. Rep. Solid State Phys., Univ. of Tokyo, Ser. A.,* No. 1635, pp. 1–13.

Ulmet, J. P., Khmou, A., Auban, P., and Bachere, L. (1986). New results on the high field magnetoresistance of (TMTSF)$_2$ClO$_4$: Oscillatory effects and phase transitions, *Solid State Commun.* **58**, 735.

Underhill, A. E., Tonge, J. S., Clemenson, P. I., Wang, H. H., and Williams, J. M. (1985). New conductors based on metal complexes of sulfur chelate ligands, *Mol. Cryst. Liq. Cryst.* **125**, 439.

Van Duyne, R. P., Cape, T. W., Suchanski, M. R., and Siedle, A. R. (1986). Determination of the extent of charge transfer in partially oxidized derivatives of tetrathiafulvalene and tetracyanoquinodimethane by resonance Raman spectroscopy, *J. Phys. Chem.* **90**, 739.

Van Smaalen, S., Kommandeur, J., and Conwell, E. M. (1986). Contributions of one- and two-phonon processes to the resistivity of tetrathiafulvalene–tetracyanoquinodimethane, *Phys. Rev. B: Condens. Matter* **33**, 5378.

Veith, H., Heidmann, C.-P., Gross, F., Lerf, A., Andres, K., and Schweitzer, D. (1985). Observation of the Meissner effect in the high-T_c (pressure) phase of the organic superconductor β-(BEDT-TTF)$_2$I$_3$, *Solid State Commun.* **56**, 1015.

Wallis, J. D., Karrer, A., and Dunitz, J. D. (1986). Chiral metals? A chiral substrate for organic conductors and superconductors, *Helv. Chim. Acta* **69**, 69.

Wang, H. H., and Williams, J. M. (1986). Ambient pressure superconducting synthetic metals β-(BEDT-TTF)$_2$X, X = I$_3^-$, IBr$_2^-$, *Inorg. Synth.,* (accepted for publication).

Wang, H. H., Reed., P. E., and Williams, J. M. (1986). Perdeuterio bis-(ethylenedithio)tetrathiafulvalene, *Synth. Met.* **14**, 165.

Weber, A., Endres, H., Keller, H. J., Gogu, E., Heinen, I., Bender, K., and Schweitzer, D. (1985). Preparation, structure and physical properties of BEDT–TTF nitrates, *Z. Naturforsch. B: Anorg. Chem., Org. Chem.* **40**, 1658.

Whangbo, M.-H., Williams, J. M., Leung, P. C. W., Beno, M. A., Emge, T. J., and Wang, H. H. (1985). Role of the intermolecular interactions in the two-dimensional ambient-pressure organic superconductors β-(ET)$_2$I$_3$ and β-(ET)$_2$IBr$_2$, *Inorg. Chem* **24**, 3500.

Williams, J. M., Wang, H. H., Beno, M. A., Geiser, U., Firestone, M. A., Webb, K. S., Nuñez, L., Crabtree, G. W., and Carlson, K. D. (1985). Ambient pressure superconductivity at 5 K in the organic system: β-(BEDT-TTF)$_2$AuI$_2$, *Physica B (Amsterdam)* **135**, 520.

Williams, J. M., Beno, M. A., Wang, H. H., Geiser, U., Emge, T. J., Leung. P. C. W., Crabtree, G. W., and Carlson, K. D. (1986). Exotic organic superconductors based on BEDT–TTF and the prospects of raising T_c's, *Physica B (Amsterdam)* **136**, 371.

Williams, K. A., Nowak, M. J., Dormann, E., and Wudl, F. (1986). Organic conductors based on diradicals: The benzobisdithiazole (BBDT) system, *Synth. Met.* **14**, 233.

Wolmershäuser, G., Schnauber, M., Wilhelm, T., and Sutcliffe, L. H. (1986). The benzobisdithiazole system in its various oxidation states: A new building block for organic conductors, *Synth. Met.* **14**, 239.

Wu, P., Mori, T., Enoki, T., Imaeda, K., Saito, G., and Inokuchi, H. (1986). Crystal structure and physical properties of (TTM-TTF)I$_{2.47}$, *Bull. Chem. Soc. Jpn.* **59**, 127.

Wu, P., Saito, G., Imaeda, K., Shi, Z., Mori, T., Enoki, T., and Inokuchi, H. (1986). Uncapped alkylthio substituted tetrathiafulvalenes (TTC$_n$-TTF) and their charge transfer complexes, *Chem. Lett. Chem. Soc. Jpn.*, p. 441.

Yakushi, K., Kuroda, H., Ikemoto, I., Kobayashi, K., Honda, M., Katayama, C., Tanaka, J. (1985). The crystal structure of a new charge-transfer for salt of hexamethylene–tetratellurafulvalene, (HMTTeF)$_4$(PF$_6$)$_2$, *Chem. Lett.*, p. 419.

Yakushi, K., Aratani, S., Kikuchi, K., Tajima, H., and Kuroda, H. (1986). Structure and optical properties of (TMTTF)$_2$IO$_4$. The role of coulomb interactions, *Bull. Chem. Soc. Jpn,* **59**, 363.

Yamaji, K. (1986). Theory of field-induced spin-density-wave in Bechgaard salts, *Synth. Met.* **13**, 29.

Yamaji, K. (1986). Pressure- and anion-dependences of the SDW phase transition in the Bechgaard salts, *J. Phys. Soc. Jpn.* **55**, 860.

Yamashita, Y., Hagiya, K., Saito, G., and Mukai, T. (1986). Novel tetracyanoqoinodimethane anion radical salts of heterocyclic cations containing a tropylium ion skeleton, *Chem. Lett. Chem. Soc. Jpn.*, p. 537.

Yui, K., Aso, Y., Otsubo, T., and Ogura, F. (1986). Syntheses and properties of binaphtho [1,8-de]-1,3-dithiin-2-yledene and its selenium analogue, *Chem. Lett.*, p. 551.

CHAPTER 3.   CONDUCTIVE ORGANIC POLYMERS

Aldissi, M. (1986). Polyacetylene block copolymers, *Synth. Met.* **13**, 87.

Aldissi, M., and Bishop, A. R. (1986). Electrochemistry of polyacetylene block copolymers, *Synth. Met.* **14**, 13.

Audebert, P., and Bidan, G. (1986). Electrochemical behavior and some characteristics of conducting polymers issued from the autoxidation of some bromopyrroles, *Synth. Met.* **15**, 9.

Audebert, P., and Bidan, G. (1986). Comparison of carbon paste electrochemistry of polypyrroles prepared by chemical and electrochemical oxidation paths. Some characteristics of the chemically prepared polyhalopyrroles, *Synth. Met.* **14**, 71.

Audenaert, M., Rachdi, F., and Bernier, P. (1986). DC conductivity versus temperature of a metallic complex of (CH)$_x$ with potassium, *Synth. Met.* **15**, 91.

Baeriswyl, D., and Maki, K. (1985). Electron correlations in polyacetylene, *Phys. Rev. B: Condens. Matter* **31**, 6633.

Batail, P., Ouchab, L., Halet, J. F., Padiov, J., Lequan, M., and Lequan, R. M. (1985). Slipped versus zig-zag TCNQ stack upon steric control in conducting 1:4 diphosphonium–TCNQ salts, *Synth. Met.* **10**, 415.

Baughman, R. H., Kohler, B. E., Levy, I. J., and Spangler, C. (1985). The crystal structure of trans, *trans*-1,3,5,7-octatetraene as a model for fully ordered *trans*- polyacetylene, *Synth. Met.* **11**, 37.

Bechtold, G., Genzel, L., and Roth, S. (1985). Far infrared of the lattice modes in trans and cis polyacetylene, *Solid State Commun.* **53**, 1.

Beck, F., and Krohn, H. (1986). The role of solvate acid in the electrochemical behavior of graphite intercalation compounds, *Synth. Met.* **14**, 137.

Begin, D., Billaud, D., and Goulon, C. (1985). Electrochemical oxidation of $(CH)_x$ in sulfuric acid, *Synth. Met.* **11**, 29.

Béniere, F., and Pekker, S. (1986). Distribution profiles of iron in $FeCl_3$-doped polyacetylene films, *Solid State Commun.* **57**, 835.

Bidan, G., and Guglielmi, M. (1986). Electrochemical characteristics of some N-(hydroxylalkyl) and N-(tosylalkyl) pyrroles; electrochemical behavior of the corresponding polymers, *Synth. Met.* **15**, 49.

Billaud, D., Haleem, M. A., and Begin, D. (1986). Interaction of HNO_3 with trans-polyacetylene $(CH)_x$: A kinetic study, *Synth. Met.* **14**, 225.

Bott, D. C., Brown, C. S., Chai, C. K., Walker, N. S., Feast, W. J., Foot, P. J. S., Calvert, P. D., Billingham, N. C., and Friend, R. H. (1986). Durham polyacetylene: Preparation and properties of the unoriented material, *Synth. Met.* **14**, 245.

Brédas, J. L., and Baughman, R. H. (1985). Organic polymers based on aromatic rings (polyparaphenylene, polypyrrole, polythiophene): Evolution of the electronic properties as a function of the torsion angle between adjacent rings, *J. Chem. Phys.* **83**, 1323.

Brédas, J. L., Thémans, B., Andre, J. M., Heeger, A. J., and Wudl, F. (1985). Geometric and electronic structures of isothianaphthene and thieno (3,4-c) thiophene: A theoretical study, *Synth. Met.* **11**, 343.

Brédas, J. L., Dory, M., and Andre, J. M. (1985). Long polyene and polydiacetylene oligomers: Pariser–Parr–Pople investigation of the geometric and electronic structures in the first B_u excited state, *J. Chem. Phy.* **83**, 5242.

Buhks, E., and Hodge, I. M. (1985). Charge transfer in polypyrrole, *J. Chem. Phys.* **83**, 5981.

Burzynski, R., Prasad, P. N., Bruckenstein, S., and Sharkey, J. W. (1985). Infrared study of electrochemically prepared homo and mixed polymer films of azulene, *Synth. Met.* **11**, 293.

Chen, J., Heeger, A. J., and Wudl, F. (1986). Confined soliton pairs (bipolarons) in polythiophene: *in-situ* magnetic resonance measurements, *Solid State Commun.* **58**, 251.

Chen, Y. C., Akagi, K., and Shirakawa, H. (1986). Anisotropic electrical conductivity of partially oriented polyacetylene doped with iodine: Effects of draw ratio and dopant concentration, *Synth. Met.* **14**, 173.

Cline, J. F., Thomann, H., Kim, H., Morrobel-Sosa, A., Dalton, L. R., and Hoffman, B. M. (1985). Electron nuclear double resonance spectra of cis-rich and trans-rich polyacetylene between 1.9–4.2 K, *Phys. Rev. B: Condens. Matter* **31**, 1605.

Colaneri, N., Kobayashi, M., Heeger, A. J., and Wudl, F. (1986). Electrochemical and opto-electrochemical properties of poly(isothianaphthene), *Synth. Met.* **14**, 45.

Conwell, E. M. (1985). Statistics for polymers whose excitations are polarons, bipolarons, electrons and holes, *Synth. Met.* **11**, 21.

Danieli, R., Taliani, C., Zamboni, R., Giro, G., Biserni, M., Mastragostino, M., and Testoni, A. (1986). Optical, electrical and electrochemical characterization of electrosynthesized polythieno (3,2-b)-thiophene, *Synth. Met.* **13**, 325.

Dian, G., Barbey, G., and Decroix, B. (1986). Electrochemical systhesis of polythiophenes and polyselenophenes, *Synth. Met.* **13**, 281.

Dorsinville, R., Tubino, R., Krimchansky, S., Alfano, R. R., Birman, J. E., Bolegnesi,

A., Destri, S., Castellani, M., and Porzio, W. (1985). Infrared-photoinduced-absorption studies in soluble trans-polyacetylene, *Phys. Rev. B: Condens. Matter* **32**, 3377.

Druy, M. A., Rubner, M. F., and Walsh, S. P. (1986). An experimental approach toward the synthesis and characterization of environmentally stable conducting polymers, *Synth. Met.* **13**, 207.

Elsenbaumer, R. L., Delannoy, P., Miller, G. G., Forbes, C. E., Murthy, N. S., Eckhardt, H., and Baughman, R. H. (1985). Thermal enchancement of the electrical conductivities of alkali metal-doped polyacetylene complexes, *Synth. Met.* **11**, 251.

Erlandsson, R., Inganäs, O., Lundström, I., and Saleneck, W. R. (1985). XPS and electrical characterization of BF_4^- doped polypyrrole exposed to oxygen, *Synth. Met.* **10**, 303.

Faulques, E., Rzepka, E., Lefrant, S., Mulazzi, E., Brivio, G. P., and Leising, G. (1986). Polarized resonant Raman spectra of fully oriented trans-polyacetylene: experiments and theory, *Phys. Rev. B: Condens. Matter* **33**, 8622.

Friend, R. H., and Giles, J. R. M. (1985). An examination of the optical absorption spectrum of poly(p-phenylene) sulfide films when exposed to arsenic pentafluoride: Comparison with poly(p-phenylene), *Synth. Met.* **10**, 377.

Friend, R. H., Bardley, D. D. C., Pereira, C. M., Townsend, P. D., Bott, D. C., and Williams, K. P. J. (1986). Conformational defects in Durham–Route polyacetylene, *Synth. Met.* **13**, 101.

Fujimoto, H., Kamiya, K., Tanaka, J., and Tanaka, M. (1985). Optical constants of polyacetylene, *Synth. Met.* **10**, 367.

Fukutome, H., and Takahashi, A. (1986). Polyelectrolytic theory of doped polyacetylene, *Synth. Met.* **13**, 135.

Funt, B. L., and Lowen, S. W. (1985). Mechanistic studies of the electropolymerization of 2,2'-bithiophene and of pyrrole to form conducting polymers, *Synth. Met.* **11**, 129.

Galuzzi, F., Schwarz, M., and Pedretti, V. (1985). On the photovoltaic spectral response of polyacetylene films, *Synth. Met.* **10**, 205.

Genies, E. M., Syed, A. A., and Salmon, M. (1985). Electrochemical study of some chemically prepared para-substituted poly-n-phenylpyrroles, *Synth. Met.* **11**, 353.

Guillaud, G., Boudjema, B., Gamoudi, M., Ranaive-Harisoa, R., Maitrot, M., Andre, J. J., Francois, G., and Mathis, C. (1985). Role of oxygen in trans-polyacetylene-metal contacts, *Synth. Met.* **10**, 397.

Hahn, S. J., Gajda, W. J., Vogelhut, P. O., and Zeller, M. V. (1986). Auger and infrared study of polypyrrole films: Evidence of chemical changes during electrochemical deposition and aging in air, *Synth. Met.* **14**, 89.

Hasegawa, S., Kaniya, K., Tanaka, J., and Tanaka, M. (1986). Far-infrared spectra and electronic structure of polypyrrole perchlorate, *Synth. Met.* **14**, 97.

Hicks, J. C., and Blaisdell, G. A. (1985). Lattice vibrations in polyacetylene, *Phys. Rev. B: Condens. Matter* **31**, 919.

Howard, I. A., and Conwell, E. M. (1985). Binding energy of a charged soliton to a dopant ion in trans-polyacetylene, *Synth. Met.* **10**, 297.

Hyodo, K., and MacDiarmid, A. G. (1985). Effect of sulfate ion on the electrochemical polymerization of pyrrole and N-methylpyrrole, *Synth. Met.* **11**, 167.

Inganäs, O., Liedberg, B., Chang-Ru, W., and Wynberg, H. (1985). A new route to polythiophene and copolymers of thiophene and pyrrole, *Synth. Met.* **11**, 239.

Ishida, K., Nagamine, K., Matsuzaki, T., Kuno, Y., Yamazaki, T., Torikai, E., Shirakawa, H., and Brewer, J. M. (1985). Diffusion properties of the Muon-produced solition in trans-polyacetylene, *Phys. Rev. Lett.* **55**, 2009.

Jow, T. R., Jen, K. Y., Elsenbaumer, R. L., Shacklette, L. W., Angelopoulos, M., and

Cava, M. P. (1986). Electrochemical studies of fused-thiophene systems, *Synth. Met.* **14**, 53.

Kaner, R. B., and MacDiarmid, A. G. (1986). Reversible electrochemical reduction of polyacetylene, $(CH)_x$, *Synth. Met.* **14**, 3.

Kitani, A., Kaya, M., and Sasaki, K. (1986). Performance study of aqueous polyaniline batteries, *J. Electrochem. Soc.* **133**, 1069.

Kivelson, S., and Heeger, A. J. (1985). First-order transition to a metallic state in polyacetylene, *Phys. Rev. Lett.* **55**, 308.

Lefrant, S., Faulques, E., Brivio, G. P., and Mulazzi, E. (1985). Resonant Raman scattering of partially isomerized and doped polyacetylene: An application of the conjugation length distribution model, *Solid State Commun.* **53**, 583.

Leising, G., Faulques, E., and Lefront, S. (1985). Polarized resonance Raman spectroscopy of fully-oriented crystalline trans-$(CH)_x$, *Synth. Met.* **11**, 123.

LeMehaute, A., Perichaud, A., Bernier, P., Van Hoang, T., and Guyot, A. (1985). Charge transfer complexes between sulfur and polyacetylene-like materials, *Synth. Met.* **11**, 373.

Macconnachie, A., Dianoux, A. J., Shirakawa, H., and Tasumi, M. (1986). Incoherent inelastic neutron scattering from polyacetylenes in the 3500–400 cm^{-1} region, *Synth. Met.* **14**, 323.

Matsumura, K., Takahashi, A., and Tsukamoto, J. (1985). Structure and electrical conductivity of graphite fibers prepared by pyrolysis of cyanoacetylene, *Synth. Met.* **11**, 9.

Mazumdar, S., and Campbell, D. K. (1986). Bond alternation in the infinite polyene: Effect of long-range coulomb interactions, *Synth. Met.* **13**, 163.

Mele, E. J., and Hicks, J. C. (1986). Lattice dynamics in the continuum model for polyacetylene, *Synth. Met.* **13**, 149.

Meyer, W. H., Kiess, H., Benggeli, B., Meier, E., and Harbeke, G. (1985). Polypyrrole for use in information storage, *Synth. Met.* **10**, 255.

Mishima, A., and Kimura, M. (1985). Superconductivity of the quasi-one-dimensional semiconductor-polyacene, *Synth. Met.* **11**, 75.

Mohammadi, A., Hasan, M. A., Liedberg. B., Lundström, I., and Salaneck, W. R. (1986). Chemical vapor deposition (CVD) of conducting polymers: polypyrrole, *Synth. Met.* **14**, 189.

Moraes, F., Davidov, D., Kobayashi, M., Chung, T. C., Chen, J., Heeger, A. J., and Wudl, F. (1985). Doped poly(thiophene) electron spin resonance determination of the magnetic susceptibility, *Synth. Met.* **10**, 169.

Moraes, F., Chen, J., Chung, T.-C., and Heeger, A. J. (1985). First-order transition to a novel metallic state in $[Na_y^+(CH)^{-y}]$: *in-situ* electron spin resonance during chemical and electrochemical doping, *Synth. Met.* **11**, 271.

Moraes, F., Park, Y. W., and Heeger, A. J. (1986). Soliton photogeneration in trans-polyacetylene: Light-induced electron spin resonsance, *Synth. Met.* **13**, 113.

Moses, D., Colaneri, N., and Heeger, A. J. (1986). Alkali vapor phase doping of polyacetylene, *Solid State Commun.* **58**, 535.

Nechtschein, M., Devreux, F., Genoud, F., Vieil, E., Pernaut, J. M., and Genies, E. (1986). Polarons, bipolarons, and charge interactions in polypyrrole: Physical and electrochemical approaches, *Synth. Met.* **15**, 59.

Ogasawara, M., Funahashi, K., Demura, T., Hegiwara, T., and Iwata, K. (1986). Enhancement of electrical conductivity of polypyrrole by stretching, *Synth. Met.* **14**, 61.

Ohnishi, T., Murase, I., Noguchi, T., and Hirooka, M. (1986). Highly conductive graphite film prepared for pyrolysis of poly(p-phenylene vinylene), *Synth. Met.* **14**, 207.

Philpot, S. R., Baeviswyl, D., Bishop, A. R., and Lomdahl, P. S. (1986). Statics and adiabatic dynamics of non-linear excitations in defected polyacetylene, *Synth. Met.* **13**, 129.

Prön, A., Kucharski, Z., Budroweski, C., Zagorski, M., Krichene, S., Suwalski, J., Deke, G., and Lafrant, S. (1985). Mössbauer spectroscopy studies of selected conducting polypyrroles, *J. Chem. Phys.* **83**, 5923.

Qian, R., Qiu, J., and Yan, B. (1986). Electrochemical behavior of oxidized polypyrrole in aqueous solutions, *Synth. Met.* **14**, 81.

Saleneck, W. R., Lundstrom, I., Huang, W.-S., and MacDiarmid, A. G. (1986). A two-dimensional-surface "state diagram" for polyaniline, *Synth. Met.* **13**, 291.

Sasai, M., and Fukutome, H. (1986). Coulomb effects on the carbon 1s XPS in doped polyacetylene, *Solid State Commun.* **58**, 735.

Sato, M. A., Tanaka, S., and Kaeriyama, K. (1986). Electrochemical preparation of conducting poly(3-methylthiophene): Comparison with polythiophene and poly(3-ethylthiophene), *Synth. Met.* **14**, 279.

Satoh, M., Kaneto, K., and Yoshino, K. (1986). Dependences of electrical and mechanical properties of conducting polypyrrole films on conditions of electrochemical polymerization in an aqueous medium, *Synth. Met.* **14**, 289.

Saxena, A., and Gunton, J. D. (1986). Quasi-one-dimensional devices utilizing two conducting polymers, *Synth. Met.* **15**, 23.

Schmeltzer, D., and Ohana, I. (1985). The amplitude mode in trans-$(CH)_x$: A scaling approach, *J. Phys. C* **18**, L687.

Shacklette, L. W., and Toth, J. E. (1985). Phase transformations and ordering in polyacetylene, *Phys. Rev. B: Condens. Matter* **32**, 5892.

Shoo-Lin, M. (1986). The electrochemical properties and stability of polyacetylene in organic electrolyte solutions, *Synth. Met.* **14**, 19.

Skotheim, T. A. (1986). Interfaces between electronically and ionically conducting polymers: Applications to ultra-high-vacuum electrochemistry and photoelectrochemistry, *Synth. Met.* **14**, 31.

Soderholm, L., Mathis, C., Francois, B., and Friedt, J. M. (1985). The preparation, characterization and physical properties of $SbCl_5$-doped polyacetylene, *Synth. Met.* **10**, 261.

Stezowksi, J. J., Stigler, R. D., and Karl, N. (1986). Crystal structure and charge transfer energies of complexes of the donor biphenylene with the acceptors TCNB and PMDA, *J. Chem. Phys.* **84**, 5162.

Sum, V., Fesser, K., and Büttner, H. (1986). Electronic correlations in a model of polyacetylene, *Synth. Met.* **13**, 173.

Sun, X., Chen, L., and Yu, E. X. (1985). Vibrational mode of polaron in trans-$(CH)_x$, *Solid State Commun.* **53**, 973.

Talaat, H., Bucaro, J. A., Huang, W. S., and MacDiarmid, A. G. (1985). Photoacoustic detection of plasmon surface polaritons in heavily doped polyacetylene films, *Synth. Met.* **10**, 245.

Tanaka, J., Kamiya, K., Shimizu, M., and Tanaka, M. (1986). Optical and electrical conductivities of iodine-doped polyacetylene, *Synth. Met.* **13**, 177.

Tanaka, K., Koike, T., and Yamabe, T. (1985). Electronic structures of poly(vinylene sulfide) and its fluorinated and methylated derivatives, *Synth. Met.* **11**, 221.

Tanaka, K., Shichiri, T., and Yamabe, T. (1986). Electronic properties of mislinked polypyrrole and polythiophene, *Synth. Met.* **14**, 271.

Tasumi, M., Harada, I., Takeuchi, H., Shirakawa, H., Suzuki, S., Maconnachie, A., and Diamoux, A. J. (1985). Incoherent, inelastic neutron scattering from polyacetylene in the low-frequency region, *Synth. Met.* **10**, 293.

Thomann, H., Dalton, L. R., Grabowski, M., and Clarke, T. C. (1985). Direct observation of coulomb correlation effects in polyacetylene, *Phys. Rev. B: Condens. Matter* **31**, 3141.

Tober, R. L., Ferraris, J. P., and Glosser, R. (1986). Isomerization-induced evolution of piezoreflectance structures in polyacetylene films, *Phys. Rev. B: Condens. Matter* **33**, 8768.

Tokito, S., Tsutsui, T., and Saito, G. (1985). Morphology and conductivity of polypyrrole containing halogen counter anions, *Chem. Lett. J. Chem. Soc. Jpn.,* p. 531.

Vicente, R., Ribas, J., Cassoux, P., and Valade, L. (1986). Synthesis, characterization and properties of highly conducting organometallic polymers derived from the ethylene tetrathiolate anion, *Synth. Met.* **13**, 265.

Wang, B., Tang, J., and Wang. F. (1986). The effect of anions of supporting electrolyte on the electrochemical polymerization of aniline and the properties of polyaniline, *Synth. Met.* **13**, 329.

Wanqun, W., Mammone, R. J., and MacDiarmid, A. G. (1985). Stability and electrochemistry of polyacetylene in aqueous media, *Synth. Met.* **10**, 235.

Yacoby, Y., and Roth, S. (1986). Resonance Raman scattering of polyacetylene doped with bromine, iodine and AsF₅ and compensated by ammonia, *Synth. Met.* **13**, 299.

Youjiang, G., and Lu, Y. (1986). Bipolaron interpretation of resonance Raman scattering spectra of cis-polyacetylene, *Solid State Commun.* **58**, 407.

Yu, L., Shen, A., and Yang, M. (1985). The kinetics and mechanism of polymerization of acetylene by rare earth coordination catalysts, *Synth. Met.* **11**, 53.

Yumoto, Y., and Yoshimura, S. (1986). Synthesis and electrical properties of a new conducting polythiophene prepared by electrochemical polymerization of α-terthienyl, *Synth. Met.* **13**, 185.

Yun, M. S., and Yoshiro, K. (1985). Doping effect on carrier mobility in poly-*p*-phenylsulfide, *J. Appl. Phys.* **58**, 1950.

CHAPTER 4. KROGMANN SALTS—PARTIALLY OXIDIZED PLATINUM-CHAIN METALS AND RELATED MATERIALS

Apostol, M., and Baldea, I. (1985). Incommensurate pinning mechanism in KCP, *Solid State Commun.* **53**, 687.

Kato, R., Kobayashi, H., Kobayashi, A., and Sasaki, Y. (1985). Crystal structure of a new molecular conductor (DBTTF) [Ni(dmit)₂], *Chem. Lett. Chem. Soc. Jpn.,* p. 131.

Kobayashi, H., Kato, R., Kobayashi, A., and Sasaki, Y. (1985). The crystal structure of [BEDT–TTF] [Ni(dmit)₂]. A new route to design of organic conductors, *Chem. Lett. Chem. Soc. Jpn.,* p. 191.

Kobayashi, H., Kato, R., Kobayashi, A., and Sasaki, Y. (1985). The crystal and electronic structures of a new molecular conductor (TMTSF) [Ni(dmit)₂], *Chem. Lett. Chem. Soc. Jpn.,* p. 535.

Tanino, H., Takahashi, K., Oyanagi, H., Koshizuka, N., Kato, K., Yamashita, M., and Kobayashi, K. (1985). Pressure study of quasi-one-dimensional platinum complexes, *Abst. High Pressure Conf., 10th, Jpn, Oct. 21–23, Tokushima.*

Vincente, R., Ribas, J., Cassoux, P., and Sourisseau, C. (1986). Synthesis and resonance Raman study of conductive iodinated materials derived from M(dmit)(bipy) complexes, *Synth. Met.* **15**, 79.

Yamamoto, Y., Takahashi, K., and Yamazaki, M. (1985). Electrochemical preparation of isocyanide complexes of non-valent platinum containing a metal–metal bond, *Chem. Lett. Chem. Soc. Jpn.,* p. 201.

CHAPTER 5. TRANSITION ELEMENT–MACROCYCLIC LIGAND COMPLEXES

Djurado, D., Hamwi, A., Cousseins, J. C., Bidar, H., Fabre, C., and Berthet, G. (1985). Study of chemically (AsF₅) and electrochemically doped aluminum polyfluorophthalocyanine (PcAlF)$_n$, *Synth. Met.* **11**, 109.

Inabe, T., Marks, T. J., Burton, R. L., Lyding, J. W., McCarthy, W. J., Kannewurf, C. R., Reisner, G. M., and Herbstein, F. H. (1985). Highly conductive metallophthalocyanine assemblies, structure, charge transport and anisotropy in the metalfree molecular metal H₂(Pc)I, *Solid State Commun.* **54**, 501.

Inabe, T., Liang, W.-B., Lomax, J. F., Nakamura, S., Lyding, J. W., McCarthy, W. J., Carr, S. H., Kannewurf, C. R., and Marks, T. J. (1986). Phthalocyanine-based electrically conductive, processible molecular/macromolecular hybrid materials, *Synth. Met.* **13**, 129.

Kutzler, F. W., and Ellis, D. E. (1986). A comparison of the one-dimensional band structures of Ni tetrabenzo-porphyrin and phthalocyanine conducting polymers, *J. Chem. Phys.* **84**, 1033.

Lang, G., Boso, B., Erler, B. S., and Reed., C. A. (1986). Spin-coupling in ferric metalloporphyrin radical cation complexes: Mössbauer and susceptibility studies, *J. Chem. Phys.* **84**, 2998.

Le Moigne, J., and Even, R. (1985). Spectroscopic properties and conductivity of thin films of partially reduced metallo-phthalocyanines, *J. Chem. Phys.* **83**, 6472.

Pietro, W. J., Marks, T. J., and Ratner, M. A. (1985). Resistivity mechanisms in phthalocyanine-based linear-chain and polymeric conductors: Variation of bandwidth with geometry, *J. Am. Chem. Soc.* **107**, 5387.

Toscano, P. J., and Marks. T. J. (1986). Electrically conductive metallomacrocyclic assemblies. High resolution solid state NMR spectroscopy as a probe of local architecture and electronic structure in phthalocyanine molecular and macromolecular metals, *J. Am. Chem. Soc.* **108**, 437.

Wynne, K. J., Zacharides, A. E., Inabe, T., and Marks, T. J. (1985). Conducting, high modulus, molecular-macromolecular composites: Mechanical properties of oriented doped-phthalocyanine/kevlar fibers, *Poly. Commun.* **26**, 162.

CHAPTER 6. MISCELLANEOUS CONDUCTORS

Araya, K., Mukoh, A., Narahara, T., and Shirakawa, H. (1986). Highly conductive graphite film prepared in a liquid crystal solvent, *Synth. Met.* **14**, 199.

Bittner, D. N., and Bretz, M. (1985). Heat capacity of antimony pentachloride in intercalated graphite, *Phys. Rev. B: Condens. Matter* **31**, 1060.

Blonsky, P. M., Shriver, D. F., Austin, P., and Allcock, H. R. (1984). Polyphosphazene solid electrolytes, *J. Am. Chem. Soc.* **106**, 6854.

Blonsky, P. M., Shriver, D. F., Austin, P., and Allcock, H. R. (1986). Complex formation and ionic conductivity of polyphosphazene solid electrolytes, *Solid States Ionics* **18 & 19**, 258.

Blumenfeld, A. L., Isaev, Y. V., and Novikov, Y. N. (1985). Molecular motions in the ternary graphite intercalation compounds with potassium and methylbenzenes, *Synth. Met.* **10**, 193.

Flandrois, S., and Herran, J. (1986). Battery electrodes based on metal–chloride–graphite intercalation compounds, *Synth. Met.* **14**, 103.

Franco, H., Godfin, H., and Thoulouze, D. (1985). Ferromagnetic instability of ^{3}He layers adsorbed on grafoil, *Phys. Rev. B: Condens. Matter* **31**, 1699.

Hardy, L. C., and Shriver, D. F. (1986). Poly(ethylene oxide)–sodium polyiodide conductors: Characterization, electrical conductivity and photoresponse, *J. Am. Chem. Soc.* **108**, 2887.

Joos, B., and Duesberg, M. S. (1985). Dislocation energies in rare-gas monolayers on graphite, *Phys. Rev. Lett.* **55**, 1997.

Koike, Y., Morita, S., Nakanomyo, T., and Fukasi, T. (1985). Weak localization in graphite, *J. Phys. Soc. Jpn.* **64**, 713.

Nayak, K., Bhaumik, D., and Mark, J. E. (1986). Band structure analysis of a bis(oxalato) platinate complex, *Synth. Met.* **14**, 309.

Prietsch, M., Wortmann, G., Karndl, G., and Schlögl, R. (1986). Mössbauer study of stage-2 FeCl$_3$-graphite, *Phys. Rev. B: Condens. Matter* **33**, 7451.

Quinton, M. F., Legrand, A. P., and Bequin, F. (1986). Structure of the intercalated layer in graphite–potassium–tetrahydrofuran compounds, *Synth. Met.* **14**, 179.

Roth, G., Chaiken, A., Enoki, T., Yeh, N. C., Dresselhaus, G., and Tedrow, P. M. (1986). Enhanced superconductivity in hydrogenated potassium–mercury–graphite intercalation compounds, *Phys. Rev B: Condens. Matter* **32**, 533.

Shioya, J., Matsubara, H., and Murakami, S. (1986). Properties of AsF$_5$ intercalated vapor-grown graphite, *Synth. Met.* **14**, 113.

Szeto, K. Y., Chen, S. T., and Dresselhaus, G. (1986). Two-dimensional spin-flop transition CoCl$_2$–graphite intercalation compounds, *Phys. Rev. B: Condens. Matter* **33**, 3453.

Tanaka, K., Koike, T., Ueda, K., Ohzeki, K., Yamabe, T., and Yata, S. (1985). Electronic structures of polyacenacene and polyphenanthrophenanthrene. Design of one-dimensional graphite, *Synth. Met.* **11**, 61.

Tsukamoto, J., Matsumura, K., Takahashi, T., and Sakoda, K. (1986). Structure and conductivity of graphite fibers prepared by pyrolysis of cyanoacetylene, *Synth. Met.* **13**, 255.

Ummat, P. K., Zaleski, H., and Datars, W. R. (1986). Intercalation of graphite by antimony tetrachloride fluoride SbCl$_4$F, *Synth. Met.* **14**, 317.

Yashima, H., Nogami, T., and Mikawa, H. (1985). Lithium batteries and electric double-layer capacitors composed of activated carbon fiber electrodes, *Synth. Met.* **10**, 229.

Yosida, Y., and Tanuma, S. (1985). De Haas-van Alphen effect of SbCl$_5$–graphite intercalation compounds I, *J. Phys. Soc. Jpn.* **54**, 701.

Yosida, Y., and Tanuma, S. (1985). De Haas-van Alphen effect of SbCl$_5$–graphite intercalation compounds II, *J. Phys. Soc. Jpn.* **54**, 707.

Yosida, Y., and Tanuma, S. (1985). Induced Torque of SbCl$_3$–graphite intercalation compounds, *J. Phys. Soc. Jpn.* **54**, 650.

Yosida, Y., Tanuma, S., and Iye, Y. (1985). Magnetic susceptibility of SbCl$_5$–graphite intercalation compounds, *J. Phys. Soc. Jpn.* **54**, 2635.

Zaleski, H., Ummat, P. K., and Datars, W. R. (1985). Preparation and electronic properties of graphite hexafluoroantimonate intercalation compound, *Synth. Met.* **11**, 183.

Zhang, Q. M., Kim, H. K., and Chan, M. H. W. (1986). Nonwetting growth and cluster formation of CF$_4$ on graphite, *Phys. Rev. B*, **34**, 2056.

NOTE ADDED IN PROOF: HIGH SUPERCONDUCTING TRANSITION TEMPERATURES

Major breakthroughs have recently been made in increasing the superconducting transition temperatures (T_c) of certain metal oxide materials. Studies of compounds in the La–Ba–Cu–O system have led to T_c's with values from 30–40 K. This represents a substantial increase in T_c over the highest previously recorded T_c for Nb_3Ge of 23.2 K (see Chapter 1). These systems also have structures different from the A-15 structures (see Chapter 1) that had previously yielded the highest T_c materials. These new materials possess the K_2NiF_4 type layered perovskite structure.

In 1985, Michel et al. [C. Michel, I. Er-Rakho, and B. Raveau, *Mater. Res. Bull.* **20**, 667 (1985)] reported the synthesis of a new oxygen deficient perovskite, $BaLa_4Cu_5O_{13.41}$, which was interpreted as containing copper in a mixed valence state and was reportedly a very good metallic conductor (to 100 K). Bednorz and Müller in 1986 [J. G. Bednorz and K. A. Müller, *Z. Phys. B: Condens. Matter* **64**, 139 (1986)] prepared metallic, oxygen-deficient compounds in the La–Ba–Cu–O system with a composition of $La_{5-x}Ba_xCu_5O_{5(3-y)}$. They found three phases in samples when $x = 1$ and 0.75, $y > 0$. One phase consisted of a perovskite-like, mixed valent copper compound. The materials showed a linear decrease in resistance upon cooling, then a quasi-logarithmic increase, and finally an abrupt decrease of up to three orders of magnitude with an onset of superconductivity at 30 K and a zero resistance state at ~13 K. Unfortunately, the authors were unable to specify which phase was superconductive.

Subsequently, high T_c values were reported by Uchida et al. [S. Uchida, H. Takagi, K. Kitazawa, and S. Tanaka, *J. Appl. Phys. Jpn.* **26**, 21 (1987)] and Takagi et al. [H. Takagi, S. Uchida, K. Kitazawa, and S. Tanaka, *J. Appl. Phys. Jpn.*, submitted for publication Dec. 9, 1986] for the La–Ba–

Cu–oxide systems with onset superconducting temperatures in excess of 30 K. They were able to identify the superconducting phase as being a $(La, Ba)_2CuO_{4-y}$ material. Magnetic susceptibility and resistivity measurements indicated a high onset T_c of 32 K, with a zero resistance state achieved at ~22 K. This phase has a K_2NiF_4 (tetragonal, space group I4/*mmm*) layered perovskite structure. The authors also suggest that the mixed valence state of the copper atoms plays an important role in causing this material to have a high T_c, as was originally suggested by the work of Bednorz *et al.*

Another compound in the Ba–La–Cu–O system, $La_{1.35}Ba_{0.15}CuO_4$ has now been characterized [J. D. Jorgensen, D. G. Hinks, D. W. Capone, Z. Zhang, H.-B. Schuttler, and M. B. Brodsky, *Phys. Rev. Lett.*, submitted January 1987], which possesses a high T_c with an onset at ~33 K measured by resistivity and magnetic susceptibility. In agreement with the Japanese work, the structure was found to be I4/*mmm* (tetragonal, layered perovskite), which is depicted in Fig. 1. The onset of the drop in resistivity

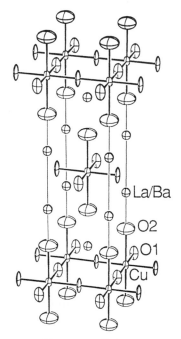

Fig. 1. Tetragonal structure of $La_{1.85}Ba_{0.15}CuO_4$. The atoms shown comprise one unit cell except that the full oxygen coordination spheres around the copper atoms at the corners are shown for clarity. [Taken from J. D. Jorgensen, D. G. Hinks, D. W. Capone II, Z. Zhang, H.-B. Schuttler, and M. B. Brodsky, *Phys. Rev. Lett.*, submitted January 1987.]

occurs at ~33 K, and the resistance reaches zero below ~19 K. A diamagnetic state also begins at 33 K. Thus, bulk superconductivity was
indicated for the compound. Barium is believed to stabilize the metallic
and superconducting structure, compared to the parent compound,
La_2CuO_4, and prevent a Peierls $2k_f$ instability from occurring. The insertion of the smaller Sr^{2+} ion for Ba^{2+} in the system also stabilizes the
superconducting phase, thought to be $La_{2-x}Sr_xCuO_4$, with a T_c of 34.7 K
[D. W. Capone, D. G. Hinks, J. D. Jorgensen, and Z. Zhang, *Appl. Phys.
Lett.* **50**, 543, 1987]. The strontium-doped samples demonstrated a zero
resistance at $T > 30$ K. The structure of the strontium-doped compound
is presently under study.

Cava and co-workers have also prepared an alloy phase consisting of
$La_{2-x}Sr_xCuO_4$ where $x \leq 0.3$ [see R. J. Cava, R. B. van Dover, B. Batlogg,
and E. A. Rietman, *Phys. Rev. Lett.* **58**, 408, 1987]. When $x = 0.2$ differences were observed between oxygen-annealed and argon-annealed samples. The former sample showed a resistance midpoint shifted to 36.2 K,
whereas the latter sample showed an outset temperature of 38.5 K.
The $La_{2-x}Sr_xCuO_4$ system has a K_2NiF_4 structure with planes of CuO_6
octahedra exclusively sharing corners separated by (La, Sr)–O layers
whereby lanthanum and strontium are ninefold coordinated to oxygen.
The copper–oxygen bonding is very distorted and copper assumes a
planar fourfold coordination to oxygen. C. W. Chu, P. H. Hor, R. L.
Meng, L. Gao, Z. G. Huang, and Y. Q. Wang, *Phys. Rev. Lett.* **58**, 405
(1987), have also found high T_c values for materials having compositions
of $La_{1-x}Ba_xCuO_{3-y}$, with $x = 0.20$ or 0.15 and y undetermined. Onset
superconducting temperatures above 40.2 K were obtained when 13 kbar
pressure was applied (T_c at ambient pressure was 32 K). All samples were
multiphased, consisting of defected perovskite structure (>60%), K_2NiF_4
(~30%), and an unidentified phase (<10%).

The official New China News Agency has reported that scientists at the
Chinese Academy of Sciences in Beijing have produced superconductivity
in similar materials at 70 K [see Science in Science/Technology Concen-
trates, *Chem. Eng. News,* Jan. 12, 1987, p. 20]. This report has not been
verified.

These major developments are somewhat preliminary and precautions
will have to be taken in the future to assure that the preparations result in
homogeneous materials and that measurements are made on single-phase
systems. Nevertheless, these results appear to be very promising and
considerable interest has been generated in the hopes of obtaining higher
T_c systems. The remarkable increase in T_c's lends credence to the notion
that superconductors with even higher T_c's can be prepared. Figure 2
shows the increase in T_c with time for several inorganic superconductors.

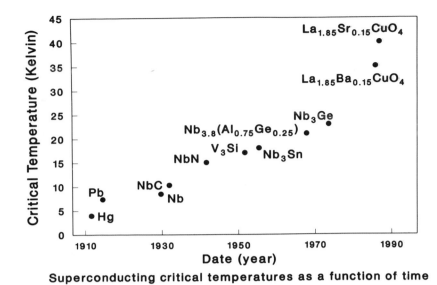

Superconducting critical temperatures as a function of time

Fig. 2. Superconducting critical temperatures as a function of time for several inorganic superconductors.

INDEX